T0189320

Logic, Epistemology, and the Unity of Science

Volume 43

Logic, Epistemology, and the Unity of Science aims to reconsider the question of the unity of science in light of recent developments in logic. At present, no single logical, semantical or methodological framework dominates the philosophy of science. However, the editors of this series believe that formal techniques like, for example, independence friendly logic, dialogical logics, multimodal logics, game theoretic semantics and linear logics, have the potential to cast new light on basic issues in the discussion of the unity of science.

This series provides a venue where philosophers and logicians can apply specific technical insights to fundamental philosophical problems. While the series is open to a wide variety of perspectives, including the study and analysis of argumentation and the critical discussion of the relationship between logic and the philosophy of science, the aim is to provide an integrated picture of the scientific enterprise in all its diversity.

More information about this series at http://www.springer.com/series/6936

Hassan Tahiri
Editor

The Philosophers and Mathematics

Festschrift for Roshdi Rashed

Springer

Editor
Hassan Tahiri
Center for Philosophy of Science
University of Lisbon
Lisbon, Portugal

and

MSH, Nord-Pas-de Calais
Université Charles de Gaulle Lille 3
Villeneuve-d'Ascq, France

ISSN 2214-9775 ISSN 2214-9783 (electronic)
Logic, Epistemology, and the Unity of Science
ISBN 978-3-030-06712-0 ISBN 978-3-319-93733-5 (eBook)
https://doi.org/10.1007/978-3-319-93733-5

Softcover re-print of the Hardcover 1st edition 2018

This Springer imprint is published by the registered company Springer Nature Switzerland AG
The registered company address is: Gewerbestrasse 11, 6330 Cham, Switzerland

Science is universal. This is not a postulate, but a basic feature which defines scientific knowledge itself. A scientific result, of whatever kind, can only be fully communicable and provable by stringent arguments. But this epistemic universality is not at all separate from the living history of human beings and from institutions. That is to say that this universality is not an immediate given of the consciousness, but rather reveals itself through a lengthy and bold conceptual process. This work organises itself along the lines of scientific traditions in which human beings and institutions are active. But these people and these institutions arise from a value-based system.

Roshdi Rashed

Several years ago, Prof. Kokiti Hara, a leading historian of mathematics in Japan, once stated his observation to me: "The most productive field of history of mathematics for the past quarter century was without doubt the history of Arabic mathematics. He added to this statement: "Thanks mainly to Mr. Rashed."

Sasaki Chikara
Editor of Historia Scientiarum

Acknowledgements

We would like to express our gratitude to our strong team of collaborators who shared with us their expertise by thoroughly and rigorously reviewing the contributions to the volume: Mark van Atten (Université de Paris 1), John Burgess (Princeton University), Giovanna Cifoletti (Centre A. Koyré, EHESS Paris), Michel Crubellier (Université de Lille 3), Sébastien Gandon (Université Blaise Pascal, Clermont-Ferrand), Richard Heck (Brown University), Gerhard Heinzmann (Université de Nancy), Leon Horsten (University of Bristol), Alberto Naibo (Université de Paris 1), Franco Oliveira (FCUL, Lisbon), Klaus Volkert (Cologne University), Heinrich Wansing (Bochum University). Our special thanks to Springer for publishing this special volume to mark this memorable scientific event.

About the book

During the nineteenth century and the first half of the twentieth century, mathematics was dominating the scientific and philosophical scene because of the richness and radicality of research programmes designed by great mathematicians to tackle problems regarding the foundations of their own discipline. The multiple different approaches and lively discussions that were generated had led to the transformation of mathematics and with it other major scientific disciplines like logic and philosophy. While the links between logic and mathematics opened a new chapter in their history, the interaction between philosophy and mathematics goes back to antiquity. A long historical relationship that motivated Jules Villemin to write: "the history of mathematics and philosophy shows that a renewal of the methods of the former has, each time, repercussions on the latter" (in La philosophie de l'algèbre, p. 4). An object of intrigues, the first scientific discipline, has always fascinated philosophers since antiquity because of the nature of its object, the rigour of its methods, its widespread applications, its deep relation with the real world and its ability to reinvent and propel itself. Mathematics has very often been the favoured interlocutor of the philosophers and a major source of inspiration in the formation of their system in an attempt to find in it consistence and permanence. How is mathematics viewed by the philosophers during the critical periods of its development? To what extent did their conceptions contribute to its progress? What role did it play in the formation of their doctrine? And what impact did its remarkable expansion have on the philosophical systems they erected? The purpose of this volume, which is the outcome of the international conference that was held in Lisbon in 2014 in honour of Roshdi Rashed,[1] is to examine the distinctive

[1]The dedication conference was organised in the context of the research project "Argumentation and Scientific Change: A case Study of How Ibn al-Haytham's *al-Shukūk* Changed the Course of Astronomy Forever" sponsored by the FCT, Fundação para a Ciência e a Tecnologia.

historical relationship between mathematics and philosophy and in particular the attitude of some philosophers towards mathematics throughout history both in those who recognised its relevance to their thought and in those who ignored it.

Contents

Roshdi Rashed: His Research Career, Awards and Publications

Research Career

Emeritus Research Director (distinguished class) at the National Center for Scientific Research (CNRS—France).
Director of the Center for the History of Arabic and Medieval Sciences and Philosophy until 2001.
Director of the Doctoral School in Epistemology and History of Sciences, Denis Diderot (Paris VII) University, until 2001.
Honorary Professor at the University of Tokyo.
Emeritus Professor at the University of Mansourah (Egypt).
Founder (1984) and first Director (till May 93) of the CNRS research team REHSEIS (Research in Epistemology and History of Sciences and Scientific Institutions).
President of the International Society of History of Arabic and Islamic Sciences and Philosophy (SIHSPAI).
Member of the French National Committee for the History and Philosophy of Sciences (Académie des Sciences).
Head Editor of Arabic Sciences and Philosophy: A Historical Journal, Cambridge University Press.
Head Editor of the collection « Histoire des Sciences arabes » (Beyrouth).
Head Editor of the collection « Sciences dans l'histoire », Blanchard.
Member of the Scientific Comittee of Revue de Synthèse (France), Bollettino di storia delle scienze matematiche (Italy); ИСТОРИКО-МАТЕМАТИЧЕСКИЕ ИССЛЕДОВАНИЯ (Moscou), Islamic Studies (Pakistan), Le Journal Scientifique Libanais (Beyrouth).

Awards and Honorary Functions

1977
Bronze Medal of the CNRS, for the Arithmetics of Diophantus.
1983
Member of the International Academy of History of Sciences. [1997-Vice-President].

1986
Member of the Academy of Arabic Language (Damascus).
1989
"Chevalier de la Légion d'Honneur", awarded by the President of the French Republic, on the occasion of the 50th anniversary of the CNRS.
1989
Member of the Academy of Arabic Language (Cairo).
1990
Alexandre Koyré Medal, awarded by the International Academy of History of Sciences, in recognition of all of his work.
1990
Third World Academy of Sciences History of Science Prize and Medal, in recognition of his works in the history of optics.
1990
Medal of the Organization of the Islamic Conference Research Center for Islamic History, Art and Culture (IRCICA), in recognition of his contributions to scholarship in the History of Islamic Culture.
1991
Member of the Third World Academy of Sciences (Mathematics section).
1998
World Prize for the best book of research in Islamology, awarded by the President of the Islamic Republic of Iran, for History of Arabic Sciences.
1999
Prize and Medal from Kuwait Foundation for the Advancement of Sciences, given by Emir of Kuwait, for his works on the history of geometry.
1999
Avicenna Gold Medal from Federico Mayor, General Director of UNESCO (United Nations Educational, Scientific and Cultural Organization), "for his contribution to recognition of Islamic culture as a part of universal scientific heritage and for promoting the dialog among different cultures".
2001
Medal from CNRS (National Scientific Research Center) for his research activities and his contribution to the international reputation of the CNRS.
2002
Member of the Royal Academy of Belgium
2004
Medal of the "Institut du Monde Arabe" (Paris), awarded by Denis Bauchard, President, in recognition of his works in the history of Arabic Sciences (15 June 2004).
2005
Medal of the Tunisian Academy "Bayt al-Hikma" and of the UNESCO Chair in Philosophy, awarded on the occasion of a colloquium (Bayt al-Hikma-UNESCO) devoted to his works (9–10 December 2005).

2007

King Faisal International Prize (Islamic Studies, Contribution to Pure or Applied Sciences) awarded by Prince Sultan Ibn Abd Al-Aziz Al-Saud, Crown Prince and Deputy Prime Minister, in Riyad on 15 April.

2010

Prize Doha "capitale culturelle arabe 2010".

2011

World Prize for the Book of the Year of the Islamic Republic of Iran, for Thâbit ibn Qurra. Philosophy in Ninth-Century Baghdad.

2012

Member of the Tunisian Academy "Bayt al-Hikma".

2015–16

Prize and Medal from the Sultan Bin Ali Al Owais Cultural Foundation (Dubai), in the field of "Humanities and Futures Studies", in recognition of his innovative studies on Arab mathematics.

2015–16

Sheikh Zayed Prize and Medal (Abu Dhabi) in the field of "Arab Culture in Other Languages" awarded for the book Angles et grandeur : D'Euclide à Kamāl al-Dīn al-Fārisī, Berlin, New York : Walter de Gruyter, 2015.

2016

Member of the Academy of Arabic Language (Jordan).

Kenneth O. May Prize in the History of Mathematics for 2017, awarded by the Executive Committee of the International Commission for the History of Mathematics.

Fields of Research

History of mathematics and their applications.
The classical theory of numbers.
History of optics.
History of mathematical and scientific instruments.
Applications of mathematics in social sciences.
Philosophy of mathematics and sciences.

Publications
Books

1. *Introduction à l'Histoire des Sciences* (co-author).
    - Vol. 1. *Eléments et instruments*, Paris: Hachette, 1971.
    - Vol. 2. *Objet et méthodes. Exemples*, Paris: Hachette, 1972.
2. *Al-Bāhir en Algèbre d'As-Samaw'al* (in collaboration with S. Ahmad), Damascus: University Press of Damascus, 1972, 347 p.
3. *Condorcet: Mathématique et Société*, Collection «Savoir», Paris: Hermann, 1974, 218 p.
    - Spanish translation, 1990.

4. *L'Art de l'Algèbre de Diophante* (in Arabic), Cairo: National Library, 1975, 253 p.
5. *L'Œuvre algébrique d'al-Khayyām* (in collaboration with A. Djebbar), Aleppo: University Press of Aleppo, 1981, 336 p.
6. *Entre Arithmétique et Algèbre. Recherches sur l'histoire des mathématiques arabes*, Collection «Sciences et philosophie arabes—Études et reprises», Paris: Les Belles Lettres, 1984, 321 p.
   - Arabic translation: Beirut: Markaz Dirāsat al-Wahda al-'arabiyya, 1989.
   - English translation: Kluwer, Boston Studies in Philosophy of Science, 1994.
   - Japanese translation: Tokyo University Press, 2004.
7. *Diophante: Les Arithmétiques, Livre IV*, vol. 3, «Collection des Universités de France», Paris: Les Belles Lettres, 1984, 487 p.
8. *Diophante: Les Arithmétiques, Livres V, VI, VII*, vol 4, «Collection des Universités de France», Paris: Les Belles Lettres, 1984, 451 p.
9. Jean Itard, *Essais d'Histoire des Mathématiques*, collected and presented by R. Rashed, Paris, Blanchard, 1984, 384 p.
10. *Etudes sur Avicenne*. Directed by J. Jolivet and R. Rashed. Collection «Sciences et philosophie arabes - Études et reprises», Paris: Les Belles Lettres, 1984, 151 p.
11. *Sharaf al-Din al-Tusi, Œuvres mathématiques. Algèbre et Géométrie au XIIe siècle*, vol. I. Collection «Sciences et philosophie arabes—textes et études», Paris: Les Belles Lettres, 1986. 480 p.
    - Arabic translation, Beirut: Centre for Arab Unity Studies, 1998.
12. *Sharaf al-Din al-Tusi, Œuvres mathématiques. Algèbre et Géométrie au XIIe siècle*, vol. II. Collection «Sciences et philosophie arabes—textes et études», Paris: Les Belles Lettres, 1986. 470 p.
    - Arabic translation, Beirut: Centre for Arab Unity Studies, 1998.
13. *Sciences à l'époque de la Révolution française. Recherches historiques*, Works of REHSEIS research team, edited by R. Rashed, Paris: Blanchard, 1988, 474 p.
14. *Mathématiques et Philosophie de l'Antiquité à l'Âge classique*. Études en hommage à Jules Vuillemin, edited by R. Rashed, Paris: éditions du C.N.R.S, 1991, 315 p.
15. *Optique et mathématiques: Recherches sur l'histoire de la pensée scientifique en arabe*, Variorum reprints, Aldershot, 1992, 310 p.
16. *Géométrie et dioptrique au Xe siècle: Ibn Sahl, al-Qūhī et Ibn al-Haytham*, Paris: Les Belles Lettres, 1993, 705 p.
    - Arabic translation, Beirut: Centre for Arab Unity Studies, 1996; second edition 2002.
17. *Les Mathématiques infinitésimales du IXe au XIe siècle*, Vol. II: *Ibn al-Haytham*, London: al-Furqan Islamic Heritage Foundation, 1993, 586 p.

- Arabic translation: Beirut, Centre for Arab Unity Studies, 2011.
- English translation: *Ibn al-Haytham and Analytical Mathematics*. A history of Arabic sciences and mathematics, vol. II, Culture and Civilization in the Middle East, London: Centre for Arab Unity Studies, Routledge, 2013, XIV–448 p.

18. *Les Mathématiques infinitésimales du IXe au XIe siècle*, Vol. I: *Fondateurs et commentateurs: Banū Mūsā, Thābit ibn Qurra, Ibn Sinān, al-Khāzin, al-Qūhī, Ibn al-Samḥ, Ibn Hūd*, London: al-Furqan Islamic Heritage Foundation, 1996, 1125 p.
    - Arabic translation: Beirut, Centre for Arab Unity Studies, 2011.
    - English translation: *Founding Figures and Commentators in Arabic Mathematics*. A history of Arabic sciences and mathematics, vol. I, Culture and Civilization in the Middle East, London: Centre for Arab Unity Studies, Routledge, 2012, XXIII–808 p.
19. *Œuvres philosophiques et scientifiques d'al-Kindī*, Vol. I: *L'Optique et la Catoptrique d'al-Kindī*, Leiden: E.J. Brill, 1997, 790 p.
20. *Descartes et le Moyen Âge*, ed. J. Biard and R. Rashed, Paris: Vrin, 1997, 425 p.
21. *Encyclopedia of the History of Arabic Science* (editor and co-author), London and New York, Routledge, 1996, 3 vol., 1105 p.
    - Vol. 1: Astronomy—Theoretical and applied
    - Vol. 2: Mathematics and the physical sciences
    - Vol. 3: Technology, alchemy and the life sciences
    - French translation: *Histoire des sciences arabes*, 3 vol., Paris: Le Seuil, 1997.
    - Arabic translation: *Mawsu'a Tārīkh al-'ulūm al-'arabiyya*, 3 vol., Beirut: Centre for Arab Unity Studies, 1997.
    - Persian translation, in press, Teheran.
    - Polish translation, *Historia nauki Arabskiej*, 3 vol., Varsovie, Dialog, 2000–2001.
22. *Œuvres philosophiques et scientifiques d'al-Kindī*, Vol. II: *Métaphysique et Cosmologie*, (with J. Jolivet), Leiden: E.J. Brill, 1998, XIII–243 p.
23. *Pierre Fermat: La théorie des nombres*, Texts translated by P. Tannery, introduced and commented by R. Rashed, Ch. Houzel et G. Christol, Paris, Blanchard, 1999, 512 p.
24. *Les Doctrines de la science de l'antiquité à l'âge classique*, ed. R. Rashed and J. Biard, Leuven: Peeters, 1999, 272 p.
25. *Al-Khayyām mathématicien*, in collaboration with B. Vahabzadeh, Paris: Librairie Blanchard, 1999, 438 p.
    - English Version: *Omar Khayyām. The Mathematician*, Persian Heritage Series no. 40, New York: Bibliotheca Persica Press, 2000, 268 p. (without the Arabic texts).

- Arabic translation: *Riyāḍiyyāt 'Umar al-Khayyām*, Silsilat Tārīkh al-'ulūm 'inda al-'Arab 7, Beirut: Centre for Arab Unity Studies, 2005.

26. *Les Catoptriciens grecs.* I: *Les miroirs ardents*, edition, translation and commentary, Collection des Universités de France, published under the patronage of the Association Guillaume Budé, Paris: Les Belles Lettres, 2000, 577 p.
27. *Ibrāhīm ibn Sinān. Logique et géométrie au Xe siècle*, in collaboration with Hélène Bellosta, Leiden: E.J. Brill, 2000, XI–809 p.
28. *Les Mathématiques infinitésimales du IXe au XIe siècle,* vol. III: *Ibn al-Haytham. Théorie des coniques, constructions géométriques et géométrie pratique*, London: al-Furqan, 2000, XXIII–1034 p.

    - English translation: *Ibn al-Haytham's Theory of Conics, Geometrical Constructions and Practical Geometry.* A History of Arabic Sciences and Mathematics, vol. 3, Culture and Civilization in the Middle East, London, Centre for Arab Unity Studies, Routledge, 2013.

29. *Les Mathématiques infinitésimales du IX au XI siècle*, vol. IV: *Méthodes géométriques, transformations ponctuelles et philosophie des mathématiques*, London: al-Furqan, 2002, XIII–1064–VII p.

    - Arabic translation: Beirut, Centre for Arab Unity Studies, 2011.
    - English translation: *Ibn al-Haytham's Geometrical Methods and the Philosophy of Mathematics*, A History of Arabic Sciences and Mathematics, vol. 5, Culture and Civilization in the Middle East, London, Centre for Arab Unity Studies, Routledge, 2016, xx–664 p.

30. *Storia della scienza*, vol. III: *La civilta islamica* (scientific direction and co-author), Enciclopedia Italiana, Roma, 2002, XX–941 p.
31. *Recherche et enseignement des mathématiques au IXe siècle. Le recueil de propositions géométriques de Na'īm ibn Mūsā*, in collaboration with Christian Houzel, Les Cahiers du Mideo, 2, Louvain-Paris: Peeters, 2004.
32. *Maïmonide, philosophe et savant (1138–1204)*, edited by Tony Lévy and Roshdi Rashed, Ancient and Classical Sciences and Philosophy, Leuven, Peeters, 2004, XI–477 p.
33. *Œuvre mathématique d'al-Sijzī.* Volume I: *Géométrie des coniques et théorie des nombres au Xe siècle*, Les Cahiers du Mideo, 3, Louvain-Paris: Peeters, 2004, 541 p.
34. *Klasik Avrupali Modernitenin Icadi ve Islam'da Bilim*, (Turkish translation by Bekir S. Gür), Ankara, Kadim Yayinlari, 2005, 360 p.
35. *Geometry and Dioptrics in Classical Islam*, London, al-Furqān, 2005, XIII–1178–VI p.
36. *Philosophie des mathématiques et théorie de la connaissance. L'Œuvre de Jules Vuillemin*, ed. R. Rashed and P. Pellegrin, Collection Sciences dans l'histoire, Paris, Librairie A. Blanchard, 2005, XIII–393 p.

37. *En histoire des sciences. Études philosophiques*, Tunisian Academy "Beït al-Hikma" and UNESCO chair in Philosophy, Carthage, 2005, French-Arabic 165–100 p.
38. *Les Mathématiques infinitésimales du IXe au XIe siècle*. Vol. V: *Ibn al-Haytham: Géométrie sphérique et astronomie*, London: al-Furqan Islamic Heritage Foundation, 2006, xiv–972–v p.
    - Arabic translation: Beirut, Centre for Arab Unity Studies, 2011.
    - English translation: *Ibn al-Haytham. New Spherical Geometry and Astronomy*, A History of Arabic Sciences and Mathematics, vol. 4, London, Centre for Arab Unity Studies, Routledge, collection Culture and Civilization in the Middle East, 2014, 642 p.
39. *Al-Khwārizmī: Le commencement de l'algèbre*, Paris, Librairie A. Blanchard, 2007, viii–386 p.
    - *Al-Khwārizmī: The Beginnings of Algebra*, «History of Science and Philosophy in Classical Islam», London, Saqi Books, 2009.
    - Arabic translation by Nicolas Farès, Silsilat Tārīkh al-ʻulūm ʻinda al-ʻArab 11, Beirut, Centre for Arab Unity Studies, 2010.
40. *Apollonius: Les Coniques*, tome 1.1: *Livre I*, commentaire historique et mathématique, édition et traduction du texte arabe, Berlin/New York: Walter deGruyter, 2008, xv–666 p.
41. *Apollonius : Les Coniques*, tome 2.2: *Livre IV*, commentaire historique et mathématique, édition et traduction du texte arabe, Berlin/New York, Walter de Gruyter, 2009, xii–319 p.
42. *Apollonius: Les Coniques*, tome 3: *Livre V*, commentaire historique et mathématique, édition et traduction du texte arabe, Berlin/New York: Walter de Gruyter, 2008, xv–550 p.
43. *Apollonius: Les Coniques*, tome 4: *Livres VI et VII*, commentaire historique et mathématique, édition et traduction du texte arabe, Scientia Graeco-Arabica, vol. 1.4, Berlin/New York: Walter de Gruyter, 2009, xi–572 p.
44. *Thābit ibn Qurra. Science and Philosophy in Ninth-Century Baghdad* (editor and co-author), Scientia Graeco-Arabica, vol. 4, Berlin/New York: Walter de Gruyter, 2009, x–790 p.
45. Apollonius de Perge, *La section des droites selon des rapports*, commentaire historique et mathématique, édition et traduction du texte arabe par Roshdi Rashed et Hélène Bellosta, Scientia Graeco-Arabica, vol. 2, Berlin/New York : Walter de Gruyter, 2009, viii–493 p.
46. *Apollonius: Les Coniques*, tome 2.1: *Livres II et III*, commentaire historique et mathématique, édition et traduction du texte arabe, Berlin/New York: Walter de Gruyter, 2010, xv–682 p.
47. *Dirāsāt fī tārīkh al-ʻulūm al-ʻarabiyya wa-falsafatihā*, Silsilat Tārīkh al-ʻulūm ʻinda al-ʻArab 12, Beyrouth: Markaz Dirāsat al-Waḥda al-ʻArabiyya, 2011.

48. *D'al-Khwārizmī à Descartes. Études sur l'histoire des mathématiques classiques*, Paris: Hermann, 2011, 795 p.
49. *Abū Kāmil : Algèbre et analyse diophantienne*, Berlin/New York: Walter de Gruyter, 2012, XV–819 p.
50. *Les courbes: Études sur l'histoire d'un concept* (co-editor and co-author), Collection Sciences dans l'histoire, Paris: Librairie A. Blanchard, 2013, VIII–245 p.
51. *Les* Arithmétiques *de Diophante: Lecture historique et mathématique* (in collaboration with Christian Houzel), Berlin, New York: Walter de Gruyter, 2013, IX–629 p.
52. *Histoire de l'analyse diophantienne classique: D'Abū Kāmil à Fermat*, Berlin, New York: Walter de Gruyter, 2013, X-349 p.
53. *Ibn al-Haytham. New Spherical Geometry and Astronomy*, A History of Arabic Sciences and Mathematics, vol. 4, London, Centre for Arab Unity Studies, Routledge, collection Culture and Civilization in the Middle East, 2014, 642 p.
54. *Classical Mathematics from al-Khwārizmī to Descartes*, Culture and Civilization in the Middle East, Londres: Centre for Arab Unity Studies, Routledge, 2014, 749 p.
55. *Dirāsāt fī tārīkh 'ilm al-kalām wa-al-falsafa* (editor and co-author), Silsilat Dirāsāt tārīkhiyya fī al-falsafa wa-al-'ulūm fi al-ḥaḍāra al-arabiyya al-islāmiyya 1, Beirut: Centre for Arab Unity Studies/al-Tafāhum, 2014, 483 p. (in Arabic).
56. *Angles et grandeur: D'Euclide à Kamāl al-Dīn al-Fārisī*, Berlin, New York: Walter de Gruyter, 2015, vi–706 p.
57. *Dirāsāt fī falsafa Abī Naṣr al-Fārābī* (editor), Silsilat Dirāsāt tārīkhiyya fī al-falsafa wa-al-'ulūm fi al-ḥaḍāra al-arabiyya al-islāmiyya 2, Beirut: Centre for Arab Unity Studies/al-Tafāhum, 2015, x–383 p. (in Arabic).
58. *Nūr al-Dalāla li-Fakhr al-Dīn al-Khilāṭī: al-jabr al-ḥisābī fī al-qarn al-thālith 'ashar*, Markaz Ḥasan b. Muḥammad li-al-Dirasāt al-tārīkhiyya, Qatar, 2016, 103 p. (in Arabic).
59. *Bayna al-falsafa wa-al-riyāḍiyāt: min Ibn Sīnā ilā Kamāl al-Dīn al-Fārisī* (editor and co-author), Silsilat Dirāsāt tārīkhiyya fī al-falsafa wa-al-'ulūm fi al-ḥaḍāra al-arabiyya al-islāmiyya 3, Beirut: Centre for Arab Unity Studies/al-Tafāhum, 2016, 624 p. (in Arabic).
60. *Algebra. Origini e sviluppi tra mondo arabo e mondo latino* (co-auteurs: Laura Catastini, Franco Ghione), Roma, Carocci editore & Frecce, 2016, 223 p.
61. *Lexique historique de la langue scientifique arabe* (editor), Hildesheim, W. Georg Olms, 2017.
62. Menelaus' Spherics: Early Translation and al-Māhānī / al-Harawī's Version (in collaboration with A. Papadopoulos), edition, translation and commentary, Berlin, De Gruyter, 2017, X–877 p.
63. Light-Based Science: Technology and Sustainable Development, the Legacy of Ibn al-Haytham, Azzedine Boudrioua, Roshdi Rashed, Vasudevan Lakshminarayanan (eds.), CRC Press, Taylor and Francis Group, Boca Raton, London, New York, 2018.

Forthcoming

1. *Les Mathématiques infinitésimales du IXe au XIe siècle. Vol. VI: Ibn al-Haytham Problèmes de fondements et commentaires des* Éléments *d'Euclide*, London: al-Furqan Islamic Heritage Foundation.

Articles

1. «Le discours de la lumière d'Ibn al-Haytham (Alhazen)», *Revue d'Histoire des Sciences,* 21 (1968), p. 197–224.
2. «Optique géométrique et doctrine optique chez Ibn al-Haytham», *Archive for History of Exact Sciences*, 6.4 (1970), p. 271–298.
3. «Le modèle de la sphère transparente et l'explication de l'arc-en-ciel: Ibn al-Haytham, al-Fārisī», *Revue d'Histoire des Sciences*, 23 (1970), p. 109–140.
4. «L'introduction de la mathématique du probable dans la science sociale», in *Actes du XIIe Congrès d'Histoire des Sciences*, vol. 9, Paris: Blanchard, 1971, p. 55–59.
5. «Islam (Les expressions)—Les sciences dans le monde musulman», in *Encyclopaedia Universalis*, Paris, 1971.
6. «La mathématisation des doctrines informes dans la science sociale», in *La mathématisation des doctrines informes*, sous la direction de G. Canguilhem, Paris: Hermann, 1972, p. 73–105.
7. «Idéologie et mathématique: l'exemple du vote au XVIIIe siècle», publication de la Faculté des Arts et des Sciences, Montréal, 1972.
8. «L'induction mathématique: Al-Karajī, As-Samaw'al», *Archive for History of Exact Sciences*, 9 (1972), p. 1–21.
9. «Modernisme et tradition», *Al-Katib* (1972), p. 35–47.
10. «Kamāl al-Dīn al-Fārisī», in *Dictionary of Scientific Biography*, vol. 7, New York: Scribner, 1973, p. 212–219.
11. «Algèbre et linguistique: l'analyse combinatoire dans la science arabe», in R. Cohen (ed.), *Boston Studies in the Philosophy of Sciences*, Reidel: Boston, 1973, p. 383–399.
12. «Al-Karajī», in *Dictionary of Scientific Biography*, vol. 7, New York: Scribner, 1973, p. 240–246.
13. «Ibrāhīm ibn Sinān», in *Dictionary of Scientific Biography*, vol. 7, New York: Scribner, 1973, p. 2–3.
14. «L'arithmétisation de l'algèbre au XIIe siècle», in *Actes du XIIIe Congrès d'Histoire des Sciences*, Moscou, 1974, p. 3–30.
15. «Résolution des équations numériques et algèbre: Sharaf al-Dīn al-Ṭūsī - Viète», *Archive for History of Exact Sciences*, 12.3 (1974), p. 244–290.
16. «Les travaux perdus de Diophante, I», *Revue d'Histoire des Sciences*, 27.2 (1974), p. 97–122.
17. «Les travaux perdus de Diophante, II», *Revue d'Histoire des Sciences*, 28.1 (1975), p. 3–30.

18. «Les recommencements de l'algèbre aux XIe et XIIe siècles», in J.E. Murdoch and E.D. Sylla (eds.), *The cultural Context of Medieval Learning*, Dordrecht: Reidel, 1975, p. 33–60.
19. «Condorcet», in *Enciclopedia Scientziati e tecnologi* (Arnoldo Mondadori, 1975).
20. Traduction française in *De Révolution en Révolution, Spécial Options*, 16 (1986), p. 34–36.
21. «Al-Bīrūnī algébriste», in *The Commemoration Volume of Biruni International Congress in Teheran*, Teheran, 1976, p. 63–74.
22. «Les fractions décimales. As-Samaw'al, al-Kāshī», in *Proceedings of the First International Symposium for the History of Arabic Science*, Aleppo, 1976, p. 169–186.
23. «Le concept de l'infini à l'époque de Rhazès», in *Actes du Colloque Rhazès*, Cairo, 1977.
24. «Lumière et vision: l'application des mathématiques dans l'optique d'Alhazen», in R. Taton (ed.), *Roemer et la vitesse de la lumière*, Paris: Vrin, 1978, p. 19–44.
25. «À propos d'une édition du texte de Dioclès sur les miroirs ardents», *Archives Internationales d'Histoire des Sciences*, 28 (1978), p. 329–334.
26. «L'extraction de la racine nième et l'invention des fractions décimales», *Archive for History of Exact Sciences*, 18.3 (1978), p. 191–243.
27. «Un problème arithmético-géométrique de Sharaf al-Din al-Tusi», *Journal for the History of Arabic Sciences*, 2.2 (1978), p. 233–254.
28. «La notion de science occidentale», in E.G. Forbes (ed.), *Human Implications of Scientific Advance*, Edinburgh, 1978, p. 45–54.
29. English translation «Science as a Western Phenomenon», *Fundamenta Scientiae*, 1 (1980), p. 7–21.
30. Arabic translation in *Al-Mustaqbal al Arabi*, 47 (1983), p. 4–19.
31. «L'analyse diophantienne au Xe siècle: l'exemple d'al-Khāzin», *Revue d'Histoire des Sciences*, 32 (1979), p. 193–222.
32. «La construction de l'heptagone régulier par Ibn al-Haytham», *Journal for the History of Arabic Science*, 3 (1979), p. 309–387.
33. «Al-Kindī» (co-author), in *Encyclopédie de l'Islam*, Leiden, 1979, p. 123–126.
34. «Ibn al-Haytham et le théorème de Wilson», *Archive for History of Exact Sciences,* 22.4 (1980), p. 305–321.
35. «Al-Kindī» (co-author) in *Dictionary of Scientific Biography,* vol. 15, New York: Scribner, 1980, p. 260–267.
36. «Remarks on the history of diophantine analysis», in *Conference on Algebra and Geometry*, Kuwait, 1981, p. 102–103.
37. «Remarques sur l'histoire de la théorie des nombres dans les mathématiques arabes», in *Proceedings of the Sixteenth International Congress of Science: Meetings on specialized Topics*, Bucarest, 1981, p. 255–261.

38. «L'Islam et l'épanouissement des sciences exactes», in *L'Islam, la philosophie et la science* (co-author), Paris: UNESCO, 1981. (English, Spanish, Arabic translations).
39. «Matériaux pour l'histoire des nombres amiables et de l'analyse combinatoire», *Journal for the History of Arabic Science*, 6 (1982), p. 209–278.
40. «Ibn al-Haytham et la mesure du paraboloïde», *Journal for the History of Arabic Sciences*, 5 (1982), p. 191–262.
41. «L'idée de l'algèbre selon al-Khwarizmī», *Fundamenta Scientiae*, 4 (1983), p. 87–100. Russian translation by B. Rozenfeld and A. Youschkevitch in *Muhammad ibn Musa al-Khwarizmī, 1200 ans*, Moscou, 1983, p. 85–108. Arabic translation in *Al-Mustaqbal al-Arabī*, Beirut, 1984. English translation in *Arab Civilization, Challenges and Responses*, edited by G.N. Atiyeh and I.M Oweiss, New York State University Press, 1988, p. 98–111.
42. «Nombres amiables, parties aliquotes et nombres figurés aux XIIIe et XIVe siècles», *Archive for History of Exact Sciences*, 28 (1983), p. 107–147.
43. «Les pratiques culturelles et l'émergence des connaissances scientifiques», *Al-Mustaqbal al-Arabī*, 68 (1984), p. 24–29.
44. English translation in *Unesco Meeting of experts on comparative philosophical studies on changes in relations between science and society*, New Delhi, 1986, p. 23–31.
45. «Diophante d'Alexandrie», in *Encyclopædia Universalis* (1985), p. 235–238.
46. «Histoire des sciences et modernisation scientifique dans les pays arabes», in *Problèmes du développement scientifique dans les pays arabes*, Beirut: *Al-Mustaqbal al-Arabi*, 1985 (in Arabic), p. 147–164.
47. «Al-Sijzī et Maïmonide: Commentaire mathématique et philosophique de la proposition II-14 des *Coniques* d'Apollonius», *Archives Internationales d'Histoire des Sciences,* no. 119, vol. 37 (1987), p. 263–296.
48. English translation: «Conceivability, Imaginability and Provability in Demonstrative Reasoning: al-Sijzī and Maimonides on II.14 of Apollonius' *Conics Sections*», *Fundamenta Scientiae*, vol. 8, no. 3/4 (1987), p. 241–256.
49. «Al-Sijzī and Maimonides: A Mathematical and Philosophical Commentary on Proposition II-14 in Apollonius' *Conic Sections*», in R.S. Cohen and H. Levine (eds), *Maimonides and the Sciences*, Kuwer Academic Publishers, 2000, p. 159–172.
50. «La périodisation des mathématiques classiques», *Revue de synthèse*, IVe S., no. 3–4 (1987), p. 349–360.
51. «Lagrange historien de Diophante», in *Sciences à l'époque de la Révolution française. Recherches historiques*. Travaux de l'équipe REHSEIS, édités par R. Rashed, Paris: Blanchard, 1988, 474 p.
52. «Ibn al-Haytham et les nombres parfaits», *Historia Mathematica*, 16 (1989), p. 343–352.
53. «Problems of the Transmission of Greek Scientific Thought into Arabic: Examples from Mathematics and Optics», *History of Science*, XXVII (1989), p. 199–209. Persan: in *Miras-e Elmi-ye Eslam va Iran* (Semiannual Journal on

the Scientific Heritage of Islam & Iran), vol. 3, no. 2, Autumn 2014 & Winter 2015, p. 109–119.

54. «Transmissions et recommencements: l'exemple de l'optique», in *Espaces et Sociétés du monde arabe*; La Documentation Française, no. 123 (1989), p. 22–26.
55. «A Pioneer in Anaclastics. Ibn Sahl on Burning Mirrors and Lenses», *Isis*, 81, 1990, p. 464–491.
56. «Al-Samaw'al, al-Bīrūnī et Brahmagupta: les méthodes d'interpolation», *Arabic Sciences and Philosophy: a Historical Journal*, 1 (1991), p. 100–160.
57. «L'analyse et la synthèse selon Ibn al-Haytham», in *Mathématiques et philosophie de l'Antiquité à l'âge classique*. Études en hommage à Jules Vuillemin, éditées par R. Rashed, Paris: éditions du CNRS, 1991, p. 131–162.
58. English translation: «Analysis and Synthesis according to Ibn al-Haytham», in C. C. Gould and R. S. Cohen (eds.), *Artifacts, Representations and Social Practice*, Kluwer Academic Publishers, 1994, p. 121–140.
59. «Science classique et science moderne à l'époque de l'expansion de la science européenne», in P. Petitjean, C. Jami and A. M. Moulin (eds.), *Science and Empires*, Boston Studies in the Philosophy of Science, Kluwer Academic Publishers, 1992, p. 19–30.
60. Portuguese translation in A. Garibaldi (ed.), *Principios*, n 27, Sao Paulo, p. 39–47.
61. «La philosophie mathématique d'Ibn al-Haytham. I: L'analyse et la synthèse», *Mélanges de l'Institut Dominicain d'Etudes Orientales du Caire*, 20, 1991, p. 31–231.
62. «Archimède et les mathématiques arabes», in *Archimede, Mito Tradizione Scienza*, a cura di Corrado Dollo, Firenze 1992, 43–61.
63. «Mathématiques traditionnelles dans les pays islamiques au XIXe siècle: l'exemple de l'Iran», in E. Ihsanoglu (éd.), *Transfer of Modern Science and Technology to the Muslim World,* Istanbul, 1992, p. 393–404.
64. «Futhitos (?) et al-Kindi sur "l'illusion lunaire"», in Goulet, Madec, O'Brien (eds.), SOFIHSMAIHTORES, «Chercheurs de sagesse», Hommage à Jean Pépin, Collection des Études Augustiniennes. Série Antiquité 131, Paris: Institut d'Études Augustiniennes, 1992, p. 533–559.
65. «Les traducteurs», in *Palerme 1070– 1492. Mosaïque de peuples, nation rebelle: la naissance violente de l'identité sicilienne*, Autrement, 1993, p. 110–119.
66. «De Constantinople à Bagdad: Anthémius de Tralles et al-Kindī», in *Actes du Colloque: La Syrie de Byzance à l'Islam* (Lyon, 1990); Damas, 1992, p. 165–170.
67. «Al-Kindī's commentary on Archimedes' "The Measurement of the Circle"», *Arabic Sciences and Philosophy*, vol. 3 (1993), p. 7–53.
68. «La philosophie mathématique d'Ibn al-Haytham. II: *Les Connus*», *Mélanges de l'Institut Dominicain d'Etudes Orientales du Caire* (*MIDEO*), 21, 1993, p. 87–275.

69. «Probabilité conditionnelle et causalité: un problème d'application des mathématiques», in J. Proust et E. Schwartz (eds), *La connaissance philosophique. Essais sur l'œuvre de Gilles Gaston Granger*, Paris: PUF, 1994, p. 271–293.
70. «Indian Mathematics in Arabic», in Ch. Sasaki, J.W. Dauben, M. Sugiura (eds.), *The Intersection of History and Mathematics*, Basel, Boston, Berlin, Birkhaüser Verlag, 1994, p. 143–148.
71. «Notes sur la version arabe des trois premiers livres des *Arithmétiques* de Diophante, et sur le problème 1. 39», *Historia Scientiarum*, 4–1 (1994), p. 39–46.
72. «Fibonacci et les mathématiques arabes», *Micrologus* II, 1994, p. 145–160.
73. Italian translation: «Fibonacci e la matematica araba», in *Federico II e le scienze*, Palermo, 1994, p. 324–337.
74. «Mathématiques arabes» of the *Encyclopédie de l'Islam*, Brill, 1994, p. 567–580. English translation: *Encyclopaedia of Islam*, Brill, 1994.
75. «Al-Yazdī et l'équation et la théorie des nombres», *Historia Scientiarum*, vol. 4–2 (1994), p. 79–101.
76. «Ibn Sahl et al-Qūhī: dioptrique et méthodes projectives au Xe siècle», in S. Garma, D. Flament, V. Navarro (eds.), *Contra los titanes de la rutina*, Madrid: CSIC, 1994, p. 9–18.
77. Articles published in Turkish in the *Islamic Encyclopædia*(Istanbul) 1994: «Mathematics», «Thābit ibn Qurra», «Ibrāhīm ibn Sinān».
78. «Recherche scientifique et modernisation en Égypte. L'exemple de 'Ali Mustafa Musharafa (1898–1950). Étude d'un type idéal», in *Entre réforme sociale et mouvement national. Identité et modernisation en Égypte (1882–1962)*, sous la direction d'A. Roussillon, CEDEJ, Le Caire, 1995.
79. Arabic translation p. 219–232.
80. «Conic Sections and Burning Mirrors: An Example of the Application of Ancient and Classical Mathematics», in K. Gavroglu *et al.* (eds.), *Physics, Philosophy and the Scientific Community*, Dordrecht, Kluwer Academic Publishers, 1995, p. 357–376.
81. «Modernité classique et science arabe», in C. Goldstein and J. Ritter (eds.), *Mathématiques en Europe*, MSH, 1996, p. 68–81.
82. Traduction portugaise in A. M.Alfonso-Goldfarb and C. A. Maia (eds), *História da ciência: o mapa do conbecimento*, Sao Paulo, 1996, p. 27–39.
83. «Les commencements des mathématiques archimédiennes en arabe: Banū Mūsā», in A. Hasnawi, M. Aouad and A. Elamrani (eds.), *Perspectives médiévales arabes et latines sur la tradition scientifique et philosophique grecque, Actes du Colloque de la SIHSPAI*, Paris/Louvain, 1996, p. 1–19.
84. Greek translation published in *Neusis*, 1995, p. 133–154.
85. English translation, «Archimedean Learning in the Middle Ages: The Banu Musa», *Historia Scientiarum*, 6–1 (1996), p. 1–16.
86. «Thābit ibn Qurra», in *Lexikon des Mittelalters*, Munich, 1996.

87. Articles published in *Encyclopedia of the History of Arabic Science* (editor and co-author), London, mars 1996, Routledge, 3 vol.
88. «Algebra», p. 349–375.
89. «Combinatorial analysis, numerical analysis, Diophantine analysis and number theory», p. 376–417
90. «Infinitesimal determinations, quadrature of lunules and isoperimetric problems», p. 418–446
91. «Geometrical optics», p. 643–671.
92. Articles «Ibn Sahl», «Ibn Sinān», «Ibn al-Haytham», «Science as a western phenomenon» (new translation), published in Helaine Selim (ed.), *Encyclopaedia of the History of Science, Technology and Medicine in Non-Western Cultures*, Dordrecht, Kluwer Academic Publishers, 1997.
93. «Le commentaire par al-Kindī de l'*Optique* d'Euclide: un traité jusqu'ici inconnu», *Arabic Sciences and Philosophy*, 7.1 (1997), p. 9–57.
94. «La *Géométrie* de Descartes et la distinction entre courbes géométriques et courbes mécaniques», in J. Biard and R. Rashed (eds.), *Descartes et le Moyen Âge*, Études de philosophie médiévale LXXV, Paris: Vrin, 1997, p. 1–22.
95. «Coniques et miroirs ardents: un exemple de l'application des mathématiques anciennes et classiques», in *Langages et philosophie. Hommage à Jean Jolivet*, Études de philosophie médiévale LXXIV, Paris: Vrin, 1997, p. 15–30.
96. «Dioclès et "Dtrūms": deux traités sur les miroirs ardents», *MIDEO*, 23 (1997), p. 1–155.
97. «L'histoire des sciences entre épistémologie et histoire», *Historia scientiarum*, 7.1 (1997), p. 1–10;
98. Japanese translation in *MISUZU*, vol. 41, no. 7 (Juillet 1999), p. 25–37.
99. «De la géométrie du regard aux mathématiques des phénomènes lumineux», text in Persan in *L'histoire des sciences en Terre d'Islam, Waqf, mirath-e jawidan*, vol. 4, n 3–4 (1996–97), p. 25–34.
100. «Mathématiques et autres sciences», *Dictionnaire de l'Islam, religion et civilisation, Encyclopædia Universalis*, Paris, 1997, p. 537–561.
101. Articles published in Japanese (*Japanese Dictionary of History of Sciences*): «Arabic Mathematics», «Arabic science», Tokyo, 1998.
102. «Hawla tārīkh al-'ulūm al-'arabiyya» (in Arabic), *al-Mustaqbal al-Arabī*, 231 (mai 1998), p. 19–29.
103. «Al-'ulūm al-'arabiyya bayna naẓariya al-ma'rifa wa-al-tārīkh» (in Arabic), *Bulletin d'Études Orientales*, Tome L, 1998, IFEAD, Damas, p. 223–232.
104. «Al-Qūhī *vs.* Aristotle: On motion», *Arabic Sciences and Philosophy*, 9.1, 1999, p. 7– 24.
105. French Version: «Al-Qūhī contre Aristote : sur le mouvement», *Oriens-Occidens Sciences, mathématiques et philosophie de l'Antiquité à l'Âge classique*, 2 (1998), p. 95–117.
106. «Nasha'at al-lugha al-'arabiyya al-'ilmiyya wa-taṭawwuruhā» (in Arabic), *al-Mawsim al-thaqafī al-sādis 'ashar*, Amman, 1998, p. 121–138.

107. «Combinatoire et métaphysique: Ibn Sīnā, al-Ṭūsī et al-Halabī», in R. Rashed and J. Biard (eds.), *Les Doctrines de la science de l'antiquité à l'âge classique*, Leuven: Peeters, 1999, p. 61–86.
108. German translation «Kombinatorik und Metaphysik: Ibn Sīnā, aṭ-Ṭūsī und Ḥalabī», dans Rüdiger Thiele (Hrg.), *Mathesis, Festschrift siebzigsten Geburtstag von Matthias Schramm*, Berlin, Diepholz, 2000), p. 37–54.
109. English translation «Metaphysics and Mathematics in Classical Islamic Culture: Avicenna and his Successors", in *God, Life, and the Cosmos*, p. 151–171.
110. «Sur une construction du miroir parabolique par Abu al-Wafā' al-Buzjānī» (with Otto Neugebauer), *Arabic Sciences and Philosophy*, 9.2, 1999, p. 261–277.
111. «Ibn al-Haytham, mathématicien de l'époque fatimide», in *L'Égypte fatimide. Son art et son histoire*, Actes du colloque organisé à Paris les 28, 29 et 30 mai 1998, sous la direction de Marianne Barrucand, Paris, Presses de l'Université de Paris-Sorbonne, 1999, p. 527–536.
112. «Turath al-fikr wa-turath al-nass: Makhtutat al-'ilm al-'arabiyya», in *Tahqiq makhtutat al-'ulum fi al-turath al-islami*, Actes du 4e Congrès de Furqan Islamic Heritage Foundation—29–30 novembre 1998, London, 1998, p. 29–76;
113. English version «Conceptual Tradition and Textual Tradition: Arabic Manuscripts on Science», in Y. Ibish (ed.), *Editing Islamic Manuscripts on Science,* Proceedings of the Fourth Conference of al-Furqan Islamic Heritage Foundation (London 29th–30th November 1997), London, al-Furqan, 1999, p. 15–51.
114. «Fermat et les débuts modernes de l'analyse diophantienne», *Historia Scientiarum*, vol. 9–1, 1999, p. 3–16.
115. «De la géométrie du regard aux mathématiques des phénomènes lumineux», in G. Vescovini (ed.), *Filosofia e scienza classica, arabo-latina medievale e l'età moderna*, Fédération Internationale des Instituts d'Études Médiévales (FIDEM), Textes et études du Moyen Âge, 11, Louvain-la-Neuve, 1999, p. 43–59.
116. «Analyse diophantienne», «Analyse et synthèse», «Isopérimètre», in *Dictionnaire d'histoire et philosophie des sciences*, sous la direction de Dominique Lecourt, Paris : PUF, 1999, resp. p. 45–47, p. 47–49, p. 550–552.
117. «The Invention of Classical Scientific Modernity», *Revista Latinoamericana de Historia de las Ciencias y la Tecnología*, vol. 12, núm. 2, mayo-agosto de 1999, p. 135–147.
118. «Ibn Sahl et al-Qūhī: Les projections. Addenda & Corrigenda», *Arabic Sciences and Philosophy*, vol. 10.1, 2000, p. 79–100.
119. «Thābit b. Kurra», *Encyclopédie de l'Islam*, p. 459–460.
120. «Astronomie et mathématiques anciennes et classiques», *Épistémologiques* (Revue internationale Paris/Sao Paulo): *Cosmologie et philosophie, hommage à Jacques Merleau-Ponty*, vol. I (1-2), janvier-juin 2000, p. 89–100.

121. «Fermat and Algebraic Geometry», *Historia Scientiarum*, 11.1, 2001, p. 24–47.
122. «Al-Quhi: From Meteorology to Astronomy», *Arabic Sciences and Philosophy*, 11.2, 2001, p. 157–204.
123. «Scienze "esatte" dal greco all'arabo: transmissione e traduzione», in *I Greci Storia Cultura Arte Società*, 3. *I Greci oltre le Grecia*, a cura di Salvatore Settis, Torino, Giulio Einaudi Editore, 2001, p.705–740.
124. «Diofanto di alessandria», *Storia della scienza*, vol. I: *La scienza antica*, *Enciclopedia Italiana*, 2001, p. 800–805.
125. «Al-bu'd al-'ilmî fî turâth al-thaqâfî al-'arabî bayna târîkh wa nazariya al-ma'rifa», Damascus, Institut Français de Damas.
126. «Transmission et innovation: l'exemple du miroir parabolique», dans *4000 ans d'histoire des mathématiques: les mathématiques dans la longue durée*, Actes du treizième colloque Inter-IREM d'Histoire et d'Epistémologie des mathématiques, IREM de Rennes, les 6-7-8 mai 2000, IREM de Rennes, 2002, p. 57–77; short version in S.M.Razaullah Ansari, *Science and Technology in the Islamic World, Proceedings of the XXth International Congress of History of Science* (Liège, 20–26 July 1997), vol. XXI, coll. De Diversis Artibus, Turnhout, Brepols, 2002, p. 101–108.
127. Articles in *Storia della scienza*, vol. III: *La civilta islamica*, *Enciclopedia Italiana*, Roma, 2002: «Dal Greco all'Arabo: trasmissione et traduzione», p. 31–49; «Algebra e linguistica, gli inizi dell'analisi combinatoria» p. 86–93; «Le tradizioni matematiche», p. 322–326; «Gli Archimedei e i problemi infinitesimali», p. 360–385; «Le tradizioni sulle coniche e l'inizio delle ricerche sulle proiezioni (co-author)», p. 385–402; «Tracciato continuo delle coniche e classificazione delle curve», p. 402–431; «Aritmetiche euclidea, neopitagorica e diofantea: nuovi metodi in teoria dei numeri», p. 448–457; «L'algebra e il suo ruolo unificante», p. 457–471; «I metodi algoritmici», p. 472–483; «Filosofia della matematica», p. 483–498; «Specchi ustori, anaclastica e diottrica», p. 561–579.
128. «Al-Qūhī et al-Sijzī: sur le compas parfait et le tracé continu des sections coniques», *Arabic Sciences and Philosophy*, 13.1, 2003, p. 9–44.
129. «Inaugural Lecture: History of Science and Diversity at the Beginning of the 21st Century», in Juan José Saldana (ed.), *Science and Cultural Diversity*, Proceedings of the XXIst International Congress of History of Science (Mexico City, 7–14 July 2001), Mexico, 2003, vol. I, p. 15–29.
130. «Les mathématiques de la terre», in G. Marchetti, O. Rignani et V. Sorge (eds.), *Ratio et superstitio*, Essays in Honor of Graziella Federici Vescovini, Textes et études du Moyen Âge, 24, Louvain-la-Neuve : FIDEM, 2003, p. 285–318.
131. «Philosophie et mathématiques: interactions», *Bulletin UTCP* (University of Tokyo Center for Philosophy), volume 1, 2003, p. 66–76.
132. «Fibonacci et le prolongement latin des mathématiques arabes», *Bollettino di Storia delle Scienze Matematiche*, Anno XXIII, Numero 2, Dicembre 2003, Pisa-Roma, Istituti Editoriali e Poligrafici Internaziolali, MMV, p. 55–73.

133. «Philosophie et mathématiques selon Maïmonide: Le modèle andalou de rencontre philosophique», in *Maïmonide, philosophe et savant (1138–1204)*, Études réunies par Tony Lévy et Roshdi Rashed, Ancient and Classical Sciences and Philosophy, Leuven: Peeters, 2004, p. 253–273.
134. «Thābit ibn Qurra et la théorie des parallèles» (in collaboration with Ch. Houzel, *Arabic Sciences and Philosophy*, 15.1, 2005, p. 9–55.
135. «Les premières classifications des courbes», *Physis*, XLII.1, 2005, p. 1–64.
136. «La modernité mathématique: Descartes et Fermat», in R. Rashed and P. Pellegrin (eds.), *Philosophie des mathématiques et théorie de la connaissance. L'Œuvre de Jules Vuillemin*, Collection Sciences dans l'histoire, Paris: Librairie A. Blanchard, 2005, p. 239–252.
137. «Les ovales de Descartes», *Physis*, XLII.2, 2005, pp. 325–346.
138. «The Celestial Kinematics of Ibn al-Haytham», *Arabic Sciences and Philosophy*, 17, 1 (2007), p. 7–55.
139. «The Configuration of the Universe: a Book by al-Ḥasan ibn al-Haytham?», *Revue d'Histoire des Sciences*, t. 60, no. 1, janvier-juin 2007, p. 47–63
140. «Greek into Arabic: Transmission and Translation», in James E. Montgomery (ed.), *Arabic Theology, Arabic Philosophy. From the Many to the One: Essays in Celebration of Richard M. Frank*, Orientalia Lovaniensia Analecta 152, Leuven-Paris: Peeters, 2006, p. 157–196.
141. «Arabic Versions and Reediting Apollonius' *Conics*», in *Study of the History of Mathematics*, Research Institute for Mathematical Sciences, Kyoto University, Kyoto, Avril 2007, p. 128–137.
142. «Lire les anciens textes mathématiques : le cinquième livre des *Coniques* d'Apollonius», *Bollettino di storia delle scienze matematiche*, vol. XXVII, fasc. 2, 2007, p. 265–288.
143. «L'étude mathématique du lieu», in *Oggetto e spazio. Fenomenologia dell'oggetto, forma e cosa dai secoli XIII-XIV ai post-cartesiani*, Atti del Convegno (Perugia, 8–10 settembre 2005), a cura di Graziella Federici Vescovini e Orsola Rignani, Micrologus' Library 24, Firenze, Sismel-Edizioni del Galluzzo, 2008, p. 71–79.
144. «Le concept de tangente dans les *Coniques* d'Apollonius», in *Kosmos und Zahl. Beiträge zur Mathematik- und Astronomiegeschichte, zu Alexander von Humboldt und Leibniz*, Berlin, 2008, p. 361–371.
145. «The Arab nation and indigenous acquisition of scientific knowledge (*tawṭīn al-'ilm*)», *Contemporary Arab Affairs*, vol. 1, no. 4, October 2008, p. 519–538.
146. «Thābit ibn Qurra, Scholar and Philosopher (826-901)», p. 3–13; «Thābit ibn Qurra: From Ḥarrān to Baghdad», p. 15–24; «Thābit ibn Qurra et la théorie des parallèles» (in collaboration with Ch. Houzel), p. 27–73; «Théorie des nombres amiables» (in collaboration with Ch. Houzel), p. 77–151; «Résolution géométrique des équations du second degré», p. 153–169; «Thābit ibn Qurra et l'art de la mesure», p. 173–209; «Lemmes géométriques de Thābit ibn Qurra», p. 211–253; in *Thābit ibn Qurra. Science and*

Philosophy in Ninth-Century Baghdad, Scientia Graeco-Arabica, vol. 4, Berlin/New York, Walter de Gruyter, 2009.
147. «The Philosophy of Mathematics», in Shahid Rahman, Tony Street, Hassan Tahiri (eds.), *The Unity of Science in the Arabic Tradition, Science, Logic, Epistemology and their Interactions*, vol. 11, Springer, 2008, p. 153–182.
148. «Les constructions géométriques entre géométrie et algèbre: l'Épître d'Abū al-Jūd à al-Bīrūnī», *Arabic Sciences and Philosophy*, 20.1 (2001), p. 1–51.
149. «Sur un théorème de géométrie sphérique: Théodose, Ménélaüs, Ibn 'Irāq et Ibn Hūd» (in collaboration with Mohamad al-Houjairi), *Arabic Sciences and Philosophy*, 20.2 (2001), p. 207–253.
150. «Le pseudo-al-Hasan ibn al-Haytham: sur l'asymptote», in R. Fontaine, R. Glasner, R. Leicht and G. Veltri (eds.), *Studies in the History of Culture and Science. A Tribute to Gad Freudenthal*, Leiden/Boston, Brill, 2011, p. 7–41.
151. «L'asymptote : Apollonius et ses lecteurs», *Bollettino di storia delle scienze matematiche*, vol. XXX, fasc. 2, 2010, p. 223–254.
152. «Mathématiques», in Pierre Pellegrin *et al.* (eds.), *Le savoir grec: Dictionnaire critique*, Paris: Flammarion, 2011, p. 447–469.
153. «Le concept de lieu: Ibn al-Haytham, Averroès», in Ahmad Hasnawi (ed.), *La lumière de l'intellect. La pensée scientifique et philosophique d'Averroès dans son temps*, Paris/Louvain: Peeters, 2011, p. 3–9.
154. «Founding Acts and Major Turning-Points in Arab Mathematics», in J. Z. Buchwald (ed.), *A Master of Science History: Essays in Honor of Charles Coulston Gillispie*, Archimedes 30. New Studies in the History and Philosophy of Sciences and Technology, Dordrecht/Heidelberg/London/New York, Springer, 2012, p. 253–271.
155. «L'angle de contingence : un problème de philosophie des mathématiques», *Arabic Sciences and Philosophy*, 22.1, 2012, p. 1–50.
156. «History of Science at the Beginning of the 21th Century», in Liao Yuqun *et al.* (ed.), *Multi-cultural Perspectives of the History of Science and Technology in China*, Proceedings of the 12th International Conference on the History of Science in China, Beijing, China, Science Press, 2012, p. 16–21.
157. «Etudes et travaux: Otto Neugebauer (1899–1990)» (in collaboration with Lewis Pyenson), *Revue d'histoire des sciences*, tome 65-2, juillet-décembre 2012, p. 381–394.
158. «Otto Neugebauer, historian» (in collaboration with Lewis Pyenson), *History of Science*, 1, 2012, p. 402–431.
159. «Qu'est-ce que les *Coniques* d'Apollonius?», in *Les Courbes: Études sur l'histoire d'un concept*, edited by Roshdi Rashed and Pascal Crozet, Collection Sciences dans l'histoire, Paris: Librairie A. Blanchard, 2013, p. 1–16.
160. «Descartes et l'infiniment petit», *Bollettino di storia delle scienze matematiche*, vol. XXXIII, Fasc. 1, 2013, p. 151–169.
161. «Abū Naṣr ibn 'Irāq: 'indamā kāna al-Amīr āliman (When the Prince was a scientist», *al-Tafahom*, 40, 2013, p. 145–170

162. «Philosophy and Mathematics: Interactions», *Physis*, p. 243–259.
163. «On Menelaus' Spherics III.5 in Arabic Mathematics, I : Ibn ʿIrāq» (in collaboration with Athanase Papadopoulos), *Arabic Sciences and Philosophy*, 24.1, 2014, p. 1–68.
164. «On Menelaus' Spherics III.5 in Arabic Mathematics, II: Naṣīr al-Dīn al-Ṭūsī and Ibn Abī Jarrāda» (in collaboration with Athanase Papadopoulos), *Arabic Sciences and Philosophy*, 25.1, 2015, p. 1–33.
165. «I problemi impossibili in numeri razionali e i problemi inaccessibili», *La Matematica nella Società e nella Cultura, Rivista della Unione Matematica Italiana*, Serie I, vol. VIII, Agosto 2015, p. 279–312.
166. «Al-Ḥasan ibn al-Ḥasan al-Haytham (in Arabic)», *Riwaq of History and Heritage*, n 1, Janvier 2016 (Qatar), p. 6–17.
167. «Avicenne, "Philosophe analytique" des mathématiques, *Les Études philosophiques*, April 2016-2, p. 283–306.
168. «Ibn al-Haytham's Scientific Research Programme», in M. Alamri, M. El-Gomati and M. Suhail Zubairy (eds.), *Optics in Our Time*, Springer, 2016, p. 25–39.
169. «Ptolemy, Ibn al-Haytham, and al-Farisi: the beginnings of quantitative research in optics», dans Ana Maria Cetto, Maria Teresa Josefina Pérez de Celis Herrero (eds.), Light Beyond 2015, Luz mas alla de 2015, Univ. Nacional Autonoma de Mexico, 2017.
170. «Ibn al-Haytham's scientific research program», in Rashed et alii. (eds.), Light-Based Science, CRC Press, Taylor and Francis Group, Boca Raton, London, New York, 2018, pp. 3–8.
171. «Ibn al-Haytham's Problem» (in collaboration with Pierre Coullet), in Rashed et alii. (eds.), Light-Based Science, CRC Press, Taylor and Francis Group, Boca Raton, London, New York, 2018, pp. 109–121.

Introduction: Interview with Roshdi Rashed by Hassan Tahiri

First things first, let us start with your education in your home country. You were born in the famous Egyptian capital Cairo in 1936. What was your educational training there?

The nature of my training was threefold: classical academic courses, extra-academic activities and training in science. After high school I decided to study philosophy at Cairo University, despite already having some familiarity with mathematics. At the same time, I felt the need to acquire a classical education, mainly in linguistics, with who was then the greatest linguist in Egypt, Maḥmud Shākir. I also undertook a variety of extra-university training. During my second year in philosophy at Cairo University, I decided to return to studying mathematics and it was at that time that I was allowed to join the Faculty of Science at Cairo University.

Can you briefly describe for us the intellectual and social background of your education at that time, i.e. Egypt of the 1940s and 50s?

In Egypt of the 40s and 50s, there was already a pretty active and dynamic intellectual milieu. A young outward-thinking man could not remain indifferent to the changing Egyptian environment of the time. There were very real opportunities for excellent training in many disciplines, including linguistics and science. In such an intellectual environment, a young student in Cairo could immerse himself in all the intellectual and philosophical paradigms that were gaining global momentum. Therefore, it was not necessary to be initiated or particularly gifted to participate. Philosophical training was another matter. Egyptian professors were neither the best nor the worst. They were academics who had obtained their Doctorat d'Etat at the Sorbonne, so they were equal to other professors, but none of them could be considered true philosophers. This was the situation within universities but, outside those confines, some real philosophers could be found. Back then there was a substantial difference with the current situation; some students had a good secondary education that enabled them to learn at least two languages in addition to Arabic. They could work both in English and French, for example, and even sometimes in German. Scientific teaching, such as in mathematics, was good from the 1930s to 1950s, i.e. prior to reform and the so-called ‘modernisation’. This was the academic atmosphere at that time, but there was a contradictory situation that oscillated between ultra-democracy and dictatorship. This was really influential on the formation of personality, especially those of the young people at the time. Then there was the military coup of July 1952. Personally, I was in opposition from the outset for the very simple reason that I did not think that a military coup had any capacity or competence to transform a civil society. Then, the Suez Canal was nationalised, which reopened some perspectives, but I had already left Egypt by that stage. So, that was the scenario in Egypt when I left and before coming to Paris in September 1956, i.e. shortly before the tripartite invasion of Egypte.

But why Paris in particular? And following your arrival, what were your first impressions of and steps in the French academic system?

Paris was a deliberate choice. I had been simultaneously accepted into Oxford, but I preferred to come to Paris because it was widely considered a place of freedom, where it was not necessary to join a particular system. The decision was also influenced by tradition; Egyptian intellectuals, in philosophy or other fields, were heirs to extensive training in the French tradition in Paris or even in Cairo given by motivated professors such as André Lalande and Alexandre Koyré. When I arrived in Paris, I was a little bit disappointed by the Sorbonne because, apart from some courses such as those of Georges Canguilhem, I found that the teaching of logic and the philosophy of science was not at a very advanced level. Teaching of the history of philosophy was more convincing. I also followed courses outside the Sorbonne, such as those of Merleau-Ponty and Gueroult. Despite the student overcrowding and poor conditions at the Faculty of Science, it was a great pleasure to attend the high quality courses of Godement and Cartan. Attending courses of this calibre of professor allowed me to discover types of mathematics other than simple analysis. I discovered modern or abstract algebra, number theory, and algebraic geometry, which I had to study because of my work on Diophantus. I remember luminous lessons, which were not simply intended to transmit knowledge. This teaching was really a discovery because it covered topics not taught in Cairo and on which I previously had no idea.

What was the topic of your thesis?

When I arrived in Paris, there were two professors who taught logic and philosophy of science at the Sorbonne: René Poirier and Georges Canguilhem, who had just been appointed. I prepared my Ph.D. with René Poirier and I wanted to work on a specific topic, but its first formulation was clumsy: “L’objectivité de la loi sociologique” (The Objectivity of Sociological Law). I also wanted to attend Canguilhem’s lectures. Two years later, i.e. in 1959, a friend convinced me to go to the Institute of the History of Science at rue du Four, and it is there that I started working with Canguilhem.

Can you tell us more about your doctoral training and the topic of your thesis, its motivation and what you were aiming to achieve?

I need to go back a little to my training as a young student to answer that question. The topic of *L'objectivité de la loi sociologique* was motivated by a combination of logical concerns and was the project of a young Egyptian living in an intellectual milieu where all kinds of philosophical and social doctrines could be discussed and who refused to be a Marxist. Under the influence of mathematics, the title of the thesis was later changed to "La mathématisation des doctrines informes" (The Mathematisation of Shapeless Doctrines). It was also at this time that I seriously started studying probability for my thesis with my friend Salah Ahmad who was a probabilist. As indicated by the title, the main thrust of the thesis was to examine the conditions under which mathematics could be applied to an area that has no elaborate theory and no precise or controlled concepts, either syntactically or semantically? In other words, to what extent is the application of mathematics possible in that area? This question is not similar to the Kantian question of the possibility of knowledge. For the knowledge that concerned Kant, there is the fully-developed theory based on physics. Me, I wanted to take a domain where there was no theory, such as the humanities. Social sciences were indeed a very apt field to raise this question. I was very convinced that to do a real epistemological work and not indefinitely repeat what had been done before, this type of question needed to be asked. My experience of working in sociology, social psychology and economics (I also did a degree in economics), allowed me to examine the question in its full generality from the philosophical point of view. I reviewed all or almost all the applications of mathematics in social psychology. I studied factorial analysis. This was a massive undertaking, and it was at that time that I wrote the book on Condorcet. I felt it was not enough to do this work from a simple comparative perspective, i.e. it would not be sufficient to analyse selected examples. Instead, the whole tradition in this area needed to be reconstituted, i.e. the entire tradition of the application of mathematics to social realities. That is why I went back to the 18th century and to authors like Condorcet and others. My thesis was equally a work on the history of probability, inspired by the Bernoulli brothers. I have written about 600 pages on this topic that have never been published. However, I found that mathematics was always external, i.e. though it has some effectiveness and allows certain phenomena to be examined and models to be built for some of them, it always remains external and without any real power of prediction. It was at this time that I started to turn away from this area.

You are now widely known more as a philosopher and historian of mathematics mainly in the Arabic-Islamic tradition and it is interesting, not to say surprising, to find out that your thesis has little to do with it. The intriguing question is how did you come across this tradition and its mathematicians?

One of the most interesting aspects of my doctoral thesis was the question as to whether there are similar historical situations? My investigations began with mechanics, and it was then that I read the analyses of Pierre Duhem, Annelise Maier, Marshall Clagett, Alexandre Koyré and others, with whom I had some discussions after attending their seminars. I was then working on Tartaglia and his contemporaries of the 16th century, and the study of mechanics was essential to answering the question I had always asked myself regarding *La mathématisation des doctrines informes*. I encountered references to Arab scientists while working on mechanics but, to be honest, I was not greatly interested by them and one day I just turned to optics. From that moment I became interested in Arab science as part of my research on situations similar to those I had met in the social sciences. I decided to do a complementary thesis (as required by the French academic system at the time), on Alhazen (or Ibn al-Haytham) entitled "L'optique d'Alhazen, problèmes de clivage entre histoire et préhistoire des sciences" (Alhazen's Optics, the Boundary Problem Between History and Prehistory of Science). This was the topic I proposed to Canguilhem, who accepted it immediately. It was with this double theme, "La mathématisation des doctrines informes" and "L'optique d'Alhazen", that I presented my candidacy to the CNRS.

How from Alhazen's optics did you become more heavily involved in, what seems to be a turning point in your career, the study of mathematics as such and of its history during the classical Arabic-Islamic period, and not simply as an applied discipline?

There are several reasons for this evolution. One is purely part of the logic of research and another has little to do with it. At the research level, the discovery of optics, following my work on Alhazen and al-Fārisī that was published in 1968, prompted my interest in Arab science and led me to focus on the history of mathematics. There was also the fortuitous event in Istanbul that led me to come across another Arabic mathematician. While waiting for the manuscript of Alhazen's optics at Süleymaniye's Library, I stumbled upon a reference to al-Samaw'al's book of algebra, a mathematician of the 12th century. I asked for it to find out what it was about and didn't believe what I read. I learned that his was the algebra of the 16th and 17th centuries. I asked for a microfilm with the idea of working on it, but was not at all sure that the task was for me. I had come from Canguilhem's tradition, where there was no training in the critical editing of historical texts. In fact, I didn't even know what that entailed. And for me personally, editing an Arabic text did not interest me at all since what does it mean to edit text in a language that is your mother tongue? Nevertheless, I decided to undertake this work with the collaboration of my friend Salah Ahmad who was living in Damascus. There, I invented strange editing rules, believing that Arabic being a living language, there was no need to put variants. The book of al-Samaw'al was edited like this. I was surprised when it was later reviewed by someone with very little knowledge of Arabic, who played the scientist, but published untruths on the work. I then realised the importance of the history of manuscripts. I really couldn't see the point of such work until my discovery of the book by Diophantus in the early 1970s. I began to treat Diophantus' work in much the same way, but I realised that if I continued like this I would just end up sabotaging my own work. It was only later that I actually began to understand the importance of the history of texts for their critical editing, especially when it came to unique manuscripts such as

Diophantus' text where if a word is changed (especially when it is a translation from Greek) it can have a huge impact. The scholar who helped me to become aware of the importance of these things was André Allard, who is both a Hellenist and Latinist. Together, we studied the Greek text of Diophantus and I then saw all the problems. It was at that time, quite late, that I realised the need to very seriously address the issue of critical editing. I remember that whenever I had a conference in the Institut d'Histoire des Sciences of rue du Four (I gave at least six a year), and I argued something drawn from the texts, I was asked for the evidence. Learn Arabic was my spontaneous reply! I then realised that to manage this evidence, a whole library of basic texts should be established for the domain to exist. Apart from the anecdotal element, it was this awareness of the need to get the editing work done that has prevailed. I also felt the need to find special rules for the critical editing of Arabic texts. This work began with Diophantus and I haven't stopped developing and refining standards. However, I must confess that I have always performed the hard work of collecting, editing, commenting on and publishing Arabic manuscripts without enthusiasm. I felt that it was not my job. But, in any given historical period, we do not choose the tasks that need to be tackled; each period determines its own requirements. The work needed to be done and I did it.

With the discovery of al-Samaw'al's work, it is obvious that one of the aims of your later work is to provide a new understanding of the development of classical mathematics. One of the key concepts you have identified in your investigations is that of a specific mathematical tradition, which sheds new light on Arabic mathematical practice and its great influence on wider scientific practice due to the unprecedented expansion of mathematics since al-Khwārizmī. Can you tell us more about the emergence of the little-known rich mathematical traditions that greatly opened up the scope of mathematics and mathematical practice, their interactions, transformation and extension?

With al-Samaw'al, the idea that immediately came to my mind was the arithmetisation of algebra. In other words, at some point, algebraists began to apply arithmetic operations to algebraic expressions, thereby defining the notion of polynomial and determining its conditions and examining its implications. This was something that nobody had noticed before. There is a major corollary that has far-reaching implications. In the history of algebra, there are traditions; to understand al-Samaw'al, it is necessary to study his predecessor al-Karajī, and to understand al-Karajī, it is necessary to study his predecessors, Abū Kāmil and al-Khwārizmī. There is a tradition that began with al-Khwārizmī and culminated in the arithmetisation of algebra and the development of abstract algebraic calculation. What kind of transformations has this tradition undergone? When did it end? Quite naturally I looked to Italian algebra and to the German Cossists up to Michael Stifel (who was not doing anything particularly different from Arab algebraists). We can identify a particular pattern in topics treated by these mathematicians. If we take the books of al-Karajī for example, they end with a chapter on Diophantine analysis. If we take Euler's algebra textbook, we find the same pattern. Of course I am not saying that Euler is al-Karajī. However, I am saying that either there is a tradition or there are others or there is none; each time there is a specific case to consider. We have the tradition of arithmetical algebra. We have another tradition of algebraic geometry and we can look at when and how it started, what scheme it followed, and what changes it underwent. It was this conviction that led me to discover the work of al-Khayyām's successor, Sharaf al-Dīn al-Ṭūsī. Prestigious researchers at the London India Office Library examined the manuscript of al-Ṭūsī, but nobody was able to understand its content. The key to understanding the work of the mathematicians since al-Khwārizmī is to identify these traditions in algebra, algebraic geometry, number theory, etc. However, it can happen that the epistemic features specific to a tradition can surface in other places and other cultural eras. Is this due to influences, a spontaneous generation, or an internal logic of the development of the research under consideration? All these questions are open and there is no single answer, but this community of features and scientific practices could only lead to the concept of the so-called "classical mathematics". The same can be said for al-Khāzin. I was not the only researcher interested in his work but, to really measure his contribution, I had to rewrite the history of another tradition, i.e. Diophantine integer analysis. Once I had found the lost part of Diophantus' work, I started to compare it with that of other mathematicians: al-Khāzin, al-Karajī, etc. I was then able to see the point at which a tradition both in algebra and number theory formed. Diophantine analysis then split into two elements from the tenth century; the rational Diophantine analysis that is part of algebra, and the Diophantine integer analysis that inaugurated a new tradition in number theory. Al-Khāzin is the founder of the geometrisation of algebra since, according to what we know, he was the first to solve a cubic equation by conic intersection. Others then followed; Abū al-Jūd really advanced things until al-Khayyām provided a complete articulated theory. In this tradition, the successors of al-Khayyām did not simply follow his work since they also made other advances; for example, it is certain that al-Ṭūsī brought a lot of novelty. Al-Khāzin is a very interesting mathematician since, in his work, we have a good example illustrating the kind of interactions between the various traditions. Of course, a mathematician can work in different areas belonging to different traditions; though, in the case of al-Khāzin, he did not have the same impact in both traditions.

The concept of tradition enables you to sharpen the study of the development of mathematics until the 17th century that illuminates, among other things, the link between the Arabic and European mathematical traditions. From your extensive study of the results achieved in both traditions, you have concluded that knowledge of Arabic mathematical works is necessary to understand what happened in the 17th century. In what way? And how can the comparison of two successive traditions effectively explain the actual increase in mathematical knowledge?

Because the traditions within which these works arose are spread over a long period, when we consider the novelty of these scientific works a question arises as to when these traditions stop. On the other hand, the strong similarities between the works of the mathematicians of the 11th, 12th and 13th centuries and those of the mathematicians of the 16th and 17th centuries immediately raise the question:

aren't the results obtained in the 16th century a natural continuation of the previous work conducted in the 11th and 12th centuries? These two issues are closely related, but how can we clarify them? By trying to answer this very specific question: if there is something new occurs in the development of a subject matter, where do we place it exactly? For example, what is new in Fermat's number theory? If I say its algebrisation, this does not explain Fermat's novelty because this algebrisation had already been done before him. So the question is what is specific, what is really new in Fermat's work that enabled him to go further. This novelty exists. It is the method of infinite descent. For Descartes, it is pretty much the same; if what he did was just inherited from al-Khayyām, how then do we explain some new development? We can answer that Descartes speaks of an algebraic curve, while al-Khayyām remains at the level of conics. That is why I insisted on the distinction between mechanical and geometric curves in my study on Descartes' work. It is not the place here to go into details. The question of timing matters. Why at this moment has this particular thing occurred? In that sense, knowing Arab mathematics is absolutely essential both to pose these crucial historical questions that nobody else had and to specifically localise novelty. When you know what others have done, you know if the same tradition applies. There is logic in the development of mathematics that requires further extension of the traditions at work. Anyway, for the 17th century, it is impossible to really see and recognise the influence of Arabic works, at least at the mathematical level, without knowing what happened at the time. We can ask whether Descartes had read al-Khayyam or if he knew his work through the mathematician and orientalist Jacob Golius or his son; all this is possible and does not matter too much. What I mean is that it is necessary to read al-Khayyām and to read al-Ṭūsī before tackling the study of Descartes' geometry. For it comes down to understanding what made Descartes' geometry his, otherwise we cannot understand wherein lies its novelty, since everything is not new. In another way, I cannot agree with those experts of the 17th century when they claim that Fermat's novelty is the algebrisation of number theory, given what I know about what was written before his work. So, it turns out that besides interest in the history of Arab mathematics in itself, its epistemic significance is further demonstrated by the fact that the mathematical traditions of the 16th and 17th centuries cannot be understood without knowing this mathematics.

Since the end of the 1980s, you have published an impressive series of volumes on infinitesimal mathematics. How did you come about the idea of this great research project and why did you focus so much on geometry?

By the end of the 1980s, geometry arose as the subject that immediately needed to be tackled. With my work on al-Khayyām and al-Ṭūsī, I understood that the history of algebra was needed to actually see how things were already developing in geometry; for example, determining the knowledge that mathematicians of that time had of conics and, above all, the presence of algebraic transformations that could only be suggested by geometrical transformations. Because of many problems of this kind, I found it necessary to look at the work of geometers. During my work on al-Ṭūsī, particularly on the second part of his treatise where he introduced analytical notions, I felt that it was necessary to look first at the work of infinitesimalists. It is always the same story. I start working on an author, in this case Ibn al-Haytham, and I immediately discover that he is part of a tradition and that, to understand him, all that tradition should be reconstituted. Hence, my project on Arab infinitesimal mathematics and the publication of a series of volumes. In algebra too, there were problems of geometrical construction that algebraists had tackled from the outset and tried to translate into equations; the problem of the trisection of the angle, the regular heptagon, the line of Archimedes, etc. Similarly, my research on the theory of conics. What brought that about? Is it come down to simply read Apollonius' work and apply it or was there something more to it? I was led to the study of a much more general theme, i.e. that of the theory of conics and its applications. There are two corpuses to which I have committed this type of research since the 1980s (*Mathématiques infinitésimales*, 5 volumes; *Apollonius' Conics*, 6 volumes). It was during my research on the theory of conics that I identified the emergence of another tradition, i.e. that of geometers who began to think about geometric transformations and projections. My research on Arabic geometry started quite late because, on the one hand, it arose from my work on algebra and, on the other hand, because of a prejudice. This prejudice is that of everyone. While algebra and number theory were certainly new chapters in Arab mathematics, in geometry the Arabs had to repeat what the Greeks had done; this was what had been portrayed in all mathematical textbooks and I ended up believing it. But when I started working on Fermat and got to know the works of geometers such as Thābit ibn Qurra and al-Qūhī, I realised that this geometry is not a replica of that of the Greeks. It is certainly linked to Greek geometry, but it gave rise to developments in new directions that did not previously exist. It is a constant intellectual work; when we begin to see the landscape differently prejudices fall away, almost effortlessly.

In recent years, you have begun to be more interested in the relationship between mathematics and philosophy. It is not by chance that this is the topic of our conference. Your present contribution on Ibn Sīnā is your latest work on what seems to be a still unexplored research field. Besides Ibn Sīnā, you have studied some significant examples like al-Kindī's use of the axiomatic model to prove the inconsistency of the concept of an infinite body, al-Ṭūsī's application of combinatorial analysis to solve the problem of the emanation of celestial intelligences or Ibn al-Haytham's work on analysis and synthesis. What is the idea behind this line of research and what are you aiming to show?

This is a fairly large area where I have started to make my first steps. We often read here and there a summary of the ideas of the philosophers on mathematics or those of mathematicians on their relationship to philosophy. This topic is quite poorly defined. The question should be asked differently in two respects: what is the living part of philosophy in mathematics; and is this living part the work of philosophers or mathematicians? The two issues are not independent. Where do mathematics and philosophy actually meet? I have asked myself these two questions. For the second question, two kinds of interactions can be found: mathematics has been used to solve philosophical problems; and philosophy was conceived as the method to solve mathematical problems. So I tried to find an example for each of these issues tackled either by a mathematician or a philosopher. In fact, it was often a mathematician-philosopher who dealt with such problems. This line of research began with the study of asymptotic behaviour and the use made of philosophical concepts to reflect on its paradoxical status. I then studied the metaphysical doctrine of emanation and the use made of combinatorial analysis to assess its possibilities under the constraint of certain principles. I have made some attempts to examine what concept philosophers had of mathematics; for example, what concepts they were using in their classifications of the sciences or in their ontological doctrines. My attempts are motivated by the conviction that the philosophy of that time can be reduced neither to a theory of the soul nor a theory of being, but that there was a philosophy of science as well-conceived as that of today and especially a philosophy of mathematics that is a fundamental part of philosophy. I will provide just two examples. Consider the analysis of asymptotic behaviour. This is a topic on which books of logic, mathematics and philosophy were written. The authors of those books were, for the most part, mathematicians, but sometimes also philosophers. Unfortunately some of these books are now lost. This means that the topic was widely recognised so that a mathematician or a philosopher (provided that he was informed of mathematics), could make their contribution. A topic like this was classified as a branch of mathematics, as well as a branch of philosophical teaching. The second example is that of analysis and synthesis to which mathematicians have devoted substantial treatises. It is important to attract attention to this topic as it is an interdisciplinary domain where logicians can find logic, mathematicians mathematics, and philosophers philosophy. This is what particularly kept mathematicians-philosophers away from Aristotelian logic. Mathematicians like Ibrahim ibn Sinān or Ibn al-Haytham did not feel the need to study the *Organon* of Aristotle to write about the problems of logic since they could do that from the study of logical themes generated by the development of their own subject matter such as asymptotic behaviour and analysis and synthesis. So a change of perspective took place, which was not comprehensive yet still very important. More generally, I think at that time, a new form of ontology began because of the development of mathematics and the diversity of the new disciplines. At that time, there was an attempt to develop unitary theories for the mathematical sciences that had multiplied and diverged, but we now know that there was no way to establish such theory because the algebra of the time couldn't do it. However, a change in the conception of mathematical objects was obviously brought about by algebra, but also by some new developments in geometry. When considering geometrical transformations for example, it is clear that the notion of place involved was not, and could not, be that of Aristotle. All this brings me to a hypothesis that we have developed a formal way of conceiving both mathematical objects and the unity of disciplines.

Your monumental work on the history of mathematics has enabled you to identify one of the main problems of the prevailing periodisation that sharply cuts the history of science into the three traditional periods: ancient, medieval, modern. You have argued that this "global look is necessarily deceptive, unsuited to depicting the landscape of early modern mathematics, and especially to tracing the history of its various components" and suggested instead a more accurate and refined alternative approach based on what you call the "differential" notion that makes the study of periodisation dependent on the subject matter. Can you finally sum up for us the main results of your work on this issue?

This work is the result of the efforts of me, the historian, being interested in what was written in the European medieval and classical periods and, at the same time, in what was written in the Arabic-Islamic classical period. It was necessary to clear up the deeply entrenched thesis, a postulate, according to which classical science is European in its origins and extensions. I find this postulate harmful on both historical and philosophical levels. I indeed proposed to critically examine the usual periodisation and introduce a new and more objective one that could better account for historical facts and that would give a sufficiently broad meaning to the notion of classical science. If one examines the components of "early modern mathematics", one soon notices that these chapters are far from being contemporary; each has its own history, and the inventions or discoveries are by no means simultaneous. More generally, the global landscape of mathematics in the 16th and 17th centuries appears as a composite, an edifice constituted by different elements with origins traceable to many different dates. Indeed, some topics, such as plane geometry, the geometry of conic sections, and the geometry of the sphere, go back a millennium. They take us back to Euclid, Apollonius, and Menelaus. As for algebra, the books of al-Khwārizmī, Abū Kāmil, and Fibonacci are prerequisites for anyone who wishes to deal with the authors of the 16th and 17th centuries. And in order to understand the algebraic geometry of the so-called early modern period, it is necessary to integrate into it some of the work carried out from the 11th to the 13th centuries. In algebraic geometry, the surest way to see novelty where it does not exist and overlook it when it is there is to forget all about the work of al-Khayyām and Sharaf al-Dīn al-Ṭūsī. We can easily find similar examples in the study of projections, parallel theory, spherical geometry, trigonometry, Diophantine analysis, number theory, combinatorial analysis, infinitesimal geometry, etc. And the same holds true for other mathematical sciences such as astronomy, optics and statics. As a result, I insist on two important points. The first is that all this periodisation is differential; when dealing with the history of algebra, this isn't the same as dealing with the history of mechanics or the history of optics. We can talk about classical algebra from al-Khwārizmī to Euler, and we can talk about classical kinematics from Galileo. But ought we speak of classical optics from Ptolemy or rather from Ibn al-Haytham, because of the revolution he accomplished, at least until Newton. Therefore, we should introduce a differential aspect in the periodisation of science. To return mathematics and science in general to a horizon that is truly its own, it is necessary to separate the history of science from political history by distinguishing between "conceptual" tradition and "objectal" tradition, which includes textual traditions, academic institutions and other contingent factors. This distinction seems to concretely translate the question of the place of the history of science between social history and epistemology. As an element of objectal tradition, the work of science is a material and cultural product, a product of men who live in a specific place and time. As part of the conceptual tradition, this work requires an epistemic analysis of its conceptual structure to clarify its meaning, which will enable us to delimit the very notion of tradition. This brings me to the second point, i.e. the universality of science. It is very important to understand this phenomenon given that scientific facts, even if they are the product of men, are nevertheless universal. Historically, science is the product of a society and social institutions but, epistemically, it is universal. From the epistemic point of view, mathematics has never stopped being universal. And from the historical point of view, I argue that the universality of science was realised in the Arabic-Islamic period because of its universal scientific library and its universal type of extensions. *Without the universality of science, we fall into folklore!*

July 2015
Dollon, France

Chapter 1
Analogy and Invention Some Remarks on Poincaré's Analysis Situs Papers

Claudio Bartocci

Abstract The primary role played by analogy in Henri Poincaré's work, and in particular in his "analysis situs" papers, is emphasized. Poincaré's "sixth example" (showing that Betti numbers do not suffice to classify 3-manifolds) and his construction of the homology sphere are discussed in detail.

In the chapter "L'avenir des mathématiques" of his *Science et méthode*,[1] Henri Poincaré famously defined mathematics as "the art of giving the same name to differ-

[1] I shall make reference to the following standard editions of Poincaré's works: *La science et l'hypothèse* (1902c), préface de Jules Vuillemin, Flammarion, Paris 1968 (which reproduces the text of the second edition, published in 1907); *La valeur de la science*, préface de Jules Vuillemin, Flammarion, Paris 1970; *Science et méthode*, Flammarion, Paris (1908); *Œuvres*, 11 vv., various eds., Gauthier-Villars, Paris (1916–1956). I have used the following English translations: *Science and Hypothesis*, translated by W. J. G., with a preface by J. Larmor, The Walter Scott Publishing Co., London & Newcastle-on-Tyne (1905); *The value of Science*, translated by George Bruce Halsted, Dover, New York 1958 (originally published in 1958 by Science Press in *Foundations of Science*); *Science and Method*, translated by Francis Maitland, with a preface by B. Russell, Thomas Nelson and Sons, London, Edinburgh, Dublin & New York (1914). I shall quote from these works (recently reprinted in *The Value of Science. The Essential Writings of H. Poincaré*, The Modern Library, New York 2001) by using the acronyms *SH*, *VS*, and *SM* followed by the page number. As for the paper "Analysis situs" and its five supplements, I have used the English translation provided in H. Poincaré, *Papers in Topology*, edited by John Stillwell, American Mathematical Society Chelsea Publishing, Providence (RI) & Mathematical Society, London 2010 (with slight modifications, when necessary); only the page number of the original texts will be given.

This paper is partly based on my previous paper "'Ragionare bene su figure disegnate male': la nascita della *topologia* algebrica", *Lettera Matematica*, 84–85 (2013), pp. 22–31. It is a great pleasure to thank the organizers of the International Colloquium "The Philosophers and mathematics" (Lisboa, October 29–30, 2014), in particular Hassan Tahiri. I also thank the anonymous referee for his/her helpful remarks. Finally, I am glad to take the opportunity to acknowledge my intellectual debt to professor Roshdi Rashed for his inspirational work on the history of mathematics, always carried out with admirable rigor and in a broad cultural perspective.

C. Bartocci (✉)
Università di Genova, Genova, Italy
e-mail: bartocci@dima.unige.it

H. Tahiri (ed.), *The Philosophers and Mathematics*, Logic, Epistemology, and the Unity of Science 43, https://doi.org/10.1007/978-3-319-93733-5_1

ent things."[2]This saying is far from being a mere witticism. In the first place, it aims at emphasizing the crucial importance of well-chosen words in mathematics: only a finely tuned language can enable mathematicians to establish relations between things that are different in the substance, but similar in the form, so that they can be cast in the same mold ("elles puissent [...] se couler dans le même moule"). Moreover, it is well known that, accordingly to Poincaré,

> Mathematicians do not study objects, but the relations between objects; to them it is a matter of indifference if these objects are replaced by others, provided that the relations do not change. Matter does not engage their attention, they are interested by form alone.[3]

Arguably, this conception of mathematics appears to be tightly related to the ubiquitous importance that such notions as morphism (in today's parlance) or equivalence relation (ditto) have in Poincare's oeuvre. For example, the notion of isomorphism allows mathematicians to identify groups that emerge differently in different problems: indeed, as Poincaré points out, "[w]e now know that, in a group, the matter is of little interest, that the form only is of importance".[4] Analogously, two *variétés* are equivalent from the point of view of *analysis situs* when they are "homeomorphic, that it is to say, of similar form",[5] and two plane curves defined by polynomials with integer coefficients are equivalent "if one can pass from one to the other by a birational transformation with integer or rational coefficients".[6] As for equivalence relations, Poincaré had certainly been well acquainted with the procedure of constructing the orbit space associated with the action of a discrete group on a given space since the early 1880s, when he wrote his celebrated papers on Fuchsian and

---

[2]*SM,* p. 34; "la mathématique est l'art de donner le même nom à des choses différentes", *Science et méthode*, cit., p. 29. The chapter "L'avenir des mathématiques" is based on the address prepared for the 4th International Congress of Mathematicians, held in Rome, April 6–11, 1908 (the address was read by Gaston Darboux, as Poincaré lay ill at his hotel during the whole conference). The original text, longer than that published in *Science et méthode*, appeared in *Atti del IV Congresso internazionale dei matematici*, G. Castelnuovo ed., Tipografia della Reale Accademia dei Lincei, Roma (1909), pp. 167–182, and in several journals: *Rendiconti del Circolo matematico di Palermo*, 16 (1908), pp. 152–168; *Revue générale des sciences pures et appliquées*, 19 (1909), pp. 930–939; *Scientia. Rivista di scienza*, 2nd year, 3 (1908), pp. 1–23; *Bulletin des sciences mathématiques*, 2nd ser., 32 (1908), pp. 168–190.

[3]*SH*, p. 20; "Les mathématiciens n'étudient pas des objets, mais des relations entre les objets; il leur est donc indifférent de remplacer ces objets par des autres, pourvu que les relations ne changent pas. La matière ne leur importe pas, la forme seule les intéresse", *La science et l'hypothèse*, cit., p. 49. The passage is excerpted from the chapter "La grandeur mathématique et l'expérience", which is based on the paper "Le continu mathématique", *Revue de métaphysique et de morale*, 1 (1893), pp. 26–34.

[4]*SM*, p. 35; "Nous savons maintenant que dans un groupe la matière nous intéresse peu, que c'est la forme seule qui importe", *Science et méthode*, cit., p. 30.

[5]"*[H]oméomorphes*, c'est-à-dire de forme pareille", "Analysis situs", *Journal de l'École Polytechnique*, 1 (1895), pp. 1–121 = *Œuvres*, vol. VI, pp. 193-288; the quote is at p. 199 (Poincaré's italics).

[6]"[S]i l'on peut passer de l'une à l'autre par une transformation birationnelle, *à coefficients entiers ou rationnels*", "Sur les propriétés arithmétiques des courbes algébriques", *Journal de mathématiques pures et appliquées*, 5th ser., 7 (1901b), pp. 161–233 = *Œuvres*, vol. V, pp. 483–550; the quote is at p. 484 (Poincaré's italics).

Kleinian groups. Indeed, already in the first *Supplément* to his essay presented for the prize competition announced by the *Académie des Sciences* on 1878, Poincaré was able to characterize the functions he (maybe, somewhat hastily) named *fuchsiennes* by the property of being invariant under the action of certain discrete subgroups of the group of isometries of the hyperbolic upper half-plane. Consistently, in the same supplement, he remarked:

> In fact, what is a geometry? It is the study of a *group of operations* formed by the displacements one can apply to a figure without deforming it. In Euclidean geometry the group reduces to *rotations* and *translations*. In the pseudogeometry of Lobachevsky it is more complicated.[7]

In a broader perspective, Poincaré's definition of mathematics as "the art of giving the same name to different things" reflects his conviction about the prominent role played by analogy within this discipline. Analogy is the guide that shows the way to mathematicians and steers their "groping attempts"[8] by revealing "the hidden harmony of things".[9] It lies at the very heart of the process of mathematical intuition, in a relationship of mutual dependence:

> What has taught us to know the true, profound analogies, those the eyes do not see but reason divines?
>
> It is the mathematical spirit, which disdains matter to cling only to pure form. This it is which has taught us to give the same name to things differing only in material, to call by the same name, for instance, the multiplication of quaternions and that of whole numbers.[10]

As usual in Poincaré, the methodological stance and the activity of "doing mathematics" remain always interrelated. This is especially apparent, as I shall try to argue, in the monumental memoir "Analysis situs" and its five *compléments*. Poincaré's extraordinary originality and creativity come along with his skillful and deliberate recourse to the tools of trade of a mathematician, above all that of analogy.

[7]"Qu'est-ce en effet une géométrie? C'est l'étude d'un *groupe d'opérations* formé par les déplacements que l'on peut faire subir à une figure sans la déformer. Dans la géométrie euclidienne ce groupe se réduit à des *rotations* et à des *translations*. Dans la pseudogéométrie de Lobatchewski [*sic*] il est plus compliqué.", *Trois suppléments sur la découverte des fonctions fuchsiennes,* J. Gray and S. A. Walters eds., Akademie Verlag, Berlin & Albert Blanchard, Paris (1997), p. 35 (Poincaré's italics). It is worth observing that, at that time, Poincaré was completely unaware of Felix Klein's *Erlanger programm.*

[8]*SM*, p. 29; "tâtonnements" *Science et méthode*, cit., p. 24.

[9]*VS*, p. 79; "l'harmonie cachée des choses", *La valeur de la science*, cit., p. 108.

[10]*VS*, p. 77; "Qui nous a appris à connaître les analogies véritables, profondes, celles que les yeux ne voient pas et que la raison devine? C'est l'esprit mathématique, qui dédaigne la matière pour ne s'attacher qu'à la forme pure. C'est lui qui nous a enseigné à nommer du même nom des êtres qui ne diffèrent que par la matière, à nommer du même nom par exemple la multiplication des quaternions et celle des nombres entiers.", *La valeur de la science*, cit., p. 106.

1.1 "L'inspection de La Figure Le Démontre"

In 1896 Poincaré underwent a series of anthropometric measurements and psychological and physiological tests conducted by the French "alienist" Édouard Toulouse (1867–1947). Toulouse's research work—not without connection to the ideas developed by Cesare Lombroso in his (at that time highly reputed) books *Genio e follia* (1864) and *L'uomo di genio in rapporto alla psichiatria* (1888)—was aimed at "clarifying the relationships of intellectual superiority to neuropathy".[11] Among the subjects examined by Toulouse there were some leading figures of *fin de siècle* French culture: besides Poincaré, Émile Zola and Pierre Berthelot. One of the several tests of memory devised by Toulouse consisted in observing a simple geometric figure for a few seconds and then reproducing it. The result obtained by Poincaré in this test was quite miserable: he could do not better than scribble down two clumsy attempts.[12] This is hardly surprising, since his awkwardness in drawing was almost legendary. According to Paul Appell's account, Poincaré's lack of skill in the technique called "lavis" jeopardized his own admittance to the École Polytechnique and cost him the first place in the second year final ranking.[13] Nevertheless—or rather, just because of this—figures, and in particular those "badly drawn", played an essential role in the shaping of Poincaré's geometric thinking, as he himself made it clear in the introduction to "Analysis situs":

> We know how useful geometric figures are in the theory of imaginary functions and integrals evaluated between imaginary limits, and how much we desire their assistance when we want to study, for example, functions of two complex variables.
>
> If we try to account for the nature of this assistance, figures first of all make up for the infirmity of our intellect by calling on the aid of our senses; but not only this. It is worthy repeating that geometry is the art of reasoning well from badly drawn figures; however, these figures, if they are not to deceive us, must satisfy certain conditions; the proportions may be grossly altered, but the relative positions of the different parts must not be upset.
>
> The use of figures is, above all, then, for the purpose of making known certain relations between the objects that we study, and these relations are those which occupy the branch of geometry that we have called *Analysis situs*, and which describes the relative situation of points and lines on surfaces, without consideration of their magnitude.[14]

---

[11]Édouard Toulouse, *Enquête médico-psychologique sur la superiorité intellectuelle: Henri Poincaré*, Flammarion, Paris (1910), p. 1.

[12]Toulouse, *op. cit.*, p. 66.

[13]Paul Appell, *Henri Poincaré*, Plon, Paris (1925), p. 28; for more details, see André Bellivier, *Henri Poincaré ou la vocation souveraine*, Gallimard, Paris (1956).

[14]"On sait quelle est l'utilité des figures géométriques dans la théorie des fonctions imaginaires et des intégrales prises entre des limites imaginaires, et comment on regrette leur concours quand on veut étudier, par exemple, les fonctions de deux variables complexes. Cherchons à nous rendre compte de la nature de ce concours; les figures suppléent d'abord à l'infirmité de notre esprit en appelant nos sens à son secours; mais ce n'est pas seulement cela. On a bien souvent répété que la Géométrie est l'art de bien raisonner sur des figures mal faites; encore ces figures, pour ne pas nous tromper, doivent-elles satisfaire à certaines conditions; les proportions peuvent être grossièrement altérées, mais les positions relatives des diverses parties ne doivent pas être bouleversées. L'emploi des figures a donc avant tout pour but de nous faire connaître certaines relations entre les objets de

Not only in his work on "analysis situs" (an area which today we would call algebraic and differential topology), but in the whole of his vast scientific production, Poincaré made an extensive use of figures both as useful aids to the study of certain particular cases and as key tools for proof. In this regard a paradigmatic example is provided by the *théorie des conséquents* (an early instance of the geometric construction we nowadays call "Poincaré's map") and the *théorie des cycles limites* developed in the groundbreaking "Mémoire sur les courbes définies par une équation différentielle (deuxième partie)"[15] (here, for example, one can read the phrase "l'inspection de la figure le démontre" ("the inspection of the figure proves it")[16], highly typical of Poincaré's way of reasoning).

On some occasions these two distinct functions of figures may fuse into a single one (as in the paper "Théorie des groupes fuchsiens"[17]), while in others they remain separate, if for no other reason than that the author has not yet found a way out of the maze of specific examples. This is the case of his last memoir, "Sur un théorème de géométrie",[18] that Poincaré submitted to Giovanni Battista Guccia for publication in the journal *Rendiconti del Circolo matematico di Palermo*, albeit he was sorely aware that it was *inachevé*:

> What embarrasses me is the fact that I will be forced to insert many figures, precisely because I have not yet been able to obtain a general rule, but have only accumulated the special solutions.[19]

In the "Advertissement" to his *Méchanique analytique* (1788), Lagrange notoriously declared: "On ne trouvera point de Figures dans cet Ouvrage."[20] About one century later, in Poincare's three volumes of *Les méthodes nouvelles de la mécanique céleste* (1892, 1893, 1899) figures, if not numerous,[21] are brought strategically into

nos études, et ces relations sont celles dont s'occupe une branche de la Géométrie que l'on appelée *Analysis situs*, et qui décrit la situation relative des points des lignes et des surfaces, sans aucune considération de leur grandeur.", "Analysis situs", cit., p. 194.

[15]*Journal de mathématiques pures et appliquées*, 3rd ser., 8 (1882a), pp. 251–296 = *Œuvres*, vol. I, pp. 44–84.

[16]*Ibid.*, p. 56.

[17]*Acta mathematica*, 1 (1882b), pp. 1–62 = *Œuvres*, vol. II, pp. 108–168.

[18]"Ce qui m'embarrasse, c'est que je serai obligé de mettre beaucoup de figures, justement parce que je n'ai pu arriver à une règle générale, mais que j'ai seulement accumulé les solutions particulières", "Sur un théorème de géométrie", *Rendiconti del Circolo matematico di Palermo*, 33 (1912), pp. 375–407 = *Œuvres*, vol. VI, pp. 499–538. Poincaré's "last geometric theorem" (stating that any homeomorphism of the annulus into itself that is area- and orientation-preserving and rotates the outer boundary circle clockwise and the inner boundary circle anticlockwise has at least two fixed points) was proved, in full generality and through different methods, by George David Birkhoff in (1913). Though incomplete, Poincaré's proof was basically correct, as shown by C. Golé and G. R. Hall in their paper "Poincaré's proof of Poincaré's last geometric theorem", in *Twist Mappings and Their Applications*, R. McGehee & K. R. Meyers eds., Springer-Verlag, New York (1992), pp. 135–151.

[19]*Le livre du centenaire de la naissance de Henri Poincaré, 1854–1954*, Gauthier-Villars, Paris (1955), p. 296.

[20]J.-L. Lagrange, *Méchanique analytique*, chez la veuve Desaint, Paris (1788), p. vi.

[21]Precisely, there are 4 figures in the first, 3 in the second, and 12 in the third volume of the work.

play in order to elucidate difficult points in a proof or to provide classification schemas of admissible behaviors. There is at least one case, however, in which the figure turns out to be impossible to visualize, even for the man who was likely to possess the most visionary geometric imagination at the time. This occurs towards the end of the third volume, where Poincaré studies the intersections of two curves belonging to the stable and instable manifolds (as we would say today) of a periodic solution, running up against a dynamics of chaotic type:

> Let us attempt to imagine the figure formed by these two curves and their intersections, which are infinitely many and each of which corresponds to a solution that is doubly asymptotic: these intersections form a sort of lattice, tissue or grid with an infinitely dense mesh; neither of the two curves must intersect itself, but must fold over itself in a very complicated way so as to intersect all of the meshes of the grid. The complexity of this figure, which I will not even attempt to draw, is striking. Nothing shows in a more compelling way the difficulty of the three-body problem and, more generally, of all dynamical problems having no uniform integral or whose Bohlin series are divergent.[22]

1.2 Invariants

In his four great memoirs about "les courbes définies par une équation differentielle,"[23] Poincaré adopted a "new point of view", that he himself defined "qualitative". This change in perspective enabled him to obtain results that, using a classical quantitative approach, would have been virtually impossible even to imagine. For example, in his third memoir Poincaré proved the "index theorem", according to which for any vector field X on a (closed and oriented) surface of genus g the relation

$$C - F - N = 2g - 2,$$

[22]"Que l'on cherche à se représenter la figure formée par ces deux courbes et leurs intersections en nombre infini dont chacune correspond à une solution doublement asymptotique, ces intersections forment une sorte de treillis, de tissu, de réseau à mailles infiniment serrées; chacune des deux courbes ne doit jamais se recouper elle-même, mais elle doit se replier sur elle-même d'une manière très complexe pour venir recouper une infinité de fois toutes les mailles du réseau. On sera frappé de la complexité de cette figure, que je ne cherche même pas à tracer. Rien n'est plus propre à nous donner une idée de la complication du problème des trois corps et en général de tous les problèmes de Dynamique où il n'y a pas d'intégrale uniforme et où les séries de Bohlin sont divergentes.", *Les méthodes nouvelles de la mécanique céleste. Tome III. Invariants intégraux—Solutions périodiques de deuxième genre—Solutions doublements asymptotiques*, Gauthier-Villars, Paris (1899a), p. 389.

[23]"Mémoire sur les courbes définies par une équation différentielle (première partie)", *Journal de mathématiques pures et appliquées*, 3rd ser., 7 (1881a), pp. 375–422 = *Œuvres*, vol. I, pp. 3–44; "Mémoire sur les courbes définies par une équation différentielle (deuxième partie)", cit.; "Sur les courbes définies par les équations différentielles (première partie)", *Journal de mathématiques pures et appliquées*, 4th ser., 1 (1885), pp. 167–244 = *Œuvres*, vol. I, pp. 90–161; "Sur les courbes définies par les équations différentielles (deuxième partie)", *Journal de mathématiques pures et appliquées*, 4th ser., 2 (1886), pp. 151–217 = *Œuvres*, vol. I, pp. 167–222.

is satisfied, where C is the number of saddle points of X, F the number of its foci, and N the number of its nodes.[24] In the fourth and last memoir of the series Poincaré addressed the problem of studying differential equations of higher order extending his own global theory to n-dimensional spaces. Commenting on the necessity of resorting to geometric objects of dimension greater than three, he remarked:

> Geometry, then, is just a language which can be more or less advantageous, but is no longer a representation talking to the senses. However, we may be led to use this language occasionally.[25]

This same qualitative strategy, combined with the handling of phase space as the geometric arena where the true dynamics takes places, allowed him, just a few years later, to open a breach—while describing the perturbations of periodic orbits—in the otherwise impregnable fortress of the three body problem.[26] It is therefore no coincidence (and no surprise) that in the informal definition of *analysis situs* given by Poincaré in the "Analyse de ses travaux..." we find precisely the adjective "qualitative":

> *Analysis situs* is the science that allows us to know the qualitative properties of geometric figures, not only in ordinary space but in space of more than three dimensions.[27]

While *analysis situs* in three dimensions is, according to Poincaré, "une connaissance presque intuitive" ("an almost intuitive knowledge"), enormous difficulties arise in extending its concepts to higher dimensions. In order to attempt to overcome these hurdles is thus necessary "to be profoundly convinced of the great importance of this discipline",[28] a conviction that Poincaré certainly possessed in high degree:

> As for me, all of the various paths on which I was successively engaged led me to *analysis situs*. I had need of the results of this discipline to pursue my research work on curves defined by differential equations [...] and to generalize it to higher order differential equations, in

[24]"Sur les courbes définies par les équations différentielles (première partie)", cit., p. 125. This result (known today under the name of Poincaré-Hopf theorem) had previously been stated in the short research announcement "Sur les courbes définies par une équation différentielle", *Comptes rendus de l'Académie des Sciences*, 93 (1881b), pp. 951–952 = *Œuvres,* vol. I, pp. 85–85. For some comments about Poincaré's qualitative approach to differential equations see C. Bartocci, "Introduzione", in H. Poincaré, *Geometria e caso. Scritti di matematica e fisica*, Bollati Boringhieri, Torino (1995) (repr. 2013), pp. VII–L.

[25]"La Géométrie n'est plus alors qu'un language qui peut être plus ou moins avantageux, ce n'est plus une représentation parlant aux sens. Nous pourrons néanmoins être conduits à employer quelquefois ce language.", "Sur les courbes définies par les équations différentielles (deuxième partie)", cit., p. 168.

[26]"[...] ce qui nous ces solutions périodiques si précieuses, c'est qu'elles sont, pour ainsi dire, la seule brèche par où nous puissions essayer de pénétrer dans une place jusqu'ici réputée inabordable", *Les méthodes nouvelles de la mécanique céleste. Tome I. Solutions périodiques—Non-existence des intégrales uniformes—Solutions asymptotiques*, Gauthier-Villars, Paris (1892a), p. 82.

[27]"L'Analysis situs est la science qui nous fait connaître les propriétés qualititaves des figures géométriques non seulement dans l'espace ordinaire, mais dans l'espace à plus de trois dimensions.", "Analyse de ses travaux scientifiques faite par H. Poincaré", *Acta mathematica*, 38 (1921), pp. 36–135 (posthumously published, but written down by Poincaré in 1901); the quote is at p. 100.

[28]"[Ê]tre bien persuadé de l'extrême importance de cette science", *ibid.*, p. 100.

> particular to those of the three body problem. I had need of *analysis situs* for the study of non uniform [i.e., multivalued] functions of 2 variables. I had need of it for the study of periods of multiple integrals and for the application of this study to the expansion of the perturbative function.
>
> Finally, I glimpsed in *analysis situs* a tool for tackling an important problem in group theory, namely, the search for discrete or finite groups contained in a given continuous group.[29]

But what are, in concrete terms (if ever possible), the objects that analysis situs is supposed to deal with? In analogy with the definition of geometry I have quoted above, Poincaré defines analysis situs as "the Science whose object is the study of [the] group [of homeomorphisms].[30] In other words, the crux of the question lies in being able to determine, for each *variété*, suitable "quantities" (numbers, groups, vector spaces, etc.) associated with it that remain invariant when the *variété* undergoes a differentiable transformation.

In the case of surfaces in Euclidean space, the problem of invariants had been solved by August Ferdinand Möbius in 1863. In his paper "Theorie der elementaren Verwandtschaften",[31] the German mathematician sectioned any closed and oriented surface into "primitive forms" (namely, disks, cylinders or union of cylinders) by means of parallel planes. He then showed that, between two such surfaces, it is possible to establish an "elementary correlation" (*elementar Verwandtschaft*, that is, more or less, diffeomorphism in modern terminology) if and only if they belong to the same class. Each class is identified by a nonnegative integer, g, that corresponds to the number of holes of the surface and it is related to the surface's Euler characteristic χ by the formula $2 - 2g = \chi$.

Some years earlier, Bernhard Riemann, in his *Inauguraldissertation* entitled "Grundlagen für eine allgemeine Theorie der Functionen einer veränderlichen complexen Grösse",[32] had introduced a profoundly novel geometric idea—that of Riemann surface—establishing unexpected relations between complex analysis, topology and the theory of algebraic curves. In this theory, every algebraic curve can

---

[29]"Quant à moi, toutes les voies diverses où je m'étais engagé successivement me conduisaient à l'Analysis Sitûs. J'avais besoin des données de cette science pour poursuivre mes études sur les courbes définies par les équations différentielles [...] et pour les étendre aux équations différentielles d'ordre supérieur et en particulier à celles du problème des trois corps. J'en avais besoin pour l'étude des fonctions non uniformes de 2 variables. J'en avais besoin pour l'étude des périodes des intégrales multiples et pour l'application de cette étude au développement de la fonction perturbatrice. Enfin j'entrevoyais dans l'Analysis Sitûs un moyen d'aborder un problème important de la théorie des groupes, la recherche des groupes discrets ou des groupes finis contenus dans un groupe continu donné.", *ibid*., p. 101.

[30]"[L]a Science dont l'objet est l'étude [du] groupe [des homéomorphismes]", "Analysis situs", cit., p. 198. As for the meaning of the words *variété* and *homéomorphisme*", see note 41.

[31]*Berichte über die Verhandlungen der Königlich Sächsischen Gesellschaft der Wissenschaften zu Leipzig, Mathematisch-Physische Classe*, 17 (1863), pp. 31–68 = *Gesammelte Werke*, vol. II, herausgegeben von F. Klein, Hirzel, Leipzig 1886, pp. 433–471.

[32]Reprinted in B. Riemann, *Gesammelte Mathematische Werke, Wissenschaftlicher Nachlass und Nachträge/Collected Papers*, nach der Ausgabe von H. Weber und R. Dedekind, neu herausgegeben von Raghavan Narasimhan, Springer-Verlag, Berlin-Heidelberg—Teubner, Leipzig (1990), pp. 35–77.

be associated with a Riemann surface, namely, the branched cover of the complex plane determined by the given curve: the genus of the curve (defined, for example, as the dimension of the space of holomorphic differentials) does coincide with the number $\frac{m-1}{2}$, where m is the "order of connectivity" (*Ordnung des Zusammenhangs*) of the surface (defined, instead, in topological terms).[33] Had Möbius been aware of Riemann's results,[34] he would have not failed to realize that his classifying number, multiplied by 2 and then added to 1, corresponded exactly to Riemann's "order of connectivity". Classification theorems for surfaces "equivalent" to Möbius's result were obtained, following distinct and innovative proof strategies, by Camille Jordan[35] and, successively, by William Kingdon Clifford.[36] Along a different line of research, Enrico Betti, profoundly influenced by Riemann's ideas, generalized (and modified) the notion of order of connectivity to spaces of higher dimensions: in his memoir "Sopra gli spazi di un numero qualunque di dimensioni",[37] he defined, for every n-dimensional space, $n-1$ "orders of connectivity" $p_1, \ldots, p_{n-1}$(integer numbers), which later mathematicians would call "Betti numbers". Are Betti numbers "invariants" in the same sense in which the genus of a Riemann surface is? If so, can they be used to obtain classification theorems for higher-dimensional geometric spaces analogous to the classification theorem for surfaces? These two issues—absolutely natural in the theoretical context of geometry in the early 1880s—were pinpointed with remarkable lucidity, in 1884, by Walther Dyck for the case of dimension 3:

> The object is to determine certain characteristical [*sic*] numbers for closed threedimensional spaces, analogous to those introduced by Riemann in the theory of his surfaces, so that their identity shows the possibility of its [*sic*] "one-to-one correspondence".[38]

---

[33]This result, valid for any compact Riemann surface, was proved by Riemann in his pioneering (and somewhat cryptic) paper "Theorie der Abel'schen Functionen", *Journal für die reine und angewandte Mathematik*, 54 (1857), pp. 115–155 = *Gesammelte Mathematische Werke*..., cit., pp. 88–142. For a detailed account of Riemann's geometric function theory see U. Bottazzini & J. Gray, *Hidden Harmony—Geometric Fantasies. The Rise of Complex Function Theory*, Springer, New York (2013), Chap. 5. The word "genus" was introduced by Rudolf Friedrich Alfred Clebsch in his paper "Über die Anwendung der Abelschen Functionen in der Geometrie", *Journal für die reine und angewandte Mathematik*, 63 (1864), pp. 189–243.

[34]Jean-Claude Pont (*La topologie algébrique des origines à Poincaré*, Presses Universitaires de France, Paris 1974, p. 97) argues that this was not the case.

[35]"Des contours tracés sur les surfaces", *Journal de mathématiques pures et appliquées*, 2nd ser., 11 (1866), pp. 110–130.

[36]"On the canonical form and dissection of a Riemann's surface", *Proceedings of the London Mathematical Society*, 8 (1877), pp. 292–304 = *Mathematical Papers*, edited by R. Tucker, Macmillan, London 1882 (reprinted AMS Chelsea Publishing, Providence (RI) 2007), pp. 241–254. It may be of some interest to note that, in this paper, Clifford proved what is now known as the Riemann-Hurwitz formula (the formula in itself being usually attributed to Hurwitz, who obtained it in 1893): *"an* n*-sheeted Riemann's surfaces with* w *branch-points may be transformed, without tearing, into the surface of a body with* $p = \frac{1}{2}w - n + 1$, *holes in it"* (Clifford's italics), op. cit., p. 251.

[37]*Annali di matematica pura e applicata*, 2nd ser., 4 (1871), pp. 140–158 = *Opere matematiche*, vol. II, Hoepli, Milano 1871, pp. 273–290.

[38]"On the 'Analysis situs' of threedimensional spaces", *Report of the Fifty-fourth Meeting of the British Association for the Advancement of Science held at Montreal in August and September 1884*, John Murray, London (1885), p. 648.

A similar question would be formulated, eight years later, by Poincaré in his first work explicitly devoted to the subject of *analysis situs*, a note barely four pages long presented to the *Académie des Sciences*:

> One may ask whether the Betti numbers suffice to determine a closed surface from the viewpoint of *analysis situs*; that is, whether, given two closed surfaces with || the same Betti numbers, it is possible to pass from one to the other by a continuous transformation.[39]

Unlike Dyck, Poincaré also had the answer: no. He claimed, in fact, to be able to construct a family of three-dimensional *surfaces* (the word *variété* is not employed yet) whose first Betti number can only assume values less or equal to 4, and which nonetheless contain an infinite number of non diffeomorphic surfaces. In order to prove this fact Poincaré needed to invent a new invariant: the fundamental group. The analogy with certain constructions used in the theory of Fuchsian groups provided him the guiding principle to devise the appropriate counterexample.

1.3 Poincaré's Sixth Example

The rudimentary ideas outlined in the 1892 *Comptes rendus* note were developed by Poincaré in a memoir of more than a hundred pages, "Analysis situs", published in 1895 in *Journal de l'École Polytechnique*, a masterwork that strikes even the most blasé reader with its amazing originality. Jean Dieudonné, certainly not one well-disposed toward a mathematician who represented the antithesis of the Bourbakist ideal, called it a "fascinating and exasperating paper".[40] Here I will not give a detailed account of "Analysis situs" and its five *compléments*,[41] but will rather confine myself to highlighting some key issues related to Poincaré's quest for understanding the import of invariants, with special attention paid to the role of analogy in the creative process.

After providing the definition of *variété* and *homémorphisme*[42] (henceforth, these words will be translated as "manifold" and "diffeomorphism") and introducing the

[39]"On peut se demander si les nombres de Betti suffisent pour déterminer une surface fermée au point de vue de l'Analysis situs, c'est-à-dire si, étant données deux surfaces fermées qui possèdent || mêmes nombres de Betti, on peut toujours passer de l'une à l'autre par voie de déformation continue.", "Sur l'analysis situs", *Comptes rendus de l'Académie des Sciences*, 115 (1892c), pp. 603–606 = *Œuvres*, vol. VI, pp. 189–192 (the quote is at pp. 189–190).

[40]*A History of Algebraic and Differential Topology*, Birkhäuser, Basel-Boston (1989), p. 28.

[41]Accurate accounts, each focusing on different aspects of Poincaré's work, are provided in: Jean Dieudonné, *A History of Algebraic and Differential Topology*, cit., chap. I; Erhard Scholz, *Geschichte des Mannigfaltigkeitsbegriffs von Riemann bis Poincaré*, Birkhäuser, Boston-Basel-Stuttgart (1980), chap. VII; K. S. Sarkaria, "The topological work of Henri Poincaré", in *History of Topology*, edited by I. M. James, Elsevier, Amsterdam (1999), pp. 121–167; Klaus Volkert, *Das Homöomorphismusproblem insbesondere der 3-Mannigfaltigkeiten, in der Topologie 1892–1935*, Kimé, Paris (2002); Jeremy Gray, *Henri Poincaré. A Scientific Biography*, Princeton University Press, Princeton (NJ) and London (2013), Chap. 8.

[42]In actual fact, Poincaré gives two different definitions of *variété*. According to the first definition, a *variété* is a subspace of $\mathbb{R}^n$ determined by a system of equations (satisfying suitable conditions of

notion of homology (as, in modern parlance, an equivalent relation on the free Abelian group generated by closed submanifolds), Poincaré spells out what *he* means by "order of connectivity":

> We say that the manifolds
>
> $$v_1, v_2, \ldots, v_\lambda$$
>
> which have the same number of dimensions and form part of V are *linearly independent* if they are not connected by any homology with integral coefficients.
>
> If there exist P_{m-1} *closed* manifolds of m dimensions which are linearly independent and form part of V, but not more than P_{m-1}, then we shall say that the order of connectivity of V with respect to manifolds of m dimensions is equal to P_m.[43]

In spite of the fact that these numbers are called "Betti numbers", they do not coincide—as it would emerge from the critical observations of Poul Heegaard—with the orders of connectivity defined by the Italian mathematician in his 1871 paper. In §7 of "Analysis situs", Poincaré, relying on his previous paper "Sur les résidus des intégrales doubles",[44] gives an interpretation of *his* Betti numbers as the maximum number of "periods" of multiple integrals that satisfy certain integrability conditions. This fundamental result is presented in a rather cursory way and not used in the rest of the memoir: in fact, it would remain dormant for more than thirty years, until Élie Cartan would reformulate it in the language of differential forms.[45] A rigorous (and almost complete) proof would be provided by Georges de Rham in his thesis of (1931).[46] The famous homology duality theorem is stated and "proved" in §9:

regularity) and inequalities; in modern terminology, one would call it an immersed C^1 submanifold of $\mathbb{R}^n$, possibly with boundary and with some mild singularities. According to the second broader definition, instead, the spaces to be studied are *chaînes continues* of *variétés* of the same dimension n pairwise glued together along a common smooth part of dimension n, or, more generally, *réseaux continus* of *variétés*. A *réseau continu* of *variétés* corresponds more or less to what today is called a "C^1 manifold" (with or without boundary), possibly with some singularities (for example, some of the spaces discussed by Poincaré in §15 are orbifolds). Poincaré's notion of *homéomorphisme* corresponds to that of C^1 diffeomorphism.

[43]"Nous dirons que les varietiés $v_1, v_2, \ldots, v_\lambda$ d'un même nombre de dimensions et faisant partie de V, sont *linéairement indépendentes*, si elles ne sont liées par aucune homologie à coéfficients entiers. S'il existe P_{m-1} variétés *fermées* à m dimensions faisant partie de V et linéairement indépendentes et s'il n'en existe que P_{m-1}, nous dirons que l'ordre de connexion de V par rapport aux variétés à m dimensions est égal à P_m.", "Analysis situs", cit., p. 207 (Poincaré's italics).

[44]*Acta mathematica*, 9 (1887), pp. 321–380 = *Œuvres*, vol. III, pp. 440–489.

[45]É. Cartan, "Sur le nombres de Betti des espaces de groupes clos", *Comptes rendus de l'Académie des Sciences*, 187 (1928), pp. 196–198.

[46]G. de Rham, "Sur l'analysis situs de variétés à n dimensions", *Journal de mathématiques pures et appliquées*, 9th ser., 10 (1931), pp. 115–200; the essential new tool in de Rham's proof is the notion of current. The theorem proved by de Rham can be stated in the following way: for every closed and orientable differentiable manifold of dimension n, the dimension (over $\mathbb{Z}$) of the k-th homology group with integer coefficients is equal to the dimension (over $\mathbb{R}$) of the vector space consisting of degree k closed differential forms modulo degree k exact differential forms, for all $k = 0, \ldots n$. The latter vector spaces are now (rather preposterously) called "de Rham cohomology groups". De Rham's proof cannot be considered "complete" because it takes as an assumption that every closed

"Consequently, for any closed manifold, the Betti numbers equally distant from the extremes are equal."[47] However, Poincaré's argument, which is based on the notion of the "intersection number" of two submanifolds that intersect (transversally) in a finite number of points, is convoluted and unclear, even "totally unconvincing", in the words of Dieudonné.[48] As we shall see below, it did not fail to draw the harsh criticism of Heegaard. The key ingredient to prove the claim contained in the 1892 *Comptes rendus* note is supplied by the generalization of a geometric technique that Poincaré had first employed many years earlier. In the already mentioned memoir "Théorie des groupes fuchsiens", he had constructed closed surfaces by "gluing together"[49] in a suitable way the sides of curvilinear polygons. In §§10–11 of "Analysis situs" this procedure is extended to three-dimensional manifolds:

> There is a manner of visualizing manifolds of three dimensions situated in a space of four dimensions which considerably facilitates their study.[50]

This "way of visualizing three-dimensional manifolds" consists in realizing them as quotient spaces (as we would say today) of convex polyhedra modulo equivalence relations. To begin with, Poincaré examines in detail—almost with the eye of a taxonomist of geometric forms—a series of five examples: the first is the three-dimensional torus obtained by identifying, without twisting or reflecting, the opposite faces of cube, and the last is the real projective space (inexplicably not given any name at all) constructed as quotient of a regular octahedron.

In §11 Poincaré goes on to describe another way of representing three-dimensional manifolds that "can be applied in certain cases". This construction consists in gluing together the faces of a polyhedron by using the equivalence relation induced by the action of a "properly discontinuous group". As Poincaré laconically remarks: "The analogy with the theory of Fuchsian groups is too evident to need stressing; I shall confine myself to a single example."[51]

This single example provided by Poincaré is the same which he alluded to in the note of 1892. Let us consider the transformations:

$$\begin{aligned}(x, y, z) &\to (x+1, y, z)\\ (x, y, z) &\to (x, y+1, z)\\ (x, y, z) &\to (\alpha x + \beta y, \gamma x + \delta y, z+1),\end{aligned}$$

manifold admits a cellular decomposition. Poincaré himself had attempted to prove this fact in his "Complément à l''analysis situs'", *Rendiconti del Circolo matematico di Palermo*, 13 (1899d), pp. 285-343 = *Œuvres*, vol. VI, pp. 290–337; a proof was first provided by John Henry C. Whitehead in his paper "On C^1-complexes", *Annals of Mathematics*, 2nd ser., 41 (1940), pp. 809–824.

[47]"Par conséquent, pour une variété fermée, les nombres de Betti également distants des extrêmes sont égaux.", "Analysis situs", cit., p. 228.

[48]*History of Algebraic and Differential Topology*, cit., p. 221.

[49]Poincaré himself uses the verb *coller* ("Analysis situs", cit., p. 237).

[50]"Il y a une manière de se représenter les variétés à trois dimensions situées dans l'espace à quatre dimensions, manière qui en facilite singulièrement l'étude.", "Analysis situs", cit., p. 229.

[51]"L'analogie avec la théorie des groupes fuchsiens est trop évidente pour qu'il soit nécessiare d'insister; je me bornerai à un seul exemple", *ibid.*, p. 237.

where $\alpha, \beta, \gamma, \delta$ are integers such that $\alpha\delta - \beta\gamma = 1$. This amounts to say that the matrix

$$\begin{pmatrix} \alpha & \beta \\ \gamma & \delta \end{pmatrix}$$

belongs to the modular group SL(2,$\mathbb{Z}$), a group that Poincaré knows well and since long time, precisely for the reason that it plays a role of primary importance in the theory of automorphic functions. The previous transformations can be used to glue together the faces of the cube in $\mathbb{R}^3$ having one vertex at the origin $(0, 0, 0)$ and another at the point $(1, 1, 1)$. Through the identifications prescribed by the first two *substitutions* (which are independent of z) we obtain a family of tori parameterized by the segment $0 \leq z \leq 1$, namely, a three-dimensional cylinder whose sections are two-dimensional tori. The third *substitution* identifies the two bases of this cylinder by means of the linear transformation defined by the parameters $\alpha, \beta, \gamma, \delta$. The resulting three-dimensional manifold (Poincaré's sixth-example) can be thought of as a family of two-dimensional tori parameterized by the circle (in other words, more technically, it is a torus fibration over S^1).

The strategic significance of the sixth example it is not immediately clear, but emerges gradually throughout the three sections that follow, §§12, 13, 14, which in themselves occupy some good twenty pages of the paper. In §12 Poincaré introduces the notion of fundamental group (*groupe fondamental*) of a manifold: this group is defined as the set consisting of closed paths (*contours fermés*) based at an initial point M_0 modulo the equivalence relation that identifies to zero paths that can be continuously deformed to a trivial *lacet* around M_0, equipped with the natural operations of composition and inverse. It should be remarked that all necessary ingredients for Poincaré's definition were already present in Jordan's paper "Des contours tracés sur les surfaces"[52] (where the expression "contour fermé" also appeared). Of course the fundamental group is not conceived by Poincaré as an abstract group, but as a group of *substitutions* acting on a certain set of functions F that are not supposed to be "uniform" on the given manifold:

> When the point M leaves its initial position M_0 and returns to that position afterll traversing an arbitrary path, it may happen that the functions F do not return to their initial values.[53]

In today's terminology, one would say that the functions F are not defined on the manifold itself, but on its universal cover. Poincaré comes back to his sixth example in §§13–14. After some computations involving the so-called *contours fondamentaux* (i.e. the generators of the fundamental group, which, in this case, is SL(2,$\mathbb{Z}$)) and a quite tortuous argument, he manages to prove that it is possible to find an infinite

[52] Cit.

[53] "Lorsque le point M, partant de sa position initiale M_0, reviendra à cette position, après avoirll parcoru un chemin quelconque, il pourra se faire que les fonctions F ne reviennent pas à leurs valeurs primitives.", "Analysis situs", cit., pp. 239–240.

number of matrices $\begin{pmatrix} \alpha & \beta \\ \gamma & \delta \end{pmatrix}$ in SL(2,$\mathbb{Z}$)) for which the associated quotient manifolds have fundamental groups that are not isomorphic, so that the manifolds themselves are not diffeomorphic. However, his computations show that the first Betti number of all these manifolds can only assume values less or equal to 4. Poincaré therefore concludes: "*Thus for two closed manifolds to be homeomorphic, it does not suffice for them to have the same Betti numbers.*"[54]

1.4 The Homology Sphere

No doubt a difficult paper to digest, "Analysis situs" found nonetheless an attentive reader in the Danish mathematician Poul Heegaard (1871–1948). In his thesis, presented at the University of Copenhagen in (1898),[55] Heegaard provided an original description of 3-dimensional manifolds in terms of certain diagrams[56] and showed himself critical of the results obtained by Poincaré:

> [...] the theory of Riemann and Betti regarding the order of connectivity has many shortcomings and is difficult to use in the case of manifolds of dimension greater than two. Poincaré [...] attempted to complete it but, we believe, he did not succeed.[57]

In particular, Heegaard's criticisms were directed at the duality theorem: "not only is the theorem not proved; *it must be incorrect*."[58] By March 1899 Poincaré had already been able to look through the "travail très remarquable" ("very remarkable work") of his younger Danish colleague and partially agreed with the objections raised by him:

> These criticisms are partly well founded; the theorem does not hold true for the Betti numbers as defined by Betti; this follows from an example provided by Heegaard; this followed as well by an example that I had described in my Memoir. However, the theorem is true for the Betti numbers as defined by myself; I found a new proof based on the study of n-dimensional polyhedra, which I shall expound, in the next future, in a more extended memoir.[59]

---

[54]"*Pour que deux variétés fermées soient homéomorphes, il ne suffit donc pas qu'elles aient mêmes nombres de Betti.*", *ibid.*, p. 257 (Poincaré's italics).

[55]*Forstudier til en topologisk Teori for de algebraiske Fladers Sammenhaeng*, Det Nordiske Forlag, København 1898; French translation revised by the author, "Sur l''Analysis situs'", *Bulletin de la Société mathématique de France*, 44 (1916), pp. 161–242.

[56]The "Heegaard diagrams" are still in use today.

[57]*Forstudier til en topologisk Teori for de algebraiske Fladers Sammenhaeng*, cit., p. 5.

[58]*Ibid.*, p. 72 (Heegaard's italics).

[59]"Ces critiques sont en partie fondées; le théorème n'est pas vrai des nombres de Betti *tels que Betti les définit*; c'est ce qui résulte d'un exemple cité par M. Heegaard; c'est ce qui résultait d'un exemple que j'avais moi-même rencontré dans mon Mémoire. Le théorème est vrai, au contraire, des nombres de Betti tels que je les définis; j'en trouvé une démonstration qui est fondée sur la considération des polyèdres a n dimensions et que je développerai prochainement dans un Mémoire plus étendu.", "Sur les nombres de Betti", *Comptes rendus de l'Académie des Sciences*, 128 (1899c), pp. 629–630 = *Œuvres*, vol. VI, p. 289.

In his "Complément à l''analysis situs'" Poincaré provided such a proof, based on the assumption (which he also attempted to prove) that every manifold admits a cellular decomposition).[60] On a more fundamental level, one year later, in his "Second complément à l''analysis situs'"[61] Poincaré introduced the key concepts of *variétés sans torsion* and *variétés à torsion*, whose distinction is made by the inspection of the associated *tableaux d'incidence*.[62] I will pass over these developments and focus on the theorem stated by Poincaré in the very last page of his "Second complément":

> *Each polyhedron which has all its Betti numbers equal to 1 and all tableaux T_q orientable is simply connected, i.e., homeomorphic to the hypersphere.*[63]

This is the first version of the famous Poincaré conjecture: in today's language, it claims that any 3-manifold having the same homology groups with integer coefficients as the 3-sphere is homeomorphic to the 3-sphere. As matter of fact, the statement is false: the zigzag path through which Poincaré would come up with stating an "improved" conjecture appears to fit in quite well with the Lakatosian pattern of the heuristic of "proofs and refutations".[64]

Neither Poincaré nor anyone else worried about proving or disproving the truth of the concluding "theorem" of the "Second complément à l''analysis situs'" over the course of the three/four years that followed. The third and fourth supplements to "Analysis situs",[65] published in 1902, were primarily concerned with questions related to complex analysis and algebraic geometry. Here it will be enough to recall that Poincaré took up the strategy (adopted also by his friend Émile Picard[66]) of describing an algebraic surface as the total space of a family of hyperplane sec-

---

[60]See note 45.

[61]*Proceedings of the London Mathematical Society*, 32 (1900), pp. 277–308 = *Œuvres*, vol. VI, pp. 338–370.

[62]A *variété à torsion* is, in modern terminology, a manifold whose homology groups with integer coefficients contain torsion elements. Given a polyhedral decomposition of a manifold M, the associated *tableau d'incidence* T_q ($0<q<\dim M$) can be tought of as a matrix whose columns correspond to the oriented q-polyhedra of the decomposition and whose rows correspond to the (q-1)-polyhedra; at each intersection of a row and column one places either 0, 1, or −1, depending whether, respectively, the (q-1)-polyhedron is not a face of the q-polyhedron, is a face and has the same orientation, or is a face and has the opposite orientation.

[63]"*Tout polyhèdre qui a tous ses nombres de Betti égaux à 1 et tous ses tableaux T_q bilatères est simplement connexe, c'est-à-dire homéomorphe à l'hypersphère*" (Poincaré's italics). If a polyhedron has all its "tableaux T_q orientable (*bilatères*)", then it has no torsion coefficients in its homology groups ("Second complément à l''analysis situs'", cit., p. 307); in Poincaré's terminology, a manifold is "simplement connexe" if its fundamental group is trivial.

[64]Cf. Imre Lakatos, *Proofs and Refutations. The Logic of Mathematical Discovery*, edited by J. Worrall and E. Zahar, Cambridge University Press, Cambridge (1976) (see, in particular, p. 42).

[65]"Sur certaines surfaces algébriques. Troisième complément à l''analysis situs'", *Bulletin de la Société mathématique de France*, 30 (1902a), pp. 49–70 = *Œuvres*, vol. VI, pp. 373–392; "Sur les cycles des surfaces algébriques. Quatrième complément à l''analysis situs'", *Journal de mathématiques pures et appliquées*, 5th ser., 8 (1902b), pp. 169–214 = *Œuvres*, vol. VI, pp. 397–434.

[66]See Émile Picard & Georges Simart, *Théorie des fonctions algébriques de deux variables complexes*, tome I, Gauthier-Villars, Paris (1897) (in particular, chap. IV).

tions (i.e., as a Lefschetz pencil, in today's language).[67] This approach—transposed by analogy to the realm of topology—led him, in the fifth supplement to "Analysis situs",[68] to develop a new and powerful method for studying the geometry of differentiable manifolds. Poincaré's idea—essentially the same which lies at the basis of Morse theory—consists in "slicing" a given manifold V of dimension m immersed in $\mathbb{R}^k$ by means of a one-parameter family of $(k-1)$-dimensional hypersurfaces $\varphi(x_1, x_2, \ldots, x_k) = t$. In this way, one obtains a "system" $W(t)$ composed (generally) of "a certain number of $(m-1)$-dimensional manifolds $w_1(t), w_2(t), \ldots, w_p(t)$":

> When t varies continuously from $-\infty$ to $+\infty$, the system $W(t)$ varies continuously and *generates* the manifold V. If the manifold V is closed, the manifolds $w_1(t), w_2(t), \ldots, w_p(t)$ are likewise.[69]

Next, each manifold $w_i(t)$ is associated with a point in the Euclidean 3-space; as t varies, the moving points will produce a "sort of network of lines", that Poincaré calls the *squelette* (*skeleton*) of the manifold M:

> Under these circumstances, when t varies in a continuous manner, the points representing the p manifolds $w_1(t), w_2(t), \ldots, w_p(t)$ generate p continuous lines $L_1, L_2, \ldots, L_p$, at least as long as the number p does not change. But this number can change at $t = t_0$, if one of the manifolds decomposes into two, or if, on the contrary, two manifolds merge into one. In the first case one of the lines L bifurcates, in the second case two of the lines L combine into one.[70]

It is perhaps worth recalling that Poincaré had extensively dealt with the phenomenon of bifurcation in his research work about the equilibrium figures of rotating fluid masses.[71] So, his notion of *squelette* of a manifold seems to emerge as the combined outcome of a dual analogy process: on the one hand, the slicing procedure draws inspiration from the theory of algebraic surfaces, on the other, the bifurcating

---

[67]Cf. Simon K. Donaldson, "One hundred years of manifolds topology", in *History of topology*, cit., p. 436.

[68]"Cinquième complément à l''analysis situs'", *Rendiconti del Circolo matematico di Palermo*, 18 (1904a), pp. 45–110 = *Œuvres*, vol. VI, pp. 435–498.

[69]"Quand t variera d'une manière continue de $-\infty$ à $+\infty$, le système $W(t)$ variera d'une manière continue et *engendra* la variété V. Si la variété V est fermée, les variétés $w_1(t), w_2(t), \ldots, w_p(t)$ le seront également.", *ibid.*, p. 436 (Poincaré's italics).

[70]"Dans ces conditions, quand t variera d'une manière continue, les points représentatifs des p variétés $w_1(t), w_2(t), \ldots, w_p(t)$ engendreront p lignes continues $L_1, L_2, \ldots, L_p$; du moins tant que le nombre p ne varie pas. Mais ce nombre peut varier pour $t = t_0$, si l'une des variétés se décompose en deux, ou si, au contraire, deux variétés se réunissent en une seule. Dans le premier cas l'une des lignes L se bifurque, dans le second deux des lignes L se réunissent en une seule.", *ibid.*, p. 437.

[71]See, for example, "Les formes d'équilibre d'une masse fluide en rotation", *Revue générale des sciences pures et appliquées*, 3 (1892b), pp. 809–815 = *Œuvres*, vol. VII, pp. 529–537; "Sur l'équilibre d'un fluide en rotation", *Bulletin astronomique*, 16 (1899b), pp. 161–169 = *Œuvres*, vol. VII, pp. 151–158; "Sur la stabilité de l'équilibre des figures piriformes affectées par une masse fluide en rotation", *Philosophical Transactions*, 198 A (1901a), pp. 333–373 = *Œuvres*, vol. VII, pp. 161–162.

lines are reminiscent of the bifurcating continuous families of equilibrium ellipsoids. The crucial observation, at this point, is the following:

> If we follow one of these lines, L_1, for example, described by the point representing $w_1(t)$, we see that this manifold remains homeomorphic to itself (and in such a way that two manifolds $w_1(t)$ and $w_1(t) + \varepsilon$ corresponding to neighboring points differ very little from each other) *as long as we do not pass through a value t such that* $w_1(t)$ *has a singular point.*
>
> We must then mark the lines of our network at points where the corresponding manifolds $w(t)$ have singular points. These will be the points of division which cut our lines into sections, but as long as we remain on one of these sections the corresponding manifold $w(t)$ will remain homeomorphic to itself.[72]

It is thus necessary "to study these singular points", and Poincaré does not retreat. First, he states (with no proof) the result that is called the "Morse lemma" in today's textbooks of differential topology,[73] namely that in a neighborhood of a point p in $\mathbb{R}^{m+1}$ where all first order partial derivates of an analytic functions φ vanish, there are coordinates $(y_1, y_2, \ldots, y_{m+1})$ such that

$$\varphi = \varphi(p) + \sum A_i y_i^2 - \sum B_k y_k^2$$

where the coefficients A_i, B_k are positive.[74] Next, he proves that

> if a two-dimensional V is orientable, its skeleton will not have singular points other than *culs-de-sac* and bifurcations[75]

and exploits this fact (using as well a good deal of hyperbolic geometry) to classify closed and orientable surfaces.[76] Finally, in a real tour de force, he applies his new method to three-dimensional manifolds.

As paradoxical as it may sound, the crowning achievement of the "Cinquième complément" is the disproving of the theorem stated at the end of the "Deuxième complément". Indeed, Poincaré provides an example (or rather a counterexample) of a 3-manifold, "all Betti numbers and torsion coefficients of which equal 1, but

---

[72]"Si nous suivons l'une de ces lignes L_1, par exemple, décrite par le point représentitif de $w_1(t)$, nous voyons que cette variété $w_1(t)$ reste constamment homéomorphe à elle-même (et cela de telle façon que sur les deux variétés trè voisines $w_1(t)$ et $w_1(t) + \varepsilon$, deux points correspondants diffèrent très peu l'un de l'autre) *tant que l'on ne passe pas par une valeur de t telle que* $w_1(t)$ *ait un point singulier.* Nous devons donc marquer sur les lignes de notre réseau les points qui correspondent aux variétés $w(t)$ qui ont des points singuliers. Ce seront des points de division qui partegeront nos lignes en tronçons, mais tant qu'on suivra l'un de ces tronçons, la variété $w(t)$ correspondante restera homéomorphe à elle-même.", "Cinquième complément à l''analysis situs'", cit., p. 437 (Poincaré's italics).

[73]See, for example, John Milnor, *Morse Theory*, based on lectures notes by M. Spivak and R. Wells, Princeton University Press, Princeton (NJ) (1969), p. 6.

[74]See "Cinquième complément à l''analysis situs'", cit., p. 439.

[75]"Si [...] V a deux dimensions et est bilatère, son squelette n'aura d'autre point singulier que les culs de sac et les bifurcations", *ibid.*, p. 443.

[76]Poincaré's argument follows more or less the same guidelines as the "modern" Morse theoretic proof of the classification theorem for topological surfaces; cf. M. W. Hirsch, *Differential Topology*, Springer-Verlag, New York (1976), Chap. 9.

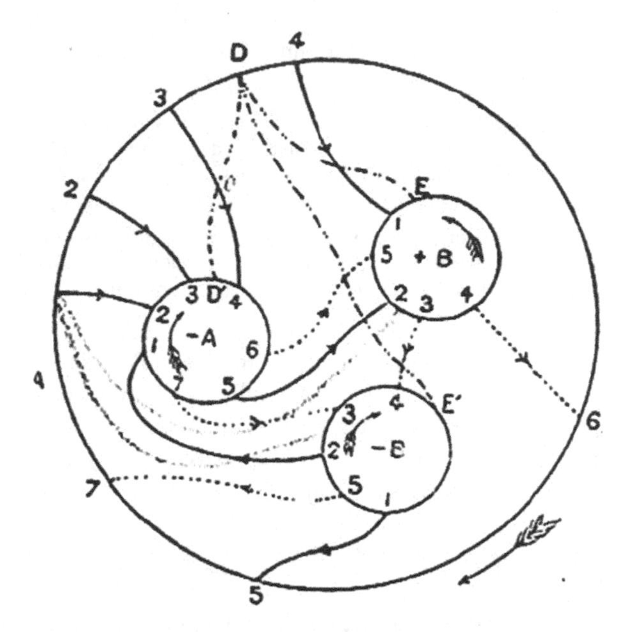

Fig. 1.1 Fig. 4, "Cinquième complément à l' 'analysis situs'", cit., p. 494

which is not simply connected".[77] The procedure followed by Poincaré consists in taking two 3-manifolds V', V'' whose boundary is a surface W of genus 2 (i.e., in today's parlance, two handle-bodies of genus 2) and pasting them together by identifying the boundary surfaces via a suitable homeomorphism, so to obtain a 3-manifold V. Since "every cycle of V is equivalent (i.e., homotopic) to a cycle of W"[78] (an essential lemma that Poincaré establishes after a lengthy argument), in order to determine "the homologies" and the fundamental group it suffices to choose a basis C_1, C_2, C_3, C_4 for the homology of W and then take into account the relations induced by the gluing homeomorphism. To this purpose, Poincaré identifies the "fundamental cycles" K_1', K_2' of V' with C_1, C_3; one has $C_1 \equiv 0$ and $C_3 \equiv 0$. Making reference to Fig. 1.1, the cycle C_1 is represented by the "conjugate circles" $-A, +A$, while C_2 is represented by the "conjugate circles" $-B, +B$. The "fundamental cycles" K_1'', K_2'' of V'' are represented by the arcs of curves running between labeled points on the perimeter of the figure; more precisely,

> The arcs which represent K_1'' are shown as unbroken lines; those which represent K_2'' are dotted. The arrows indicate the sense in which they are traversed.[79]

[77] "[D]ont tous les nombres de Betti et les coéfficients de torsion sont égaux à 1, et qui pourtant n'est pas simplement connexe", "Cinquième complément à l' 'analysis situs'", cit., p. 436.

[78] *Ibid.*, p. 490.

[79] "Les arcs qui représentent K_1'' sont en trait plein; ceux qui représentent K_2'' sont en trait pointillé. Une pointe de flèche placée sur le trait lui-même indique dans quel sens ce trait doit être parcouru.", *ibid.*, p. 494.

By expressing the cycles K_1'', K_2'' in terms of the cycles C_1, C_2, C_3, C_4, Poincaré, after some computations, gets the relations

$$-C_2 + C_4 - C_2 + C_4 \equiv 0; \quad 5C_2 \equiv 0; \quad 3C_4$$

(note that, in this case, the fundamental group is not commutative). These relations—he immediately remarks—are "the relations of the structure in which the substitutions C_2 and C_4 generate the icosahedral group".[80] He can therefore conclude that

> the fundamental group of V cannot reduce to the identical substitution, since it contains the icosahedral group as a subgroup.[81]

On the other hand, Poincaré shows that the first Betti number of V is equal to 0, so that V is what we today call a "homology sphere".

How did Poincaré arrive at producing his example? This is, of course, an unanswerable question: we can only hazard some guesses. We can safely assume that Poincaré was not unaware of Felix Klein's book *Vorlesungen über das Ikosaeder und die Auflösung der Gleichungen vom fünften Grade* (Teubner, Leipzig 1884), where the icosahedral group was described in detail.[82] However, there is evidence that Poincaré had no inkling of what today is called the Hurewicz theorem (namely, the fact that the Abelianization of the fundamental group is the first homology group with integer coefficients): in fact, he did not exploit the property of the icosahedral group of being a perfect group,[83] and instead explicitly computed both the "homology equivalences" and the "homotopy equivalences" of the cycles C_1, C_2, C_3, C_4.

[80] "[L]es relations de structure qui ont lieu entre les deux substitutions C_2 et C_4 qui engendrent le *groupe icosaédrique*", *ibid.*, p. 498 (see notes 81 and 82).

[81] "Le groupe fondamental de V ne saurait se réduire à la substitution identique, puisqu'il contient comme sous-groupe le groupe icosaédrique.", *ibid.*

[82] Klein gives a description of the icosahedral group as the group generated by the operation S, T with the relations $S^5 = 1$, $T^2 = 1$, $(ST)^3 = 1$ (in multiplicative notation; see *Vorlesungen über das Ikosaeder...*, cit., p.41); by letting $C_2 = S$ and $C_4 = ST$, one obtains Poincaré's relations (in additive notation). The same relations had been previously derived by William Rowan Hamilton in his papers "Memorandum respecting a new system of roots of unity", *The London, Edinburgh and Dublin Philosophical Magazine and Journal of Science*, 4th ser., 12 (1856), p. 446 and "Account of the icosian calculus", *Proceedings of the Royal Irish Academy*, 6 (1858), pp. 415–416.

[83] Let us briefly review a possible realization of Poincaré's homology sphere using the language and tools of differential topology. The icosahedral group I is the symmetry group of the icosahedron; it is isomorphic to the group of even permutations of 5 elements, i.e., the alternating group A_5. The group I is of order 60, simple (i.e., it does not contain any proper normal subgroup) and perfect (i.e., $[I, I] = I$); it admits the presentation $\langle x, y | (xy)^2 = x^3 = y^5 = 1 \rangle$. Since the isometries of the icosahedron are proper rotations, there is a natural immersion $I \subset \mathrm{SO}(3)$. Let us consider the double cover $\mathrm{SU}(2) \to \mathrm{SO}(3)$ and denote by I^* the inverse image of I. The group I^*—called the binary icosahedral group—is of order 120 and perfect, but no longer simple (its center is the group of the two elements $\{1, -1\}$); it admits the presentation $\langle x, y | (xy)^2 = x^3 = y^5 \rangle$. Now $\mathrm{SU}(2)$ is homeomorphic to the three-dimensional sphere S^3 and it can be shown that the quotient space $\mathrm{SU}(2)/I^*$ is smooth. The fundamental group of $\mathrm{SU}(2)/I^*$ is, of course, I^*; by the Hurewicz theorem, its first homology group is the abelianization of I^*, namely $I^*/[I^*, I^*] = 0$; moreover, one can prove that $H_2(\mathrm{SU}(2)/I^*, \mathbb{Z}) = 0$ (this follows from the fact that I^* is superperfect). In

Moreover, it seems certain that he believed that his example was only one among many possible others ("[...] nous nous bornerons à donner un example"), while we know that he stumbled over the *only possible* example.[84] In conclusion, we can plausibly suppose that Poincaré essentially proceeded by trial and error, perhaps keeping in his mind (or in the back of his mind) that the icosahedral group could have been a workable algebraic object. Figure 1.1 above is drawn well (at least in its printed version), but behind it lie, most probably, dozens of geometric experiments on figures drawn badly.[85]

As it almost universally known, the "Cinquième complement" ends with the statement of what will be later called the "Poincaré conjecture", which is but a mere question in its original formulation:

> Is it possible for the fundamental group of V to reduce to the identity without V being simply connected?[86]

Even a cursory review of the various attempts to solve this problem it until the solution provided by Grigori Y. Perelman in 2003[87] "nous entraînerait"—I can say in Poincaré's words—"trop loin".

References

Appell, P. (1925). *Henri Poincaré*. Paris: Plon.

Bartocci, C. (1995). Introduzione. In H. Poincaré (Ed.), *Geometria e caso. Scritti di matematica e fisica* (pp. VII–L). Torino: Bollati Boringhieri (repr. 2013).

Bellivier, A. (1956). *Henri Poincaré ou la vocation souveraine*. Paris: Gallimard.

---

conclusion, the manifold $\mathrm{SU}(2)/I^*$ is an integral homology sphere, actually homeomorphic to Poincaré's example V. For further details see R. C. Kirby & M. G. Scharlemann, "Eight faces of the Poincaré homology sphere", in *Geometric Topology. Proceedings of the 1977 Georgia Topology Conference*, edited by J. C. Cantrell, Academic Press, New York (1979), pp. 113–146; N. Saviliev, *Invariants for Homology 3-Spheres*, Encyclopedia of Mathematical Sciences, vol. 140, Springer-Verlag, Berlin-Heidelberg (2002).

[84]This result was proved by Michel A. Kervaire in his paper "Smooth homology spheres and their fundamental groups", *Transactions of the American Mathematical Society*, 144 (1969), pp. 67–72: if M is a 3-dimensional manifold such that $H_*(M) = H_*(\mathrm{S}^3)$ and its fundamental group Γ is finite, then either $\Gamma = \{1\}$ or else $\Gamma = I^*$.

[85]"It is clear that in order to arrive at his example of a nonsimply-connected homology sphere [...] Poincaré must have done a good deal of experimentation with Heegaard diagrams, of genus 2, and presumably of higher order too", C. McA. Gordon, "3-dimensional topoogy", in *History of Topology*, cit., pp. 449–489 (the quote is at p. 462).

[86]"Est-il possible que le groupe fondamental de V se réduise à la substitution identique, et que pourtant V ne soit pas simplement connexe?", "Cinquième complément à l''analysis situs'", cit., p. 498.

[87]See, for example, J. Stillwell, "Poincaré and the early history of 3-manifolds", *Bulletin of the American Mathematical Society*, new ser., 49 (2012), pp. 555–576; C. Rourke, "La congettura di Poincaré", in *La matematica. II. Problemi e teoremi*, edited by C. Bartocci, Einaudi, Torino (2008), pp. 731–763. It should be pointed out that Perelman solved a more general problem than the Poincaré conjecture, namely Thusrston's geometrization conjecture.

Betti, E. (1871). "Sopra gli spazi di un numero qualunque di dimensioni", *Annali di matematica pura e applicata*, 2nd ser., *4*, pp. 140–158. Also in *Opere matematiche*, vol. II, Hoepli, Milano, pp. 273–290.

Birkhoff, G. D. (1913). "Proof of Poincaré's last geometric theorem", *Transactions of the American Mathematical Society*, *14*(1913), pp. 14–22. Also in *Collected Mathematical Papers*, vol. I (pp. 673–811). Dover, New York.

Bottazzini, U., & Gray, J. (2013), *Hidden harmony – Geometric fantasies. The rise of complex function theory*. New York: Springer.

Cartan, É. (1928). Sur le nombres de Betti des espaces de groupes clos. *Comptes rendus de l'Académie des Sciences, 187,* 196–198.

Clebsch, A. (1864). Über die Anwendung der Abelschen Functionen in der Geometrie. *Journal für die reine und angewandte Mathematik, 63,* 189–243.

Clifford, W. K. (1877). On the canonical form and dissection of a Riemann's surface. *Proceedings of the London Mathematical Society*, *8*, 292–304. Also in R. Tucker (Ed.), *Mathematical papers* (pp. 241–254). Macmillan, London 1882 (repr. AMS Chelsea Publishing, Providence (RI) 2007).

De Rham, G. (1931) Sur l'analysis situs de variétés à n dimensions. *Journal de mathématiques pures et appliquées*, 9th ser., *10*, 115–200.

Dieudonné, J. (1989). *A history of algebraic and differential topology*. Basel-Boston: Birkhäuser.

Donaldson, S. K. (1999). One hundred years of manifolds topology. In I. M. James (Ed.), *History of Topology* (pp. 435–447). Amsterdam: Elsevier.

Dyck, W. F. A. (1885) On the 'Analysis situs' of threedimensional spaces. *Report of the fifty-fourth meeting of the British association for the advancement of science held at montreal in August and September 1884*, John Murray, London, p. 648.

Golé, C., & Hall, G.R. (1992). Poincaré's proof of Poincaré's last geometric theorem. In R. McGehee & K. R. Meyers (Eds.), *Twist mappings and their applications* (pp. 135–151). New York: Springer.

Gordon, C. M. A. (1999). 3-dimensional topology. In I. M. James (Ed.), *History of topology* (pp. 449–489). Amsterdam: Elsevier.

Gray, J. (2013). *Henri Poincaré: A scientific biography*. Princeton (NJ) and London: Princeton University Press.

Hamilton, W. R. (1856a). Memorandum respecting a new system of roots of unity. *The London, Edinburgh and Dublin Philosophical Magazine and Journal of Science*, 4th ser., *12*, 446.

Hamilton, W. R. (1856b). Account of the icosian calculus. *Proceedings of the Royal Irish Academy, 6*(1858), 415–416.

Heegaard, P. (1898). *Forstudier til en topologisk Teori for de algebraiske Fladers Sammenhaeng*, Det Nordiske Forlag, København (French translation revised by the author, "Sur l''Analysis situs'", *Bulletin de la Société mathématique de France*, *44*(1916), 161–242).

Hirsch, M. W. (1976). *Differential topology*. New York: Springer.

Hurwitz, A. (1893). Über algebraische Gebilde mit eindeutigen Transformationen in sich. *Mathematische Annalen 41*, 403–442. Also in *Mathematische Werke*, herausgegeben von der Abteilung fur Mathematik und Physik der eidgenossischen technischen Hochschule in Zurich, vol. 1, Birkhäuser, Basel 1932 (repr.1962), pp. 391–430.

Jordan, C. (1866). "Des contours tracés sur les surfaces", *Journal de mathématiques pures et appliquées*, 2nd ser., *11*, 110–130.

Kervaire, M. A. (1969). Smooth homology spheres and their fundamental groups. *Transactions of the American Mathematical Society, 144,* 67–72.

Kirby, R. C., & Scharlemann, M. G. (1979) Eight faces of the Poincaré homology sphere. In J. C. Cantrell (Ed.), *Geometric topology. Proceedings of the 1977 Georgia Topology Conference* (pp. 113–146). New York: Academic Press.

Klein, F. (1884). *Vorlesungen über das Ikosaeder und die Auflösung der Gleichungen vom fünften Grade*. Leipzig: Teubner.

Lagrange, J. –L. (1788). *Méchanique analytique*, Chez la veuve Desaint, Paris.

Lakatos, I. (1976). *Proofs and refutations*. In J. Worrall & E. Zahar (Eds.), *The logic of mathematical discovery*. Cambridge: Cambridge University Press.

Le livre du centenaire de la naissance de Henri Poincaré, 1854–1954, Gauthier-Villars, Paris (1955).
Milnor, J. (1969). *Morse Theory, based on lectures notes by M. Spivak and R. Wells*. Princeton (NJ): Princeton University Press.
Möbius, A. F. (1863). "Theorie der elementaren Verwandtschaften", *Berichte über die Verhandlungen der Königlich Sächsischen Gesellschaft der Wissenschaften zu Leipzig*, Mathematisch-Physische Classe, 17, pp. 31–68. Also in Gesammelte Werke, vol. II, herausgegeben von F. Klein, Hirzel, Leipzig 1886, pp. 433–471.
Picard, É., & Simart, G. (1897). *Théorie des fonctions algébriques de deux variables complexes, tome I*. Paris: Gauthier-Villars.
Poincaré, H. (1881). Mémoire sur les courbes définies par une équation différentielle (première partie). *Journal de mathématiques pures et appliquées*, 3rd ser., *7*, 375–422. Also in *Œuvres*, vol. I, pp. 3–44.
Poincaré. (1881). "Sur les courbes définies par une équation différentielle", *Comptes rendus de l'Académie des Sciences*, *93*, 951–952. Also in *Œuvres*, vol. I, pp. 85–85.
Poincaré. (1882a). Mémoire sur les courbes définies par une équation différentielle (deuxième partie). *Journal de mathématiques pures et appliquées*, 3rd ser., *8*, 251–296. Also in *Œuvres*, vol. I, pp. 44–84.
Poincaré. (1882b). Théorie des groupes fuchsiens. *Acta Mathematica*, *1*, 1–62. Also in *Œuvres*, vol. II, pp. 108–168.
Poincaré. (1885). Sur les courbes définies par les équations différentielles (première partie). *Journal de mathématiques pures et appliquées*, 4th ser., *1*, 167–244. Also in *Œuvres*, vol. I, pp. 90–161.
Poincaré. (1886). Sur les courbes définies par les équations différentielles (deuxième partie). *Journal de mathématiques pures et appliquées*, 4th ser., *2*, 151–217. Also in *Œuvres*, vol. I, pp. 167–222.
Poincaré. (1887). Sur les résidus des intégrales doubles. *Acta Mathematica*, *9*, 321–380. Also in *Œuvres*, vol. III, pp. 440–489.
Poincaré. (1892a). *Les méthodes nouvelles de la mécanique céleste. Tome I. Solutions périodiques – Non-existence des intégrales uniformes –Solutions asymptotiques*, Gauthier-Villars, Paris.
Poincaré. (1892b). Les formes d'équilibre d'une masse fluide en rotation. *Revue générale des sciences pures et appliquées*, *3*, 809–815. Also in *Œuvres*, vol. VII, pp. 529–537.
Poincaré. (1892c). Sur l'analysis situs. *Comptes rendus de l'Académie des Sciences*, *115*, 603–606. Also in *Œuvres*, vol. VI, pp. 189–192.
Poincaré, (1893). Le continu mathématique. *Revue de métaphysique et de morale, 1*, 26–34.
Poincaré. (1895). Analysis situs. *Journal de l'École Polytechnique*, *1*, 1–121. Also in *Œuvres*, vol. VI, pp. 193–288.
Poincaré. (1899a). *Les méthodes nouvelles de la mécanique céleste. Tome III. Invariants intégraux – Solutions périodiques de deuxième genre – Solutions doublements asymptotiques*, Gauthier-Villars, Paris.
Poincaré. (1899b). Sur l'équilibre d'un fluide en rotation. *Bulletin astronomique*, *16*, 161–169. Also in *Œuvres*, vol. VII, pp. 151–158.
Poincaré. (1899c). Sur les nombres de Betti. *Comptes rendus de l'Académie des Sciences*, *128*, 629–630. Also in *Œuvres*, vol. VI, p. 289.
Poincaré. (1899d). Complément à l''analysis situs'. *Rendiconti del Circolo matematico di Palermo*, *13*, 285–343. Also in *Œuvres*, vol. VI, pp. 290–337.
Poincaré. (1900). Second complement à l''analysis situs'. *Proceedings of the London Mathematical Society*, *32*, 277–308. Also in *Œuvres*, vol. VI, pp. 338–370.
Poincaré. (1901a). Sur la stabilité de l'équilibre des figures piriformes affectées par une masse fluide en rotation. *Philosophical Transactions*, *198 A*, 333–373. Also in *Œuvres*, vol. VII, pp. 161–162.
Poincaré. (1901b). Sur les propriétés arithmétiques des courbes algébriques. *Journal de mathématiques pures et appliquées*, 5th ser., *7*, 161–233. Also in *Œuvres*, vol. V, pp. 483–550.
Poincaré. (1902a). Sur certaines surfaces algébriques. Troisième complément à l''analysis situs'. *Bulletin de la Société mathématique de France*, *30*, 49–70. Also in *Œuvres*, vol. VI, pp. 373–392.

Poincaré. (1902b). Sur les cycles des surfaces algébriques. Quatrième complément à l''analysis situs'. *Journal de mathématiques pures et appliquées*, 5th ser., *8*, 169–214. Also in *Œuvres*, vol. VI, pp. 397–434.

Poincaré. (1902c). *La science et l'hypothèse*, préface de Jules Vuillemin, Flammarion, Paris.

Poincaré. (1904a). Cinquième complément à l''analysis situs'. *Rendiconti del Circolo matematico di Palermo*, *18*, 45–110. Also in *Œuvres*, vol. VI, pp. 435–498.

Poincaré. (1904b). *La valeur de la science*, préface de Jules Vuillemin, Flammarion, Paris.

Poincaré. (1905). *Science and Hypothesis* (W. J. G., with a preface by J. Larmor, Trans.). The Walter Scott Publishing Co., London & Newcastle-on-Tyne.

Poincaré. (1908). *Science et méthode*. Paris: Flammarion.

Poincaré. (1909). "L'avenir des mathématiques", in *Atti del IV Congresso internazionale dei matematici*, a cura di G. Castelnuovo, Tipografia della Reale Accademia dei Lincei, Roma, pp. 167–182 (also in *Rendiconti del Circolo matematico di Palermo*, *16*(1908), pp. 152–168; *Revue générale des sciences pures et appliquées*, *19*(1909), pp. 930–939; *Scientia. Rivista di scienza*, 2nd year, *3*(1908), pp. 1–23; *Bulletin des sciences mathématiques*, 2nd ser., *32*(1908), pp. 168–190).

Poincaré. (1912). Sur un théorème de géométrie. *Rendiconti del Circolo matematico di Palermo*, *33*, 375–407. Also in *Œuvres*, vol. VI, pp. 499–538.

Poincaré. (1958). *The value of science,* (G. B. Halsted Trans.). Dover, New York (orig. publ. in 1913 by Science Press in *Foundations of Science*).

Poincaré. (1914). *Science and Method* (Francis Maitland, with a preface by B. Russell Trans.). London, Edinburgh, Dublin & New York: Thomas Nelson and Sons (repr. Thoemmes Press, Bristol 1996).

Poincaré. (1916–1956). *Œuvres*, 11 vv., various eds., Gauthier-Villars, Paris.

Poincaré. (1921). Analyse de ses travaux scientifiques faite par H. Poincaré. *Acta Mathematica*, *38*, 36–135.

Poincaré. (1997). *Trois suppléments sur la découverte des fonctions fuchsiennes,* J. Gray and S. A. Walters (Eds.). Berlin: Akademie Verlag & Paris: Albert Blanchard.

Pont, J.-C. (1974). *La topologie algébrique des origines à Poincaré*. Paris: Presses Universitaires de France.

Riemann, B. (1857). Theorie der Abel'schen Functionen. *Journal für die reine und angewandte Mathematik*, *54*, 115–155. Also in Gesammelte Mathematische Werke..., pp. 88–142.

Riemann, B. (1990). *Gesammelte Mathematische Werke, Wissenschaftlicher Nachlass und Nachträge/Collected Papers*, nach der Ausgabe von H. Weber und R Dedekind, neu herausgegeben von Raghavan Narasimhan, Springer, Berlin-Heidelberg – Teubner, Leipzig (repr. 2014).

Rourke, C. (2008). La congettura di Poincaré. In C. Bartocci (Ed.) *La matematica. II. Problemi e teoremi* (pp. 731–763). Einaudi, Torino.

Sakaria, K. S. (1999). The topological work of Henri Poincaré. In I. M. James (Ed.), *History of topology* (pp. 121–167). Amsterdam: Elsevier.

Saviliev, N. (2002). Invariants for homology 3-spheres. *Encyclopedia of mathematical sciences*, vol. 140. Berlin-Heidelberg: Springer.

Scholz, E. (1980). *Geschichte des Mannigfaltigkeitsbegriffs von Riemann bis Poincaré*. Boston-Basel-Stuttgart: Birkhäuser.

Stillwell, J. (2012). Poincaré and the early history of 3-manifolds. *Bulletin of the American Mathematical Society*, new ser., *49*, 555–576.

Toulouse, É. (1910). *Enquête médico-psychologique sur la superiorité intellectuelle: Henri Poincaré*. Paris: Flammarion.

Volkert, K. (2002). *Das Homöomorphismusproblem insbesondere der 3-Mannigfaltigkeiten, in der Topologie 1892-1935*, Kimé, Paris.

Whitehead, J. H. C. (1940). On C^1-complexes. *Annals of Mathematics*, 2nd ser., *41*, 809–824.

Chapter 2
Scientific Philosophy and Philosophical Science

Hourya Benis Sinaceur

Abstract Philosophical systems have developed for centuries, but only in the nineteenth century did the notion of scientific philosophy emerge. This notion presented two dimensions in the early twentieth century. One dimension arose from scientists' concern with conceptual foundations for their disciplines, while another arose from philosophers' appetite for more rigorous philosophy. In the current paper, I will focus on David Hilbert's construct of "critical mathematics" and Edmund Husserl' and Jules Vuillemin's systematic philosophy. All these three thinkers integrated Kant's legacy with the axiomatic method. However, they did so in different ways, with Hilbert's goal being the opposite of that of Husserl or Vuillemin. Specifically, I will show how the scientism of Hilbert's mathematical epistemology aimed at shattering the ambition of philosophy to submit mathematical practices and problems to philosophy's own principles and methods, be they transcendental or metaphysical. On the other hand, phenomenology promoted the idea of a non-exact philosophical rigour and highlighted the need of a point of view encompassing positive sciences, ontology, and ethical values in connection with the dominant category of sense/meaning, and Jules Vuillemin built on from the inseparability of thought—scientific or philosophical—from the metaphysics of free will and choice.

2.1 Introduction

The rapid evolution, in many and varied ways, of the axiomatic approach in the 19th Century, coupled with the renewal of logic to which Gottlob Frege gave a decisive boost, shook ancient philosophical certainty concerning the status of and mutual relations between fundamental concepts such as intuition, concept, experience, object, subject and consciousness. At the same time the prestige of science triggered the revival of the idea of "philosophy as science". Physical sciences, mathematics and

H. Benis Sinaceur (✉)
Institut d'Histoire et Philosophie des Sciences et des Techniques (IHPST),
Université Paris 1 Panthéon-Sorbonne, CNRS, ENS Ulm, Paris, France
e-mail: houryabenis@yahoo.fr

H. Tahiri (ed.), *The Philosophers and Mathematics*, Logic, Epistemology, and the Unity of Science 43, https://doi.org/10.1007/978-3-319-93733-5_2

mathematical logic shaped the philosophical requirement for rigour, although there is neither test nor proof for philosophical assumptions.

The idea of philosophy as science firmly establishes itself in the first third of the 20th century. Most of its proponents share the desire to counteract the influence of Hegel's system that the author had presented as philosophical science and which had, in the 1830s, been a dominant doctrine of Berlin University and the Prussian State. But they do not have a common vision of what should replace it. The idea of philosophy as science is not univocally determined. It includes distinct elements of varying composition, borrowing from both the philosophical tradition and new scientific methods. Without being exhaustive, I will mention three or four of these compositions: the critical mathematics of Hilbert and Leonard Nelson, the scientific philosophy of Husserl and the systematic philosophy of Jules Vuillemin. Drawing from the common source of demand for rigour combining critical philosophy and the axiomatic method, the authors of these compositions develop very different, even antithetical, designs from the intersection of philosophy and science.

2.2 Science Enters Its Critique Phase: Hilbert, Hessenberg, Nelson

The first design consists of "critical" science, namely the adoption of the Kantian perspective for characterising the axiomatic renewal of science. Unlike dogmatism that takes its principles for granted, critical philosophy justifies its own principles. In this sense, a critical attitude itself appears to bring about rigorous standards for science. Reflexive reasoning effectively exercises control over its capacity and determines the conditions of its exercise. The axiomatic approach is credited with the ability to perform such a reflection back onto scientific objects and procedures to accurately delineate the extent of their validity. Exploring results previously acquired in different areas of mathematics, the axiomatic approach reflexively establishes objects of a new kind, namely structures, and it sets new standards of truth. Truth is no longer confused with evidence imposing itself on the so-called immediate knowledge (commonly known as intuition). The axiomatic approach organises scientific propositions into self-regulating systems of deduction based on axioms *set down as assumed truths.*[1]

From Descartes to Husserl, philosophy first conceived reflexiveness as self-conscious reflection. With Kant, self-consciousness abandons the certainty of cogito and the divine guarantee of clear ideas for the human tribunal of critique: the subject looks back on his own actions to first submit them to analysis with a view to revealing the a priori conditions for possibility of knowledge of objects. Pure con-

[1]Cf. The Frege-Hilbert correspondence (published by I. Angelelli in Gottlob Frege's *Kleine Schriften*, Darmstadt, Wissenschatmiche Buchgesellschaft, 1967): to Frege, who advocates the intuitive origin of geometrical axioms, Hilbert responds "Sobald ich habe ein Axiom gesetzt, ist es wahr und vorhanden" ("As soon as I have posed an axiom, it is true and available").

sciousness, native and immutable, is a condition of empirical consciousness; it is a priori and necessary condition of both experience and objects of experience. Kant calls this pure consciousness "transcendental apperception." It is the relationship to this apperception that constitutes the *form* of all understanding of the object.

In the interpretation made by scientists, reflexiveness usually abandons self-consciousness, whether pure or merely empirical, so as to vindicate only the critical ingredient, and even then in a manner little in keeping with its Kantian origins. David Hilbert is well known for placing under the banner of Critique the two essential branches of his foundations of science programme[2]: the axiomatic method and proof theory. He therefore contributed to feeding, or even starting, discussions focused on whether and to what extent the framework of *Critique of Pure Reason* still has a legitimate claim to provide us with foundations of science after the scientific and epistemological revolutions of non-Euclidean geometry, relativity theory, and quantum physics.

2.2.1 The Axiomatic Method and Intuition

Hilbert's *Foundations of Geometry* (1899) presents and classifies the axioms of geometry in order to show various combinations generating different geometries: Euclidean geometry, Cartesian algebraic geometry, non-Archimedean geometry, projective geometry, etc. In his introduction Hilbert first highlights the famous words of Kant: "all human knowledge begins with intuitions, proceeds from thence to concepts, and ends with ideas".[3] Then he presents his axiomatic ordering as an "analysis of our intuition of space". Describing what the mathematician's intuition is, or what it consists of, is a theme that runs through Hilbert's whole work. Indeed, allegiance of the Göttingian mathematician to the Königsbergian philosopher is not merely decorative or transient and is not limited to this highlight, which indicates that Hilbert intentionally placed his axiomatic ordering of geometric propositions within Kant's perspective. The conclusion, the last section of which I will quote, explains what Hilbert meant by the analysis of the intuition of space.

> The present work is a critical investigation of the principles [Prinzipien] of geometry. In this investigation the guiding precept [der leitende Grundsatz] is to examine each question so as to prove outright if the answer is *possible* [my emphasis] when some limited means are imposed in advance. [...]
>
> In modern mathematics the question of the *impossibility* [Hilbert's emphasis] of certain solutions or problems plays a leading role, and the attempts to answer such questions have often given the opportunity to discover new and fruitful areas of research. Examples of this include the demonstration by Abel of the impossibility of solving by radicals the 5th degree

[2]When Hilbert was born in Königsberg in 1862, Kant had been dead for nearly 60 years (1804), but the considerable prestige of the thinker of the Enlightenment was far from being extinguished.

[3]*Critique of Pure Reason* (*CPR*), Transcendental Logic, Transcendental Dialectic, Appendix A702/B730. Another well-known phrase: "If all our knowledge begins with experience, it does not follow that it derives all from the experience."

equation, the discovery of the impossibility of proving the parallel axiom, and the theorems of Hermite and Lindemann on the impossibility of constructing by algebraic means the numbers e and π.

The precept by which one must always examine the principles of the *possibility* [my emphasis] of proof is closely related to the requirement for "purity" of methods in proof, which in recent times has been considered of the highest importance by many mathematicians. This requirement is basically nothing more than a *subjective* [my emphasis] version [Fassung] of the precept followed here. Indeed our present investigation attempts to explain generally what axioms, assumptions or auxiliary means are necessary to establish the truth of an elementary geometric proposition, and all that remains to be gauged is which method of proof is preferable in each case from the adopted point of view.[4]

Certainly, Hilbert's guiding precept is that of conditions of possibility. But, and this is essential, the precept is played out on a case by case basis and in each case it is circumscribed by both the problem to be solved and the limited resources previously allowed for the solution. Obviously those conditions of possibility are by no means universal and necessary, unconditional and unchanging principles of experience. Hilbert limits himself to the specific experience of mathematics, and furthermore to particular experiences concerning the demonstration of definite propositions in definite situations. Moreover, in mathematics, *proving* the *im*possibility of a solution does not establish a higher absolute domain, comparable to the Kantian realm of things-in-themselves, and does not close the door to exploring *other* possibilities by changing the way of formulating the problem and the means of proof.
However, the parallel between the axiomatic approach and Kant's critical enterprise remains fixed in Hilbert's mind. In 1917, in "Axiomatisches Denken" (p. 148),[5] Hilbert says that critical examination [die kritische Prüfung] of certain proofs leads to new axioms being formed from more general and fundamental propositions than those previously held as such. This axiomatic deepening, also characterised as "proof critique" [Beweiskritik], represents the first stage of the critique of mathematical reason, which overturns the dogmatism of established evidence and practices.[6] In 1922 Hilbert links again the axiomatic method to Critique, saying on this, and to my knowledge only occasion, that axiomatising is nothing other than thinking in the light of consciousness [mit Bewußtsein denken],[7] but Hilbert added that the most important thing is the *mathematical resolution* of questions of theory of knowledge posed by the axiomatic method. He then presents the work of Dedekind and Frege on arithmetic as the inauguration of "modern critique of Analysis" (p. 162). In 1930

[4] *Grundlagen der Geometrie*, 10. Auflage, Stuttgart, B.G. Teubner, 1968, Schlusβwort, 124–125.

[5] *Mathematische Annalen* 78 (1918), 405–415, *Gesammelte Abhandlungen* III, Berlin, Springer, 1935, 146–156.

[6] In 1904, 5 years after the publication of Hilbert's *Foundations of Geometry*, the philosopher Leonard Nelson, whose habilitation and career Hilbert supervised and facilitated, would explicitly set out the programme for "transferring Critique to the axiomatic systems of mathematics in order to constitute a specific scientific discipline: critical mathematics"; cited by Volker Peckhaus in *Hilbertprogramm und Kritische Philosophie. Das Göttinger Modell interdisziplinärer Zusammenarbeit zwischen Mathematik und Philosophie*, Göttingen, Vandenhoeck und Ruprecht, 1990, p. 158. I return to critical mathematics below.

[7] Neubegründung der Mathematik, *Gesammelte Abhandlungen* III, p. 161.

in "Naturerkennen und Logik"[8] he briefly examines the Kantian a priori (to which I return below). The recourse to critical reason is constant, therefore.

Let us examine whether and to what extent this recourse is legitimate given Hilbert's actual mathematical practice. According to Hilbert, showing which geometric theorems are logically derivable from a definite set of axioms, i.e. showing how different geometry systems are each related to a definite conjunction of axioms expressing the necessary and sufficient conditions for developing the whole system, thus showing the need for a deductive link between principles and consequences, is "the analysis of our intuition of space". Then this analysis displays different concepts or systems of space. To demonstrate the *logical* compatibility or deducibility between geometric propositions, to distinguish between assumed propositions (axioms) and demonstrated propositions (theorems), to ask whether an axiom, given the other axioms simultaneously admitted, is removable or indispensable[9] is to gain in mathematics the rigour acquired in philosophy by *critical attitude*.

The explanation seems Kantian, for Hilbert uses Kant's terminology: 'critique', 'condition of possibility', and 'intuition'. But the terminology can mislead, as notable philosophers have been. Hilbert's good faith is not in question, but one has to look closer the text of *The Foundations of Geometry*. First and foremost it is clear that it is the *logical analysis* of the *objective* links of dependence between mathematical statements that is charged with taking on a critical attitude. This way is actually closer to the objectivist spirit of Bolzano, Dedekind, and Frege than to Kant's subjectivism. Hilbert's epistemological effort consists precisely in replacing the Kantian "subjectivist version" with an objectivist version or, if you will, of restoring the rights of formal logic over transcendental logic. Additionally Hilbert's understanding of 'formal logic' in the *Foundations of Geometry* and other works does not coincide with Kant's definition of formal logic. After Frege's *Begriffschrift* (1879) 'formal logic' had a definitely different meaning than before. Let me explain in detail my arguments.

1. The scope of a geometrical proposition, its meaning/significance [Bedeutung], as Hilbert says, is shown by a set of variations governing the connection between axioms and theorems. The investigation concerns the conditions of *validity* of certain *mathematical content* according to different settings. So the possibility in question is material in as much as it concerns a formal logical structure.
2. Accordingly, the conditions in question operate *locally*. The very possibility of varying the choice of axioms (with or without the parallel axiom, with or without the Archimedes' axiom, etc.) depending on the type of geometry that one wants to construct, itself attests to their regional (not universal) character and their relative

[8]*Naturwissenschaften* **18**, 959–963; in *Gesammelte Abhandlungen* III, 378–387.

[9]Notable examples: Pascal's theorem cannot be proven in the simultaneous absence of the axioms of congruence and the Archimedes' axiom; Desargues' theorem is provable in space from the axioms of incidence, but in the plane it is necessary to add the 5 axioms of congruence; if we add just one more axiom which negates the existence of points outside the plane, we cannot construct projective geometry unless you also add the theorem of Desargues as an axiom, demonstrating the constitutive role thereof within the construction of planar projective geometry.

necessity. To counter dogmatism is not to deny the universal, but to *contextualise* it. Indeed, Hilbert stresses that the axiomatic method does not change only the content but also the modality of our mathematical beliefs, and therefore dissolves dogmatism by explaining the logical connections between mathematical propositions. In this context Hilbert speaks of "necessary relativism",[10] which in this case, one should add, has nothing to do with Kant's demarcation between absolute things-in-themselves and knowable phenomena. Axiomatic relativism stems from the many systems corresponding to the same web of mathematical propositions and the many a priori possible interpretations for the same system. Dissolving dogmatism comes down here to abandon the idea of absolute truth of mathematical propositions in favour of the idea of truth relating to axiomatic systems.

3. Moreover, considering those desired conditions as axioms, which formally express the properties or relationships deemed fundamental, is completely different from designating an *empty form* of relationship between our understanding and things that appear to us only qua objects of experience or phenomena. In fact, Hilbert's description and advocacy of the axiomatic method relates to scientific practice which singles out sets of primitive propositions as the basis for proving theorems, it does not concern the theory of knowledge in general. Besides, philosophy plays only a peripheral role in *The Foundations of Geometry*, yet it is useful for piquing the interest of philosophers and bringing them into the mathematical school.
4. Finally, proposing an "analysis of our intuition of space" is to state that space is an object of our intuition and thus an intuitive datum—real or conceptual. Yet for Kant space is not a datum but the *form* of sensory data provided by perception. It is "the *subjective* condition of sensitivity under which alone external intuition is possible for us".[11] And besides, Kant distinguishes between sensory intuition (empirical intuition) and pure intuition,[12] the latter conditioning the former. Space is pure intuition (and not a pure concept), it is a formal a priori condition of experience, "the basic form of all external sensation".

> *Space is not something objective and real*, [non aliquid objectivi et realis]; nor a substance, nor an accident, nor a relation; instead, it is *subjective* and ideal, and originates from the mind's nature in accord with a fixed law [natura mentis stabili lege profiscens] as a scheme for coordinating everything sensed externally.[13]

[10]Neubegründung der Mathematik, *Gesammelte Abhandlungen* III, p. 169.

[11]*CPR*, Transcendental Aesthetic I, § 2, A26/B42 (I have highlighted 'subjective').

[12]"Space and time are pure forms [of perception], sensation in general its matter. We can cognize only the former a priori, i.e., prior to all actual perception, and they are therefore called pure intuition; the latter, however, is that in our knowledge that is responsible for its being called a posteriori knowledge, i.e., empirical intuition. The former adheres to our sensibility absolutely necessarily, whatever sort of sensations we may have". *CPR*, Transcendental Aesthetic I, § 8, A42-43/B59-60.

[13]Dissertation de 1770, Paris, Vrin, 1951, p. 55 (Kant's emphasis), English W. J. Eckoff, Columbia College, 1894, p. 65 (https://archive.org/details/cu31924029022329). *CPR*, Transcendental Aesthetic I, § 2, A28/B44: "We maintain [...] the empirical reality of space in regard to all possible external experience, although we must admit its transcendental ideality; in other words, that it is

It follows that on the requirement for the a priori representation of space rests "the apodictic certainty of all geometric principles, and the possibility of their construction a priori."[14] The properties of a triangle, for example, are constructed a priori in pure intuition. Similarly, the three-dimensional Euclidean space is pure a priori intuitive evidence. Space as a form of both experience and objects of experience is a priori and necessary representation, it is one of the principles of a priori knowledge;[15] in particular it is a necessary subjective principle of geometric propositions, which are "always apodictic, that is, *united with the consciousness of their necessity*" (emphasis added).

Thus, for Kant space as pure intuition is the principle of the axioms of geometry (Euclidean geometry, the only known then). On the contrary, for Hilbert sets of axioms are the principles of the geometry they determine. In Kantian language, we can say that Hilbert totally disregards "transcendental ideality" of space, that is, the fact that for Kant space *is nothing* from the point of view of things, their properties or relationships, and has but formal reality as a condition of possibility of phenomena, a condition belonging to the subjective constitution of the mind. In fact, the misunderstanding or confusion derives from the very meaning of the term 'space'. Kant holds that "the original representation of space is an intuition a priori, and not a concept."[16] By contrast, since Gauss', Riemann's, and Dedekind's works, geometric space is not the space of external experience, it is neither empirical intuition nor pure intuition but a body of mathematical properties, that is to say a "concept" in Dedekind' and Hilbert's wording. In *Stetigkeit und irrationale Zahlen* Dedekind clearly maintains the conceptual nature of geometric space and Hilbert will explicitly recognise that the axioms defining a geometry form a "conceptual framework" [ein Fachwerk von Begriffen][17] to formalise a structure, among several possible structures, expressing in a synthetic and coherent way experimental data collected in the real world by physical instruments. For the mathematician Hilbert, conceptual/axiomatic frameworks capture geometric intuitions, which refer less to the sensory world than to the world of scientific (physical, biological, astronomical, etc.) experiments and they do not presuppose pure a priori intuition as their formal condition of possibility.

Yet Kant's shadow continues to hang over Hilbert as over other German and non-German mathematicians (Poincaré and Brouwer in particular). In a seminar held in 1905 Hilbert presented similar material to that of his lecture at the 3rd International Congress of Mathematicians, "On the foundations of logic and arithmetic". There, preceding logical calculation, an "axiom of thought" [Axiom des Denkens] is meant

nothing, so soon as we withdraw the condition upon which the possibility of all experience depends and look upon space as something that belongs to things in themselves".

[14] *CPR*, Transcendental Aesthetic I, § 2, A25.

[15] *CPR*, Transcendental Aesthetic I, § 1, A22.

[16] "Also ist die ursprüngliche Vorstellung vom Raume Anschauung a priori, und nicht Begriff". *CPR*, Transcendental Aesthetic I, § 2, B40.

[17] 'Fachwerk' literally means half-timbering. The concepts are therefore the visible structure of the theoretical edifice. Hence, Cavaillès' and Bourbaki's insistence on the "architecture" of mathematics.

to represent "the a priori of philosophers".[18] Kant clearly constitutes the philosophical horizon of Hilbert and the Göttingen mathematicians, discussions focusing on thought, a priori, and the division between analytical judgements and synthetic a priori judgements.

2.2.2 Critique of Reason and Proof Theory

In 1917, in "Axiomatisches Denken", Hilbert once again explains the contribution of the axiomatic approach. This time he sees it as the instrument of transformation of a set of facts within a given scientific field into a *unified theory* of this field. Thus the theory of arithmetic, the Galois theory, the theory of heat, the theory of gases, the theory of money, etc. A theory is therefore a conceptual framework [Fachwerk von Begriffen] such that a concept corresponds to a particular object of the scientific field being studied and that logical relations between concepts correspond to relationships between facts within the field. At the base of the framework a few concepts and their interrelations enable the reconstruction of the entire framework, at least that is how it was envisaged before Gödel's incompleteness theorem (1931).

It is also in 1917 that Hilbert, to complete his work on the axiomatic method, sets up the project to build a new branch of mathematics, named metamathematics, whose specific object is the concept of mathematical proof. A similar approach to the physicist's theory of his technological equipment and the philosopher's critique of reason, he writes, will supply a Critique of proof [Beweiskritik]. A little later (1922) this metamathematical project will begin to be realised under the name of proof theory: the Beweiskritik becomes Beweistheorie, i.e. both method and meta-content. This crowns the whole *critique of mathematical reason* enterprise begun with the axiomatic method. An important inflection appears in this 1917 paper. Hilbert now puts the problem of non-contradiction of axioms, that of a criterion for simplicity of mathematical proof, that of the relationship between contentuality and formalism [Inhaltlichkeit und Formalismus] in logic and mathematics and that of decidability of a mathematical question by a finite number of steps in the set of "questions of the theory of knowledge with a specific mathematical coloration". Therefore, Hilbert has in mind a mathematical theory of knowledge, highlighting that the Kantian theory of knowledge can no longer prescribe the new mathematics, the new physics and the new logic. In a lecture delivered in 1919–1920 on "The role of intuition and experience", Hilbert intends to deliver a "kind of preparation for the construction of a theory of knowledge" which promises, as far as mathematics is concerned, to be a "far greater success than Kant's" (!)[19] The example of the construction of the mathematical continuum that he sets out in detail shows that Hilbert, unlike Kant, makes little distinction, or only one of degree, between intuition

[18]Cited by Peckhaus, p. 62.

[19]Hilbert, *Natur und mathematisches Erkennen*, herausgegeben von David. E. Rowe, Basel, Birkhäuser Verlag, 1992, pp. 3–4.

and perception. Indeed, intuition begins with perception and leads to a concept, which frees us from intuition as Einstein's theory shows us.[20] By speaking here on intuition Hilbert wishes above all to show that mathematics is not an empty game but "a conceptual system constructed according to an internal requirement". Echoing the view taken by F. Klein and especially R. Dedekind,[21] whose essays on numbers so impressed him, Hilbert substitutes the Kantian requirement for forms of experience with the constraint imposed by *the content of mathematical problems*. His fight against dogmatism and intuitionism leads him far from Kant: mathematical constraints are not uniquely formal, nor universal, nor unchangeable.

In 1930, in the article entitled "Naturerkennen und Logik" mentioned above, Hilbert proposed treating the old epistemological problem of the relationship between thought and experience in the light of advances in physics due to Planck, Bohr, Einstein, the Curies, Röntgen, etc. After flatly rejecting Hegel's absolute rationalism and invoking Leibniz' pre-established harmony to account for the correlation between logical axiomatics and experience, Hilbert turns to Kant, who adds, according to him, an a priori element consisting of "some knowledge of reality". Here is the text:

> In fact philosophers have argued that Kant is the classic representative by stating that in addition to logic and experience we still have some a priori knowledge. I acknowledge that certain a priori *views* [Einsichten] are required for the construction of the theoretical structure and constitute the basis of our knowledge. I believe that mathematical knowledge also rests, ultimately, on a kind of intuitive view of this nature. And even to build arithmetic we need a certain a priori intuitive attitude [eine gewisse a priori anschauliche Einstellung]. It is here, therefore, that the most fundamental thinking of Kant's theory of knowledge lies, namely the philosophical problem of establishing this a priori intuitive attitude and thereby examining the condition of possibility of conceptual knowledge and at the same time that of experience. I think this is essentially achieved by my studies on the principles *of mathematics*. The a priori is nothing more nor less than a fundamental attitude or the expression of certain indispensable preconditions [der Ausdruck für gewisse unerläßliche Vorbedingungen] for thought and experience. But we must draw differently from Kant the border between, on one hand, that which we possess a priori and, on the other, that for which experience is needed. Kant has overrated the role and scope of the a priori [...] We can say that today science has produced a safer result from the point of view expressed by Gauß and Helmholtz about the empirical nature of geometry [...] The Kantian a priori includes anthropomorphic scoria from which we must be freed; once these have been cleared out, all that remains is this a priori attitude, which is also the basis of pure mathematical knowledge: it is the sum and substance of what, in my various writings, I have characterised as finitist attitude" (emphasis added).

This text picks up on the considerations already present in the more technical articles of 1922, 1923, 1927, 1928 and 1930.[22] It shows Hilbert's interpretation of Kant's

[20]Ibid., pp. 50–51. Thus intuition is rather preliminary than a priori.

[21]Über die Einführung neuer Funktionen in die Mathematik (The introduction of new functions in mathematics), *Gesammelte mathematische Werke* III, Vieweg & Sohn, Braunschweig, 1932, 428–438, French trans. in *La création des nombres*, Paris, Vrin, 2008, 221–233.

[22]In chronological order: Neubegründung der Mathematik (reprint in *Gesammelte Abhandlungen* III, 155–177), Die logischen Grundlagen der Mathematik (*Gesammelte Abhandlungen* III, 178–191), Die Grundlagen der Mathematik (*Abhandlungen aus dem mathematischen Seminar der Universität Hamburg*, 6, 65–85), Probleme der Grundlegung der Mathematik (*Mathematis-*

pure intuition, not present in his previous defence of the axiomatic method, which was rather more a question of "analysing intuition" and gathering facts into a theory. Hilbert now recognises that building axiomatic frameworks rests on "a priori intuitive attitude" as *the* condition of possibility of any conceptual knowledge and simultaneously of all experience. However he believes that a priori is only "a fundamental attitude or the expression of certain indispensable preconditions of thought and experience", and he says later in his presentation that these preconditions may change or prove to be mere prejudice, as demonstrated by the example of the concept of absolute time, defined by Newton as an a priori datum and identical for all observers. The notion of absolute time was accepted "without critique by the philosopher of Critique", quips Hilbert, while it is refuted by Einstein's gravitational theory. For that reason Hilbert agrees now with the empiricist conception of geometry as espoused by Gauss and Helmholtz. The Kantian a priori is still, according to him, encumbered with "anthropomorphic scoria" that must be eliminated to supply an objective version. Out of this is born his finitist conception[23] of the foundations of pure mathematics, which turns a *subjective "view"* into an *objective method* for determining the conditions of acceptability of mathematical proofs. Mathematics is therefore the mediation [Vermittlung] between theory and praxis, between thought and observation. Formalism materialised as symbolic procedures renders superfluous the hypothesis of pure apperception as empty form.

It is amusing to note the aid Hilbert finds in the Hegelian term and concept of mediation, which differs from the Kantian "Verbindung" [conjunctio].[24] Evidence, if it were needed, of the pervasiveness of this Hegelian figure of reflection, which will be significant in modern philosophy.

What about the Kantian conjunctio? Conjunctio is "the highest principle in all human knowledge"; it is not given by objects and is only "an accomplishment of the understanding, which is itself nothing more than the power of conjoining a priori and of bringing the variety of given representations under the unity of [pure] apperception".[25] By contrast, Hegel's Vermittlung rejects the dualism of understanding and sensitivity and expresses the reciprocal immanence of thought and being. According to the latter perspective, the laws of thinking, that Hilbert regards as expressed by the rules of his proof theory, would give us the mathematical intelligence of reality in coincidence with reality itself. But Hilbert seems to use 'Vermittlung' by chance, he

che Annalen 102, 1–9), Die Grundlegung der elementaren Zahlenlehre (*Math. Ann.* 104, 485–494). Most of those papers are translated into French by par J. Largeault, *Intuitionisme et théorie de la démonstration*, Paris, Vrin, 1992.

[23]Finitism consists of establishing propositions involving infinity upon an analysis of finite sequences of formulas constituting proofs of the propositions in question.

[24]According to Kant, "any combination [Verbindung/conjunctio] is either composition [Zusammensetzung/compositio] or connection [Verknüpfung/nexus]". Both are synthesis of the manifold, but only the second sort is "the synthesis of a manifold, in so far as its parts do belong necessarily to each other; for example, the accident to a substance, or the effect to the cause. Consequently it is a synthesis of that which though heterogeneous is represented as connected a priori." *CRP*, Transcendental Analytic, Book II, chapter 2, section 3, B202, Footnote.

[25]*CRP* B135.

does not avail himself of Hegel's decisive contribution consisting of internal identity of being and thought. On the contrary, he stands by the externality of the traditional relationship, and even goes back from Kant to Leibniz and Galileo. Indeed, instead of speaking of subjective representation and pure apperception (Kant) he invokes Leibniz' pre-established harmony and confirms Galileo's statement: mathematics is indeed the *expression* of reality. Quite remarkably language rather than consciousness functions as mediation. One cannot reproach a great mathematician for not being a coherent philosopher, let alone for ultimately preferring language, a public and controllable vehicle, to the interiority and opacities of consciousness.

So far from being inspired by Hegel, and dismissing Kantian orthodoxy, Hilbert performs a kind of naturalisation of the mind. In effect he brings the a priori back to the linguistic sphere (a little like Quine will later): "thought", he writes, "is parallel to language and writing."[26] This naturalisation stems from his conception of arithmetic signs as constituting, in a material and visible way, the fundamentals of building formal axiomatic systems. Signs do not have a representative function they are themselves the necessary exteriority of thought, its material.[27] An equation such as $f(x)=x^2$ is both the symbolic expression of a curve and a mathematical material (a quadratic equation). Just as f and x here, mathematical signs have no predetermined content, but are the material building blocks, "concrete objects of intuitive experience [Erlebnis] preceding any thought,"[28] which nevertheless "liberate us from the subjectivism already inherent in Kronecker's intuitions and which reached a peak with [Brouwer's] intuitionism." So, Hilbert admits that intuition is rooted in perceiving sensory signs outside of the mind, but he rejects the subjective synthesis of perception. Hilbert explains that the set of formulas attacked by Brouwer for their alleged lack of content, of meaning, is actually the instrument that allows us to express the whole thought content [Gedankeninhalt] of mathematics in a uniform (standard) way so that the interconnections between the formulas of symbolic language and mathematical facts become clear. For Hilbert this ordered set of formulae not only has mathematical value but also an important philosophical meaning/significance because "it is carried out according to certain rules in which the technique of our thought is expressed." His proof theory, he says, has no other purpose than to "describe the activity of our understanding, to draw up a set[29] of rules by which our thinking actually performs."[30] Thought is a mechanism whose elements are sequences of signs connected by deductive links. In 1927 Hilbert there-

[26]Die Grundlagen der Mathematik, lecture delivered in 1927 and published in 1928 in the *Abh. Math. Sem. Hamburg* 6, pp. 79–80. French trans. in J. Largeault, *Intuitionisme et théorie de la démonstration*, Paris, Vrin, Mathesis, 1992, pp. 145–163.

[27]Neubegründung, p. 163: "In the beginning is the sign" ("Am Anfang ist das Zeichen").

[28]Die Grundlagen der Mathematik, p. 65. Hilbert writes 'Erlebnis' which indicates an empirical experience whereas the Kantian 'Erfahrung' is a synthetic unity of sensory perceptions produced by the understanding.

[29]"Protokoll", which literally means the minutes or report of proceedings.

[30]Die Grundlagen der Mathematik, p. 79: "The thrust of my proof theory is nothing but a description of the activity of our understanding, an inventory of rules under which our thought proceeds effectively. Thought is parallel to language and writing…".

fore replaces lucid thought [mit Bewußtsein denken] by technical thought: formality is not about consciousness but about language and symbol writing. Intuition is not a priori but sensory, it is not subjective but formally, that is expressly, objectified. Symbolic expressions are really an objective starting point and a material support for mathematical practice. There is no need to presuppose a native intuition of number (Brouwer) or a specific faculty of pure understanding accounting for the principle of complete induction (Poincaré's "intuition of pure number.").[31]

Intuitionism denies the autonomy of meaning from its psychological actualisation. In contrast, Hilbert, logical positivism, the Vienna Circle, model theory, and Quine deny the autonomy of meaning from its linguistic expression. Hilbert wants to save the infinite involved in mathematical abstractions through the materially expressed formality of finite sequences of signs. This is why he claims that his "finitism" is the expression of a conception of the a priori that is free from the anthropomorphic aids preserved in Kant's doctrine, and thus offers the basis for *pure* mathematical knowledge. The regulated manipulation of sequences of signs, figures or bars (see Hilbert 1905) is the mediation between thinking and observing. It is likely that Hilbert knew that J. F. Fries maintained that *Critique of Pure Reason* was a psychological or anthropological attempt to build a base for a priori knowledge.[32] What we can say with more certainty is that by "anthropomorphic scoria" Hilbert is referring to the subjective side of the Kantian reflexive perspective. The literality and formal legality of proof theory are meant to save us from this subjectivity. Thus science is substituting for philosophy of conscience, it says what the a priori consists of.

If Hilbert the mathematician entertained a real interest in philosophy it was in order to depose its supremacy rather than to recognise it as the "primary science", that which determines the conditions of intelligibility of objects of any particular science. In this respect, Hilbert certainly played a role in the philosophical rejection of philosophy as primary science and the call often issued by the proponents of historical epistemology to place it within the school of science, i.e. to learn philosophically by the practice of science. Even today proponents of the philosophy of mathematical practice intend to base philosophical insights on mathematical material.[33]

As was his wish, Hilbert's perspective has given much food for philosophical thought. Philosophers have variously used it for flatly rejecting the transcendental (Cavaillès),[34] interpreting it differently from Kant (Husserl, Vuillemin, Granger), or diminishing it on a real and empirical level (Foucault's historical a priori). With his structuralist objectivism, Hilbert's perspective contributed—in cooperation or in conflict with other elements resulting, in particular, from the promotion of philosophy

[31] Du rôle de l'intuition de la logique en mathématiques, reprint in *La valeur de la science*, chapter I, Paris, Flammarion, 1905; Sur la nature du raisonnement mathématique, reprint in *La science et l'hypothèse*, chapter I, Paris, Flammarion, 1902.

[32] Jakob Friedrich Fries, *Neue oder anthropologische Kritik der Vernunft*, 3 Band, Heidelberg, Christian Friedrich Winter, 1828–1831.

[33] CF. for example P. Mancosu (ed.), *The Philosophy of Mathematical Practice*, Oxford University Press, 2008, and P. Mancosu, *Infini, logique, géométrie*, Paris, Vrin, 2015, third part.

[34] See my work, in particular *Jean Cavaillès. Philosophie mathématique*, Paris, PUF, 1994 and *Cavaillès*, Paris, Les Belles Lettres, 2013.

of history accomplished by Hegel and from the development of human sciences (Dilthey's Geisteswissenschaften)—to the abandonment of exclusive attention to the subject (Husserl) or to the preponderant privilege of the object (Cavaillès, linguistic, philosophical structuralism).

In the following section, after a brief presentation of the philosophy of Leonard Nelson, who, as we have seen, sparked some reflections in Hilbert, I will explain more fully that of Edmund Husserl, whose ambitions for philosophy was at least as great as Hilbert's ambition for mathematics and who maintained a deeper relationship with Kantian philosophy. In the third part I will turn to Jules Vuillemin who wrote *La Philosophie de l'Algèbre*, as a dialogue with Leibniz, Kant, and Husserl in the light of modern mathematics.

2.2.3 Critical Mathematics of Gerhard Hessenberg and Leonard Nelson

Hilbert believed that the foundation of mathematics on the axiomatic method and proof theory was of great interest to philosophy. And indeed, from the philosopher's point of view the idea of considering the axiomatic approach as a critical step in the development of mathematics was very attractive, since it allowed bridges to be built between philosophy and mathematics, bridges which would stretch over and ignore the speculative Spirit of Hegel, the dialectical history of philosophy redistributed among successive figures of growing rationality, and the conquering empiricism of experimental sciences and humanities (psychophysics, statistical sociology, etc.). In fact, "critical mathematics"[35] was a rallying point for mathematicians and philosophers around Hilbert. In 1904 the mathematician Gerhard Hessenberg published a short article, Über die kritische Mathematik,[36] whose theme is more widely developed in "scientific philosophy" [wissenschaftliche Philosophie] by the philosopher Leonard Nelson.

Hilbert had initiated an exchange with Edmund Husserl who taught at Göttingen from 1901 to 1916. Husserl had already published *Philosophie der Arithmetik* (1891) and *Logische Untersuchungen* (1900–1901), which showed his strong interest in the relationship between mathematics and psychology versus logic, a relationship that was precisely the purpose of Hilbert's technical work. Husserl and Hilbert developed a close and intellectual friendship, and the latter put all his weight behind support for Husserl against the hostility of the university professors of psychology and history and for his continued employment in Göttingen, in the hope of seeing him occupy a chair of "systematic philosophy of the exact sciences".[37] The phenomenologist might

[35] See K. Herzog, "*Kritische Mathematik*"—*ihre Ursprünge und moderne Fortbildung*, Dissertation (1978) Düsseldorf.

[36] *Sitzungsberichte der Berliner mathematischen Gesellschaft* 3, 21–28.

[37] A detailed historical study, exploring many unpublished documents can be found in the excellent book by Volker Peckhaus, cited in note 6.

not have endorsed the perspective to commit to the programme, which was not his own. Testament to this are his lectures and publications during his years in Göttingen, which were far from being appropriate to Hilbert's positivistic and empiricist views: *Die Idee der Phänomenologie. Fünf Vorlesungen* (1907), *Philosophie als strenge Wissenschaft* (1911), and *Ideen zu einer reinen Phänomenologie und phänomenologische Philosophie I Allgemeine Einführung in die reine Phänomenologie* (1913). Indeed the first and third works mark a turning point of phenomenology towards transcendental idealism while the second fiercely criticises scientific positivism, which is non-philosophical by nature.

After Husserl's departure to Freiburg in 1916, Hilbert pursued his goal by supporting Nelson who, after many obstacles, was appointed professor in 1919 and put in charge of imparting lessons in the aforementioned "systematic philosophy of the exact sciences" programme.[38] In 1917 Hilbert joined the Neue Fries'sche Schule founded by Nelson in 1903. The school was a focal point for exchanges between scientists and philosophers in Göttingen: among its members there were Kurt Grelling, Richard Courant, Max Born and Paul Bernays, editor of the complete works of Nelson in 9 volumes. With the support of Gerhard Hessenberg Nelson created a journal titled *Abhandlungen der Fries'schen Schule. Neue Folge*. In the first volume Nelson defines the project to develop a philosophy "whose method is as rigorously scientific as the method of mathematics and natural sciences."[39] He argues against Hegelian "scholasticism" and its powerful influence, against historicism, against the Platonism of Schelling, and against empiricism that, he considers, was refuted definitively by Kant. The axiomatic approach is the paragon of rigor to follow. But Nelson aligned himself equally with Fries, who was a philosophical adversary of Hegel. Nelson attributes to Fries the first transfer of *Critique* to mathematics and the constitution of the "philosophy of mathematics" as an autonomous discipline in *Die mathematische Naturphilosophie*, published in 1822. Like Fries, Nelson understands mathematics as a set of a priori *synthetic propositions* and wishes to provide a critical foundation for mathematical axioms by subjecting them to transcendental deduction in the manner of Kant as revisited by Fries. Absent from Hilbert's terminology, 'consciousness', 'a priori synthetic judgements' and 'transcendentalism' come into play. Nelson distinguishes between demonstration and deduction: the former works in mathematics and science by showing the intuitions that underlie our basic judgements, the latter is the foundation, through Fries' reprise of Kant's "regressive method", of funda-

[38]Hilbert wrote: "Among the philosophers who are not primarily historians and experimental psychologists, Husserl and Nelson are the most remarkable personalities, it seems, and for me it is not a coincidence that these two found themselves on the mathematical ground of Göttingen. [...] Neither is it a coincidence that I speak out on this subject. Without me, Husserl would have been caught earlier, without me Nelson would never have been seen here in Göttingen. Göttingen is predestined for this task—a huge cultural task." (cited by Peckhaus 1990, p. 223).

[39]Die kritische Methode und das Verhältnis der Psychologie zur Philosophie. Ein Kapitel aus der Methodenlehre, *Abhandlungen der Fries'schen Schule, Neue Folge* 1 (1906), pp. 1–88. On line: http://archive.org/stream/abhandlungenderf01gtuoft/abhandlungenderf01gtuoft_djvu.txt

mental metaphysical judgements and applies equally to mathematics.[40] For Nelson deduction is the most important task of critique. Indeed he writes:

> Unlike demonstrable judgements, judgements that are only deductible are not based on intuition, that is, the immediate knowledge on which they rest is not immediate to us, but mediated by reflection, led by judgements to consciousness.[41]

And a little further on:

> A critical deduction of mathematical axioms must also be possible. This transfer of *Critique* to the axiomatic system represents a scientific discipline it its own right: the philosophy of mathematics or, by better description, critical mathematics.

For Nelson the axiomatic is but the first task of Critique; it is logical in nature. The second, real task falls within the theory of knowledge and involves the question of "origin and validity of axioms", a task whose total absence in Hilbert's work he deplores. We have seen, however, that Hilbert, probably prompted at least partly by the reproach that he could not ignore, turned from 1917 to questions of the theory of knowledge which he understood and resolved by way of the mathematical filter. For Nelson, on the contrary, it is a metaphysical deduction of axioms that achieves the aim of Hilbert's finitist programme. No doubt it was by way of obstruction of this view and reaffirmation of his own understanding that Hilbert, who ignored a priori synthetic judgements and transcendental versus metaphysical deduction, returned, in his 1930 article, to the problem of the relationship between thought and experience, *in order to assign to mathematics itself* the power to link thought and experience. Finitism is presented as intuitive a priori attitude: so a *method* of proof plays the role of Kant's *pure consciousness or apperception* which is the a priori form of any synthetic representation of objects. In other words, Hilbert maintains and reiterates the view which was at the heart of his proof theory: mathematical treatment of epistemological issues and not, like Nelson, use of axiomatic method as a model to build a metaphysical theory of knowledge. Hilbert's programme for cooperation between mathematics and philosophy did not lead in Nelson's work—nor in Husserl's, as we will now see—but to a resorption of philosophy within mathematics through a complete objectification of thought by means of symbolic processing.

---

[40]Hilbert picked up on Nelson's distinction between progressive and regressive methods in his lecture "Die Rolle der Voraussetzungen", in *Natur und mathematisches Erkennen* (David Rowe ed.), Birkhäuser, 1992, pp. 17–18. But for him the regressive method finds its most perfect expression in the axiomatic method. This means that for him transcendental deduction is advantageously replaced by mathematical proof.

[41]"Die nur deducierbaren Urteile aber haben ihren Grund nicht, wie die demonstrierbaren, in der Anschauung; d.h. die ihnen zugrunde liegende unmittelbare Erkenntnis kommt uns nicht unmittelbar, sondern nur durch *Vermittlung* der Reflexion, nur durch das Urteil zum Bewusstsein."(Die kritische Methode). I highlight the word 'Vermittlung', a Hegelian term which encroached on the author's Critique. As I have highlighted above, Hilbert also uses the term 'Vermittlung'. We will find further on, with Husserl and Vuillemin, more traces of the resonance of Hegelian thought on those very people who reject it outright.

2.3 Philosophy as Rigorous Science: Husserl

Husserl held in high esteem the work of Hilbert, at whose request he had made two presentations in 1901 to the Mathematical Society of Göttingen. In contrast he found Nelson devoid of "scientific seriousness" and disliked his programme of "Systemphilosophie" and his metaphysical deduction. Husserl intended to proceed "from below" and to develop a purely phenomenological method.[42] In fact, the two philosophers held antagonistic positions on the place of metaphysics in philosophy and the founding status of intellectual intuition. In 1908 Nelson had severely criticised categorial intuition, as it had appeared in *Logical Investigations*, and the descriptive (i.e. non-critical) nature of Husserl's analyses.[43] This played a part in the evolution of descriptive phenomenology into transcendental phenomenology, and Husserl's programme of "Philosophy as rigorous science"[44] which he conceived as a definitive eradication of the metaphysics of the things-in-themselves.

> We must eliminate any 'metaphysical' thing-in-itself and with it any Spinozist ontological metaphysics, which derives a *science of being* from a system of *pure concepts*.[45]

2.3.1 *Need for Critique*

Philosophy as rigorous science is a science, a "philosophical science" [philosophische Wissenschaft], very different from Nelson's scientific philosophy [wissenschaftliche Philosophie]. The reversal of the terms is not without significance, although Husserl also sometimes uses the term 'wissenschaftliche Philosophie', for example when he describes and critiques the worldview [Weltanschauung] philosophy. Husserl, like Nelson, considers as decisive both the critical revolution brought about by Kant, as well as the axiomatic turn of mathematics initiated by Hilbert—we will see how Vuillemin also maintained this dual allegiance. But while they share a

---

[42]I refer to the Correspondence Nelson-Hessenberg cited by Peckhaus 1990, pp. 203–204. Cf. also P. A. Varga, Ein bisher unbekanntes Portrait von Edmund Husserl, 2010, revised in Sept. 2013. http://hiw.kuleuven.be/hua/Media/mitteilungsblatt/portrait

[43]See Nelson, *Geschichte und Kritik der Erkenntnistheorie*, Part 2, Section XII entitled "Husserls phenomenological Methode und die intellektuale Anschauung", *Complete Works*, Volume II, pp. 171–177.

[44]Philosophie als strenge Wissenschaft (PSW), *Logos - Zeitschrift für Philosophie und Kultur*, Bd. 1, Tübingen 1910/11. Online http://www.gleichsatz.de/b-u-t/archiv/phenomeno/husserl_streng1.html

English trans. M. Brainard: *The New Yearbook for Phenomenology and Phenomenological Philosophy* II (2002): 249–95; online:

https://fr.scribd.com/doc/63651073/Husserl-Philosophy-as-a-Rigorous-Science-New-Translation

[45]Letter from Husserl to Dilthey cited by Q. Lauer, p. 175 of his French translation of PSW (emphasis added).

vindication of critical attitude and an appropriation of the axiomatic method, Husserl and Nelson proceed differently.

Husserl takes the critique of reason as the primary condition of philosophical scientificity. In 1903 in Göttingen, Husserl devoted a few lessons of a course on the history of philosophy to Kant. And in 1906 he wrote in his *Persönliche Aufzeichnungen*[46]:

> The general problem I have to solve for myself if I want to call myself a philosopher. I mean a *Critique of reason*.

Husserl believes that this prerequisite has been present in philosophy from Plato to Kant and Fichte while it is absent from the Hegelian system. And yet Hegel had conceived his system as *the* philosophical science totalling all knowledge. It is therefore Husserl's ambition to construct a philosophical science that combines the rigour and autonomy of exact science with the critical dimension of Kant's philosophy. While rejecting the traditional idea of *philosophical system* [Systemphilosophie],[47] "springing like Minerva fully armed from the head of a creative genius so as in later times to be preserved in the silent museum of history,"[48] Husserl is searching for a *systematic philosophy* endowed with the "seriousness of science". The "true path to a scientific theory of reason"[49] will lead to the "primary science", which Kant could not stumble upon, Husserl thinks, for lack of pushing his analysis of pure intuition far enough. The steps along the path to the primary science: epoché, reduction, eidetic variation should be *systematically* followed to answer the fundamental question, which is no longer the Kantian question of knowing *what* the conditions of possibility of experience and of knowledge of the objects of experience are, but that of knowing *how*, in concrete terms, these conditions give the conscious mind access to objectively valid concepts, how the a priori *encounters* experience, how it details and organises *structurally* this encounter. The question is the following:

> How are we to understand the fact that the "in itself" of objectivity reaches 'representation' and even 'apprehension' in knowledge, and yet eventually becomes subjective again? What does it mean to say that the object has 'being-in-itself' and is 'given' in knowledge?[50]

In the name of serious science Husserl tackles historicism and naturalism, which were then also in the sights of philosophers working in Göttingen in contact with Hilbert and willing to take into account new developments in mathematics, logic and

[46]Published by W. Biemel in *Philosophy and Phenomenological Research* XVI, 1956, p. 297.

[47]"A truly living and truly scientific philosophy can not be possible but as free philosophy, as a philosophy that questions not philosophers and their systems, but things themselves," (Hua XXVII, 198–199).

[48]The allusion to Hegel is transparent. Paradoxically, according to Husserl, building a philosophical system concurs with the "worldview philosophies" he intends to replace because they amount to subordinating thought to a particular view of the whole.

[49]*PSW*, English trans. p. 278.

[50]*Logical Investigations*, Volume II, Part 1, Investigations into Phenomenology and theory of knowledge, Introduction, English trans. J. N. Findlay, Routledge & Kegan Paul Ltd. Reprint 2001, p. 169.

physics.[51] Mathematicians and physicists reciprocally supported Husserl and Nelson against the faculties of philology, history and psychology of their university. These historical data confirm the understanding of Hilbert's theory of mathematical signs as being a linguistic objectivism rather than a psychological or cognitive naturalism.

More than the Hegelian system itself, Husserl fights the avatars of Hegel's philosophy of history, that is, the worldview philosophies, which propagate a sceptical and relativistic historicism, though contrary to the absolute Spirit of Hegel. He holds post-Hegelian historicism responsible for the weakening, indeed perversion, of the "drive towards the establishment of a rigorous philosophical science". The criticism mainly concerns the concept of human sciences [Geisteswissenschaften] introduced by Dilthey, with whom Husserl was connected in Göttingen. Husserl recognises that Dilthey rejects historical scepticism but he does not understand how one could derive decisive reasons against scepticism from Dilthey's rich analyses of the structure and typology of "worldviews".

Psychologism and positivism, which were then asserting themselves in response to Hegelianism, were even more responsible for weakening the drive for scientific philosophy in favour of sceptical relativism and naturalism. Husserl has in his sights the natural science of consciousness and ideas, that is to say experimental psychology and psychophysics which were thriving tanks to Fechner, Wundt, von Helmholtz and others.[52] In fact Husserl was reacting against his own earlier leaning. Nonetheless Husserl retains from experimental psychology the postulate of the primacy of perception and the idea of the inseparable connection between mind and body.

2.3.2 Radicalisation of Critique: From Scientific Explanation to Philosophical Understanding

The second condition of scientificity is the "radicality" of critique. By 'radical' Husserl refers to the very first origin. The question of origin has a different meaning for Husserl than for Nelson. It is not a question of metaphysical deduction, but a kind of Cartesian undertaking: not accepting anything as pre-given, not taking anything bequeathed as beginning, not being blinded by any great figure from philosophy or

[51]In 1916, Nelson wrote: "philosophy is in deep distress [...] since experimental psychologists on the one hand, and historians on the other, dispute their place ever more successfully and have begun to share the inheritance", cited by Peckhaus, p. 219.

[52]Fechner (1801–1887) and von Helmholtz (1821–1894) were lecturers at Göttingen, later followed by Georg Elias Müller (1850–1934), who succeeded Lotze in the chair of philosophy and in 1887 founded an Institute of Psychology (in the image of the one founded by Wundt in Leipzig in 1879) and in 1904 the Experimental Psychology Society. Müller was a bitter opponent of Husserl and Nelson. Husserl attended Wundt's philosophical lectures in Leipzig in the late 1870s and wrote under Carl Stumpf's supervision Über den Begriff der Zahl (Halle, 1886), which is the basis of *Philosophie der Arithmetik*.

science, starting from the problems themselves and their demands. As in science[53] we will begin by precisely formulating the problems and methods. But we must go beyond scientific positivity, beyond the positive character of the experimental sciences, natural sciences and sciences of the mind [Geisteswissenschaften], beyond mathematics and even beyond the available corpus of logic. Let me cite as examples extracts from two different texts.

> It is true that natural science is, *in its own way*, very critical. The experience of a mere individual, even when multiplied, still has very little value to it. It is in the methodical arrangement and combination of experiences, in the interplay of experience and thought which has its logically firm rules, that valid [gültig] and invalid experiences are distinguished, that each experience gains its degree of validity and that, from there, objectively valid knowledge (knowledge of nature) is worked out. Yet however much this type of critical experience may satisfy us as long as we stand *in* natural science and share its way of thinking, a totally different critique of the experience is still possible and necessary, a critique that simultaneously calls into question the whole experience as such and the way of thinking proper to experiential science.[54]

> Logic must be more than simply a positive science of logico-mathematical idealities [...]. Transcendental logic must understand *how the ideal objectualities* [ideale Gegenständlichkeiten] that originate purely in our subjective activities of judgement and knowledge, that are present *originaliter* in our field of consciousness purely as formations of our spontaneity, *acquire the being-sense of "objects"*, existing in themselves, being not contingent on actions and subjects. How does this sense "come about", how does it originate in us? And where could we get it if not from our own sense-constituting power? Can what makes sense for us receive sense ultimately from somewhere other than ourselves? This question, once posed about one sort of object, becomes immediately general: is not each and every objectivity, in every sense which is valid for us, an objectivity that is winning or has won validity in us with the sense that we ourselves have acquired for it? Accordingly, the *transcendental problem* that the *objective logic* [...] has to pose with respect to its field of ideal objectualities runs *in parallel with the transcendental problems of the sciences of reality,* namely the problems that must be raised with reference to the regions of reality to which those sciences pertain, therefore in particular with the transcendental problems concerning Nature, which were treated by Hume and by Kant.[55]

[53]"Starting from the problems themselves" was a leitmotiv of Felix Klein and Richard Dedekind. Hilbert popularised the theme in his Parisian 1900 lecture.

[54]*PSW*, English trans. slightly modified (Husserl's emphasis).

[55]*Formale und Transzendentale Logik*, § 100: "Denn nun stand man vor der Unverständlichkeit, wie *ideale Gegenständlichkeiten*, die rein in unseren subjektiven Urteils- und Erkenntnistätigkeiten entspringen, rein als Gebilde unserer Spontaneität in unserem Bewußtseinsfeld originaliter da sind, *den Seinssinn von „Objekten" gewinnen*, an sich seiend gegenüber der Zufälligkeit der Akte und Subjekte. Wie „macht" sich, wie entspringt dieser Sinn in uns selbst, und woher sollen wir ihn anders haben als aus unserer eigenen Sinn-konstituierenden Leistung; kann, was für uns Sinn hat, letztlich anders woher Sinn selbst haben als aus uns selbst? Diese Frage, einmal an einer Art von Objekten gesehen, wird sofort zur allgemeinen: ist nicht alle und jede Objektivität, mit allem Sinn, in dem sie uns je gilt, in uns selbst zur Geltung kommende oder gekommene, und das mit dem Sinn, den wir uns selbst erworben haben? Danach tritt das transzendentale Problem, das die objektive Logik in welch enger oder weiter Fassung immer in Bezug auf ihr Feld idealer Gegenständlichkeiten zu stellen hat, in Parallele zu den transzendentalen Problemen der Realitätenwissenschaften, nämlich den in Bezug auf ihre Regionen der Realitäten zu stellenden, also insbesondere den von Hume und Kant behandelten transzendentalen Problemen der Natur" (Husserl's emphasis), Halle, Max Niemeyer Verlag, 1929.

The latter text shows that Husserl aims to supply positivist scientific *explanations* with a philosophical framework for the *elucidation* of the being-sense of the phenomena that science explains. Husserl is integrating in his own way the paradigm of understanding that Dilthey opposes with the explanation. Moreover he gives a transcendental status to the sense component of objects and simultaneously does not include the hermeneutical dimension that will predominate among authors such as Hans-Georg Gadamer and Paul Ricoeur.

Husserl reproaches Kant for having kept the natural sciences as a model of rigorous science and of having removed logic from his transcendental enterprise. Is he aware of the semi-similarity between his reproach and Hegel's critique of scientific positivity?[56] Indeed, like Hegel, Husserl does not intend to subordinate philosophical thought to the "dry" abstraction of scientific truths, but, unlike Hegel, he makes the utmost of the objectivity of constituted science while assigning to philosophy the task of questioning the meaning. For contrary to the wisdom of worldviews, science, or more accurately the *Idea of science*, is not bounded by the spirit of a time [Zeitgeist], it is marked with the seal of eternity: "worldviews may conflict, science alone can decide and its decision bears the seal of eternity." Philosophy as rigorous science will even have to abandon opinions and points of view and strive for perfect objectivity which is necessary and universal to all human beings at all times. Husserl values science, not completed science but the scientific spirit, while he rejects scientism which limits the Idea of science to pre-known realisations. Phenomenology will be a new type of science and transcendental analysis will not be limited to determining the conditions of possibility of positive science.[57] Philosophical science cannot stop at the causal explanation of phenomena. Husserl deplores that naturalism is transmitted from the natural sciences to the sciences of the mind, to psychophysics and experimental psychology in particular, whose motivation is the naturalisation of consciousness, ideas and reason. Naturalists certainly challenge post-Hegelian historicism but they also fight for a worldview philosophy[58] and share the same "superstition of the fact". They interpret Ideas as facts and transform all reality "into

---

[56]See in particular *Phänomenologie des Geistes* [*Phenomenology of Mind*] on the abstract, ineffective, fixed and non-living nature of mathematical truths, French trans. Bernard Bourgeois, Paris, Vrin, 2006, pp. 86–89.

[57]I cannot go further into the characterisation of phenomenology here. Jocelyn Benoist, for example, explains that the freeing of the meaning of 'consciousness' (as the pure sphere of appearance) involves the suspension of the very purpose of any science… namely the world. Science is more interested in the a priori of actuality, and remains guided by this actuality of the world. Phenomenological reduction serves to subvert this attitude by the very questioning of the direct or indirect reference of science to the world (this 'reference' gives science its scientific meaning)." *Autour de Husserl*, Paris, Vrin, 1994, p. 175. Epoché is intended to put the world in brackets so as to reveal what knowledge, as such, means.

[58]*PSW*: "The advent of the new *worldview philosophy* [*Weltanschauungsphilosophie*] is essentially determined by the transformation of Hegel's metaphysical philosophy of history into sceptical historicism. Worldview philosophy appears to be spreading rapidly nowadays, and, incidentally, even with its mostly anti-naturalist and sometimes even anti-historical polemics, it wants nothing less than to be sceptical". English trans. modified (Husserl's emphasis). For the critique of Weltanschauungen see also *Logical Investigations*, I.

an incomprehensible mix of 'facts' from which any idea is absent." Factualism is thus the main obstacle to a more comprehensive critique of reason. Yet, it goes without saying that *empirical science* ["Tatsachenwissenschaften"] cannot provide the foundations for the *normative principles* specific to theoretical disciplines such as logic, epistemology and axiology. Natural sciences and sciences of the mind have the pre-Kantian naivety to accept the given as given, as self-evident, and have no other goal than transforming this given into an object of knowledge. However, upstream of knowledge, we must question *how* the object is given and elucidate the essential content [Wesensgehalt]. For objective knowledge of the given is not sufficient, one must grasp *the sense/meaning*[59] which emerges in it. Scientific knowledge represents the highest degree of objectivity, but it is about asking what 'objective' means. "To speak as Lotze", wrote Husserl, "*calculating* the trajectory of the world does not mean *understanding* it".[60]

What did Husserl think of Hilbert's presentation of his Beweistheorie as a positive, objective inventory of the technical rules expressing the laws of thought? One can easily guess, because for Husserl scientific objectivism equals philosophical naivety. Husserl executes a radical disconnection between scientific philosophy and positive science. One has the tendency, he observes, to conceive as rigorous science only positive sciences, and as rigorous philosophy only that based on such a science. But it is a prejudice that diverts philosophy away from its scientific future. It is not about emulating either the natural sciences or mathematics or positive logic. It is about developing an autonomous philosophical science that brings its own set of problems, methods and theories. Rigour and scientific objectivity can and should be conceived more broadly than simply in a positive way. At any rate it is the task of scientific philosophy to implement this new meaning. The concepts of science and rigour must be extended to hold not only for the enterprise of objectification of phenomena, which was as much the goal of Hilbert as of the physical and empirical sciences, but also for the genetic makeup of the meaning of phenomena. Without having to preserve "the apparatus of conclusions and proof" of the experimental or deductive disciplines and "without all indirect methods of symbolisation and mathematisation" so advocated by Hilbert, Phenomenology must assume complete responsibility for the demands of scientific rigour specific to philosophy and leave no room for opinions or points of view, be they as they might defended by geniuses.

> It lies precisely in the essence of philosophy, insofar as it returns to the ultimate origins, that its scientific work moves in spheres of direct intuition, and it is the greatest step our age has to make to see that with philosophical intuition in the right sense, the phenomenological *seizing upon essences* opens up an endless field of work and a science which, without any indirect method of symbolisation and mathematisation, without the devices of deductions

---

[59]Husserl uses 'Sinn' and 'Bedeutung' as synonymous. For a detailed account see D. Pradelle, On the notion of Sense in Phenomenology: Noematic Sense and Ideal Meaning, *Research in Phenomenology* 46, 2016, 184–204.

[60]Emphasis added. Husserl nevertheless makes a side note about the determination by Lotze of the task of philosophy: "it is essentially a worldview philosophy". Cited in Lauer's French trans, p. 179.

> and demonstrations, nevertheless obtains an abundance of the most rigorous and decisive cognitions for *all* future philosophy.[61]

Husserl further states:

> I hope nevertheless to have shown that this is not the old rationalism, which was absurd and in general unable to grasp the problems of the mind concerning us the most, and which can be renewed here. The *ratio* which is now in question is none other than the truly radical and truly universal understanding of the mind by itself, as a responsible, universal science by which *an entirely new mode of scientificity* is undertaken, where all conceivable questions, questions of being and questions of the norm, questions of so-called existence, find their place. (Emphasis added)

Thus, distancing himself from classical rationalism, where the philosopher insists on taking the mathematical or physical sciences as a model and ideal, and criticising empiricism which believes the answer to questions of law raised by the facts of experience can be found *in* the experience of facts, Husserl wishes to promote a philosophy whose rigour is of a different kind than that of the positive sciences. Against the positivity of the constituted theoretical sciences, against the platitude (factuality) of empirical science ["Tatsachenwissenschaften"] which cannot found logical, epistemological or ethical *norms*, against the depth of Romanticism which loses the sense of conceptual clarity and exactness in favour of the subjective and also secretes scepticism and relativism, Husserl argues for the purpose of a philosophy dedicated to the search for an unconditioned truth through the radical critique of the presuppositions not of this or that science (arithmetic, geometry, the theory of electromagnetism, etc.) but of the enterprise of scientific objectification itself. We must systematically challenge every fact, including the scientific fact (results, theories and methods). For the problems of being and value can neither be treated according to a positive logic nor be dissolved in the perspectivism of worldviews. Philosophy has no other given than that which it constitutes itself. It follows, in particular, that the questions of the theory of mathematical knowledge and the mathematical solution given to them by Hilbert, the scientific value of which is undeniable, fall short of or outside the proper field of philosophy and cannot claim to invalidate the specifically philosophical questioning about them. Science cannot be a substitute for philosophy. Whatever the strength of scientific data, they cannot serve as building blocks to construct a rigorous philosophy. The axiomatic method, if paradigmatically useful for analysing what a "unit of meaning" is in contrast to a unit of something,[62] does not, however, escape the need for a phenomenological reevaluation. And Beweistheorie, by the admission of its author, is the formal expression of the *techniques of thought*,

---

[61] *PSW*: last sentence (Husserl's emphasis). Husserl further asserts in Krisis that "transcendental phenomenology overcomes naturalistic objectivism and all objectivism in general." (French. trans. N. Depraz, p. 112, online: http://www.ac-grenoble.fr/PhiloSophie/wp-content/uploads/ebooks/husserl_depraz.pdf.)

[62] "Every physical thing has *its* nature—as quintessence [Inbegriff] of *what it is*: identity—by dint of being the meeting point between causal series inside a total nature and one," *PSW*, English trans. p. 268 modified slightly. In *Logical Investigations*, Husserl shows that the unity of meaning is ideal and, as such, the prototype for all "ideal objects" of science, particularly mathematical ones, for which Hilbert, following Leibniz and many others, tried to set down the rules.

and not a clarification of what thought is in its essence. "As instigating philosopher", wrote Husserl in *Cartesian Meditations*, "we do not ascribe value to any normative ideal of science; and we cannot have a normative ideal to the extent we create it ourselves." Husserl insists that mathematical standards are to be found in mathematics, logical standards in logic, ethical standards in ethics, etc. The autonomy of science as science is not in question. The purpose is not to subordinate science to philosophy, but to make philosophy an autonomous science.

2.3.3 *The Essence of Consciousness*

Philosophy is not *in principle* a "theory of science" [Wissenschaftslehre], whose purpose would be to build the foundations of science, that is to determine the principles from which one can derive valid scientific propositions or explain phenomena, a goal shared by Hilbert and Nelson who pursued it in two different ways. Philosophy has as subject the "fundamental sphere", brought about by the problem of the foundation of science but treated as such neither by scientists nor by previous philosophers; it is the sphere of the primordial givenness of the object, be the object real (physical or mental) or ideal like scientific objects. This is a long way from Hilbert's programme of an axiomatic or metamathematical foundation and far, also, from Nelson's metaphysical deduction of mathematical principles.[63] Husserl aims to discover and reveal in its very origin [Ursprung] the processes which beginning in nature engenders the theoretical interest to which the unfinished building of science responds. One must clarify "the origin of all our formal-logical and natural-logical principles and any other normative principles, and all issues intrinsically related to the correlation of the being (natural being, being of value, etc.) and consciousness."[64] Philosophical science or phenomenology views origin in a different way than in the sense of regressing to non-conditioned principles or of describing the course of a series of historical facts. It is meant neither as metaphysical deduction nor as actual history, but as a genetic constitution of meaning[65]. It is concerned with this radically originary sphere in which all objectification and all scientific practice are rooted, the sphere of "things themselves"—as totally opposed to the unknowable things-in-themselves. Things themselves are the phenomenal foundations upon which every

[63]It is also far from what we call today "metaphysics of science", or even "meta-science", understood as the fundamental analysis of concepts structuring scientific research, for example the analysis of concepts of cause, individual, disposition, etc. Thus the 5th Congress of the Philosophy of Science Society was held in June 2014 at the University of Lille 3, with the theme Meta-physics in Science and the primary objective of "considering ontological problems arising from science and, more specifically, from their discoveries, concepts, models and theories. From there, Metaphysics in Science aims to reconstruct a "scientific picture of the world." The project is one of metaphysical naturalism, revealing the ontology that results from scientific theories and practices. More broadly, metaphysics of science is interested in "any study of the links between science and metaphysics."

[64]*PSW*, English trans. p. 275 (modified).

[65]Husserl's genetic constitution is not creation but elucidation.

scientific and cultural superstructure is constructed. They are the *pure*[66] experiences of consciousness "in which the appearance of the object resides", and from where arises the dual demand for an originary relationship to the pre-scientific world and a reflexive, rational understanding of this relationship. Research must therefore be directed at determining scientifically not the *fact* of consciousness, which is the concern of experimental psychology, but the *essence* of consciousness, that is the objective structures of the givenness of the object to consciousness. Those structures are, shall we say, independent of the hypothesis of the thing-in-itself and consequently render it useless. It is a matter of studying structures, as with the axiomatic method, but applied to consciousness and here the reflexive method consists not of the axiomatic construction of philosophical concepts, but of asking *how* data enter consciousness. The answer should surpass both subjective factuality and objective factuality.

> Inquiry must be aimed at a scientific eidetic knowledge of consciousness, at what consciousness *itself*, by its *essence*, "is" in all its distinguishable formations, but at the same time at what it "signifies" [bedeutet], as well as at the different ways in which it—in accordance with the essence of these formations (now clearly, now unclearly, now presentiatingly or representiatingly, now signitively or pictorially, now simply, now mediated by thought, now in this or that attentional mode, and so on in innumerable other forms)—intends *something objectual* [Gegenständliches] and perhaps "shows" [erweist] it to be a "valid," "actual" being [Seiendes].[67]

The essence of consciousness consists of intentionality. To analyse the essence of consciousness is to discover the intentional relationship to every kind of objectualities [Gegenständlichkeiten] and with respect to every modality of this relationship. As mentioned in the above passage, essence therefore consists of a family of distinct binary relations, mathematically understood as correspondence between a source domain: the consciousness modalities and a target domain: the set of all possible objectualities. Hence there is nothing empirical about this analysis, neither in its object—*pure* experiences of consciousness—nor in its approach, which is mathematical. Husserl makes a remarkable philosophical interpretation of the mathematical concept of relation, whose particular case is the concept of function.[68] The broad definition we know today of a function as an *arbitrary correlation* between two numbers was established by Johan Peter Gustav Lejeune-Dirichlet (1805–1859)[69] and broadened to include elements of any sets by Richard Dedekind, who between 1863 and 1894 published four editions of Dirichlet's *Lectures* from the winter of 1856–57 at Göttingen University. It is more than likely that Husserl came across this famous work during the mathematical stage of his education. Husserl likewise endowed the

[66]This does not refer to empirical experiences of psychology but to theoretical experiences of phenomenology.

[67]*PSW*, English trans. p. 259 (Husserl's emphasis).

[68]In a binary relation an element of the source set (the domain) may be associated with several elements in the target domain (the range). A function is a binary relation such that every element in the domain is related to *only one* element in the range.

[69]Leonard Nelson was the great grandson of J. P. G. Lejeune-Dirichlet, who succeeded to Gauß' chair at Göttingen on his death in 1855.

concept of the invariant with a philosophical destiny. To use his language, one might say some kind of invariance characterises any mathematical ideality, whenever and wherever we might place it historically and geographically. In the 19th century, the concept of invariant was at the heart of an important trend in research, notably illustrated by Hilbert's theory of algebraic invariants. In 1888, while a young lecturer at Königsberg (and while Husserl, assistant lecturer at Halle, was working on his *Philosophy of Arithmetic*, published in 1891), Hilbert proved the existence of a finite basis for generating invariants of an algebraic form of any number n of variables. This theorem, already remarkable for its generality, reverberated like a clap of thunder in the mathematical sky. Hilbert posed the problem in Kantian terms, wondering, given an infinite system of f_i forms of a finite number of variables $x_1, x_2, \ldots, x_n$, under what conditions a finite set of forms exists in terms of which all others are expressed as linear combinations whose coefficients are rational functions of the same variables as the starting forms. He actually proved that a finite base *necessarily* exists as suggesting otherwise would lead to contradiction. Hilbert did not present an effective basis nor indicated a way of calculating one. This was a famous launch pad for lively and interminable mathematical and philosophical controversies about the validity of reductio ad absurdum and about what 'existence' means in mathematics. Kronecker, the master mathematician of Berlin decreed that there is no existence without construction. To confirm the validity of its non-constructive approach, Hilbert drew on Kronecker's theory of algebraic numbers to produce, in 1892, an algorithmic proof of his theorem. Presumably, Husserl ignored neither these discussions nor the theory of invariants from which they originated. Furthermore, as a reader and supporter of Bolzano, he knew the variation technique Bolzano applied to non-logical parts of a proposition to test its logical validity.

The concepts of relation/function and invariant very likely sustained, if not inspired Husserl's willingness to turn his back on the notion of substance, dominant from Aristotle to Descartes and Hegel, and to promote the Platonic notion of essence as disconnected from that of substance. Husserl resumes the dialogue with Kant who banished metaphysics of substance from the domain of human reflection[70] in favour of a transcendental subject, and he radicalises this banishment. For he does not speak of the essence of substance but of the essence of the phenomenon itself, and essence is not what hides behind the phenomenon (since there is nothing behind the phenomenon), but it is *the invariant of the modes in which the phenomenon enters consciousness*. The experimental method is, of course, essential, recognises Husserl, when it comes to establishing factual connections, but it presupposes what no experiment could achieve, the analysis of consciousness itself. Only phenomenology can reach the ultimate components of acts of consciousness. For the "answers to the questions of knowing how givens achieve objective ascertainment, what 'objectivity' and 'ascertainment of objectivity' mean depend on the meaning of the given itself, that is *of the meaning given to it by consciousness, in its essence*."[71]

[70]Kant understood functionally the concept of substance as being the permanence which allows change to be grasped. "All we know about substances is force", *Reflexions* 4824.

[71]*PSW* (emphasis added).

The method for accessing the systematic connections of consciousness with its immanent correlations is *eidetic seeing* [Schauen], which is unlike sensory perception, but has no other "mystical secrets" than sensory perception.[72] Indeed, since essence is nowhere except in the phenomenon itself and since the phenomenon is only as much as its correlation to consciousness, I can *see* essence *directly* with no need for "signitive representation" [signitiv vorgestellt],[73] as it is used in Hilbert's proof theory. Essence is not the being itself, but the "being-given" of the phenomenon, which is simultaneously being and sense. Originary givenness takes place in eidetic intuition, which originates in sensory intuition without being limited to it. And Husserl's sensory intuition is no longer limited to passively receiving raw data from the senses as Locke, and to a certain extent Hume, Kant and the neo-Kantians, thought. Husserl's eidetic intuition is none of the Kantian intuitions: neither sensory intuition nor pure intuition as an a priori form of sensibility. Let me stress once more that Kant's sensory intuition is not bare sensation, it is already synthesis at the sensory level. At every level, Kantian intuition indicates the *power of form* over the matter that it informs. In contrast, Husserl's eidetic intuition is not an *empty form*, but a full view of essence. It is an ability of the understanding, which is applied equally to different modalities of being, essential as much as existential. Sensory perception encompasses a sort of primitive categorial activity by which *things give themselves* in their identity. Eidetic intuition has to do with *ideal identities of the meaning* of phenomena; it does not seize upon what exists but upon its essence, turning variably towards the invariant that arises from all the correlations between noema and noesis. I perceive the thing itself and I seize eidetically upon the essence of a thing, the being of a thing, which is given to me not as a thing but as meaning.

> The seeing seizes upon the *essence as being an essence* and in no way does it posit *existence*. Accordingly, knowledge of essence is no *matter-of-fact* knowledge, and includes not the least assertive content regarding an individual (e.g. natural) existence.[74]

The "true" critique of reason is intentional constitution where both meaning and the presence of the object are led to the state of givenness.

Hilbert totally ignored the thing-in-itself and wished to nullify the exteriority of the relationship between thought and experience through semio-linguistic objectification of the form of mathematical demonstration. Hilbertian intuition is materially represented by the symbols of formal, algebraic or logical calculations. Husserl in turn also eliminated the thing-in-itself, but not consciousness—transcendental consciousness, valid for all subjects at all times. Consciousness reigns even more supremely than in Kant's conception: not only does it constitute the noematic *structure* of the object in correlation with its noetic acts, but moreover it aims and can achieve, by eidetic reduction, the essence, that is the *being-sense* of the appearance of the phenomenon. Phenomenology will go beyond noumenon-phenomenon dualism and nullify in a certain sense the reciprocal exteriority of consciousness and object, since essence is

[72] *PSW*, English trans. p. 272.
[73] *PSW*, English trans. p. 273.
[74] *PSW*, English trans. p. 272.

given within the phenomenon. As eidetic inquiry, phenomenology, wrote Husserl, is "in the genuine sense a priori inquiry, and it simultaneously takes full account of all the legitimate motifs of aprioricism".[75] The true a priori is both form and sense, and not just form, as Kant says, or principle of formal constructivity for blind thought, thought by symbols, as in Hilbert's metamathematics. Husserl promoted the concept of sense/meaning as a non-material/ideal entity, the essence of things themselves, which is independent from its linguistic expression and yet has an objective reality, not reducible to a subjective psychological experience.

2.4 Systematic Philosophy: Jules Vuillemin

2.4.1 The "Analogies of Mathematical Knowledge"

Following the animosity towards the Hegelian system, coming from diverse quarters and for diverse reasons and fed in particular by the offensives of the experimental credo of science (notably psychophysics) and positive philosophy (A. Comte), and after the spread of historical consciousness (Hegel, Dilthey), of the existential movement (Kierkegaard, fierce opponent of "empty palaces"), of deconstructive back-worlds (Nietzsche), of hermeneutics (Dilthey, Heidegger), of historic-scientific rationalism (Léon Brunschvicg and the tenants of French historic epistemology), the idea of a philosophical system was once again taken up by Jules Vuillemin.

Rejecting as strongly as Husserl the worldview perspective, Vuillemin applies analogically a method inspired by the axiomatic approach and logic to "theoretical philosophy". Contrary to psychology and history, including history of science, "theoretical philosophy" aims, he says, to determine non-contingent and objective connections, comparable to those established by logic and exact science. The idea is not to develop a philosophy which reflects on the emergence or characteristic nature of modern mathematics, nor a philosophy so well suited to it that a mathematician might agree with or endorse it,[76] but rather to use new mathematical tools to analogically build a systematic philosophy both in its content and in its architectural analysis of previous systems. Clearly presented in *La Philosophie de l'Algèbre* [*The Philosophy of Algebra*], dedicated to the mathematician Pierre Samuel, with whom he was in dialogue,[77] J. Vuillemin's persistent idea is to draw a parallel between deductive

[75] *PSW*, English trans. p. 278.

[76] Presumably conversely to the positive reception of André Weil to the philosophy of Jean Cavaillès and that of Charles Ehresmann to that of Albert Lautman.

[77] Both were teachers in Clermont-Ferrand, then in the early 1960s, at the École Normale Supérieure de Jeunes Filles. One can see the connection between Vuillemin's point of view and those, respectively, of Bourbaki and P. Samuel in Sébastien Maronne's article: Pierre Samuel et Jules Vuillemin: mathématiques et philosophie, in Thierry Lambre (éd.), *Des mathématiques en Auvergne: histoire, progrès, interactions*, t. 1, Clermont-Ferrand, Revue d'Auvergne, pp. 151–173, 2014 (I am grateful to the author for providing me with a copy of his article).

theories constructed in mathematics as sets of consequences derived from explicitly stated *axioms*, and the project of a philosophy as a coherent system of compatible propositions based on a small number of *principles*. The system is to philosophy what deductive theory is in mathematics. Hence the will to use "the analogies of mathematical knowledge to critique, reform and define, as much as possible, the Method of Theoretical Philosophy,"[78] and consequently to build a systematic history of philosophy.

"Analogies of mathematical knowledge" are meant to replace Kant's "analogies of experience" in order to achieve a "*general* critique of reason". What does Vuillemin mean by 'general'? Actually he aims to rewrite *Critique of Pure Reason* in a totally a priori manner independent as much from the sensory experience as from historical considerations. He tracks a priori conditions of possibility of knowledge freed from the limiting relationship to experience which was the fulcrum of Kant's theory. According to Kant indeed, analogies of experience are fundamental principles [Grundsätze]—of substantiality, causality, and coexistence or reciprocity—which must be presupposed in the experience of the world, and "experience is possible only through the representation of a *necessary connection* of perceptions". "An analogy of experience is ... *a rule* according to which a unity of experience may arise from perception" (emphasis added). So experience "contains in one consciousness the *synthetic unity* of the manifold of perceptions. This synthetic unity constitutes the essential in any knowledge of objects of the senses, that is, in experience as distinguished from mere intuition or sensation of the senses".[79]

Now, how should we understand Vuillemin's analogies of mathematical knowledge? That is how should we understand the role of mathematical knowledge in the reform of theoretical philosophy? If we follow Kant's general line of thought but leave aside the central status of consciousness, we would say that mathematical knowledge supplies rules for gathering and ordering diverse mathematical results or propositions in synthetic theories. Of course it's particularly about rules of the axiomatic method. And then, how should we link mathematics to philosophy? The answer is "by analogy", taking this term in a broader sense rather than in its specific use in the architecture of *Critique of Pure Reason*. Once again, according to Kant's accurate account, "analogy does not mean, as is commonly understood, an imperfect similarity between two things, but a perfect similarity of relations within two quite dissimilar things."[80] So Vuillemin's project seems quite similar to the traditional search for philosophical rigour by using *mutatis mutandis* a deductive pattern. Indeed, Vuillemin aims to elaborate a systematic philosophy whose propositions would be ordered in roughly the same way as mathematical propositions in an axiomatic theory, even if the main ordering relation would be logical compatibil-

[78] *La Philosophie de l'algèbre I. Recherches sur quelques concepts et méthodes de l'Algèbre moderne*, Paris, Presses Universitaires de France, 1962, Introduction, §2, p. 5. See also the passage extracted from Kant's *Critique of judgement* that Vuillemin quotes on p. 510.

[79] *Critique of pure Reason*, Transcendental Logic, The Analytics of Principles B 219 (emphasis added).

[80] *Prolegomena to any future Metaphysics* 58.

ity rather than strict logical deduction. And of course Vuillemin takes as guide the bare principle of the axiomatic method regardless of its embodiment in a symbolic language and its framing in a logical formal system.

There is yet another important difference between systematic philosophy and axiomatic theories. Systematic philosophy must be built on a priori principles, and those principles *are not* axioms. The analogy between theoretical philosophy and axiomatic theories concerns uniquely the respective ordering relations which state explicitly what propositions are set as primitive and what propositions are compatible with them versus are deducible from them. Philosophical a priori principles do not have the same epistemological status as mathematical axioms. Mathematical axioms are constitutive of the set of propositions derived from them, while philosophical principles have only a regulative function: they are rules for establishing an "ideal" system in accordance with the unifying and guiding role of reason's ideas. In Kant's interpretation an idea implements the necessary role for reason's demand for systematic unity, but refers to no object of experience.[81] An idea has no objective reality, it has a "transcendental subjective reality". Vuillemin holds that an "ideal" system refers to no actual system but he does not accept that it has a transcendental subjective reality, precisely because he calls transcendental subjectivism into question.

2.4.2 Abstract Algebra and Systematic Philosophy

In Chapter III of *The Philosophy of Algebra*, the study of Abel's "general method" and his proof of the impossibility of solving algebraically general equations of degree higher than the fourth degree inspires Vuillemin's aim to radicalise Kant's *Critique* through a "*general* critique of pure reason". Vuillemin quotes (pp. 208–209) Abel's prescription about general algebraic equations of the fifth degree: "One must formulate a problem in a form that always allows the possibility of solving it...Instead of trying to find a relation without knowing whether it exists or not, one must ask whether such a relation is actually possible."[82] Vuillemin comments in Kantian terms (p. 221):

> Particular proofs are real: they presuppose the principle of the possibility of experience given in sensation. General proofs concern what is possible and start from concepts alone, ignoring the limitative conditions of the senses.

There is an ambiguity here. In this comment, indeed, mathematical experience seems to rely on sensation or *empirical* intuition, whereas in Kant's *Critique* mathematical

[81] Kant distinguishes between reason and understanding. Reason does not itself provide us with concepts of objects but only orders the concepts produced by the understanding. Reason takes the understanding and its concepts as its object while the understanding, by contrast, relates to objects themselves.

[82] Niels Henrik Abel, Beweis der Unmöglichkeit der algebraischen Gleichungen von höheren Grade als dem vierten allgemein aufzulösen, *Journal für die reine und angewandte Mathematik* 1, 1826, 65–84.

experience relies on *pure* intuition. Anyway, general algebraic methods provide us with the paradigm of the purely formal exercise of reason, an exercise independent of the principle of the possibility of experience. Vuillemin's version of Kant's theory of knowledge is, in fact, very close to Dedekind's purely conceptual construction of natural numbers in *Was sind und was zollen die Zahlen?* Like Dedekind, Vuillemin rejects the Transcendental Aesthetic. He writes:

> When the critical philosopher says…that all our knowledge begins with experience, although it does not derive from it, and from the fact that in reality he mixes origin and beginning by subordinating the deduction of concepts to the principle of experience, he makes the very task he proposes impossible, proceeding as he does from individuals whose experience is at least the occasion… The limits which it [classical philosophy, otherwise known as genetic idealism] attempts to assign to the faculties of knowledge are born not *internally* of the structure of these faculties, but are suggested from the outside, by the individual objects to which it applies itself through encounter.[83]

And further on:

> Kant borrows these limits [of our knowledge] from sensory intuition, which is a faculty external to reason.[84]

An finally, to eliminate the assumption of possible experience

> is not to weaken but strengthen the Kantian objective to free itself from *extrinsic limitations* under which it was restricted by the principle of the possibility of experience.[85]

Therefore "the *general* critique of pure reason" undertakes "to define knowledge as the relationship of reason with itself, and not as a relationship between this faculty and external data."[86] From a philosophical point of view this internalisation is a legacy from Hegel's process of concept, which develops internally with no external reference. Hegelian conceptual process is not definable through a relation to something external to it. In mathematics the instrument for such an internalisation is provided by the notion of structure, which reorganises axiomatically the mathematical material. Like others philosophers (especially Jean Cavaillès[87] whose influence in France was strong), Vuillemin uses the axiomatic method as a tool to reform Kant's theory of knowledge and to achieve "the revival of the problems posed by philosophy."[88] Vuillemin endorses Hilbert's distinction between contentual axiomatic theories and

---

[83] *La Philosophie de l'Algèbre*, p. 217 (emphasis added).

[84] *La Philosophie de l'Algèbre*, p. 220. Cf. also p. 463: "The principle difficulty of Kantian doctrine is its notion of intuition".

[85] *La Philosophie de l'Algèbre*, Conclusion, pp. 474–475 (emphasis added).

[86] *Leçon inaugurale [Inaugural Lecture] au Collège de France*, p. 28.

[87] See Hourya Sinaceur, *Jean Cavaillès. Philosophie mathématique*, Paris, Presses Universitaires de France, 1994.

[88] *Mathématique et métaphysique chez Descartes*, Paris, Presses universitaires de France, 1960, p. 141. For an example of the application of the axiomatic method to philosophy see "On two cases of the application of the axiomatic approach to philosophy: Zeno of Elea's analysis of motion and Diodorus Cronus' analysis of freedom", *Fundamenta Scientiae*, 6, 1985, pp. 209–219.

formal axiomatic theories.[89] Euclid's *Elements* exemplifies contentual axiomatics, whereas Hilbert's *Foundations of Geometry* exemplifies formal axiomatic theories, that is theories where one abstracts from concrete mathematical content.[90] Let me quote Hilbert's characterisation of both kinds of deductive theories:

> Contentual axiomatics introduces its fundamental concepts by reference to known *acts of experience* and its basic principles either as *obvious facts*, which one can make clear to oneself, or as extracts from complexes of experiences, thereby expressing the belief that one is on the track of *laws of nature* and at the same time intending to support this belief through the success of the theory.[91] (emphasis added)
>
> [Formal axiomatics] cannot get a foundation through a reference to either the evident truth of its axioms or to experience; rather such a foundation can only be given when the idealization is performed, i.e. when the extrapolation through which the concept formations [Begriffs-bildungen] and the principles [Grundsätze] of the theory come to overstep the reach either *of intuitive evidence* or *of the data of experience*, is understood to be consistent.[92] (emphasis added)

As one can note, reference to external experience is what makes the difference between contentual and formal axiomatics. Consequently the concept of truth and the meaning of intuition must be changed in order to fit the new standard of idealization achieved by pure reason. Vuillemin is not completely content with Hilbert's formal axiomatics. He revives Leibniz' *logical-metaphysical* rationalism as contrasted with Kantian intuitionism. His philosophy is partly a dialogue between Kant and Leibniz, developed notably in the chapter on Galois of *The Philosophy of Algebra*. Indeed, Vuillemin uses Leibniz' philosophical concepts, suitably altered, to enlighten substitution groups. According to his philosophical reading, Galois theory furnishes a method "to construct individual elements, no longer in intuition and following flawed schemes, but in the concepts themselves, in a completely a priori and general way, now without reference to any data or owing anything to luck" (pp. 288–289). So the question: is one entitled to apply the concepts of group and structure in the philosophical method? (§ 34, pp. 292–300). Besides, *What are philosophical systems?* (Cambridge University Press, 1986) presents in a Leibnizian style a systematic and a priori classification of the various possible philosophical systems.

Vuillemin also endorses in history of philosophy and to a large extent applies in human sciences (linguistics, psychology, anthropology, etc.), the Hilbertian distinction between genetic and structural points of view.[93] He attributes the first to philosophers who ascribe priority to the mind, that is, German idealists in general and Fichte[94] in particular, and the second to Martial Gueroult, arguing that only the

[89]D. Hilbert, P. Bernays, *Grundlagen der Mathematik* I, Springer, 1934, zweite Auflage, 1968, § 1, pp. 1–8.

[90]D. Hilbert, P. Bernays, *Grundlagen der Mathematik* I, §1, p. 20.

[91]D. Hilbert, P. Bernays, *Grundlagen der Mathematik* I, §1, p. 2.

[92]D. Hilbert, P. Bernays, *Grundlagen der Mathematik* I, §1, p. 3.

[93]D. Hilbert, P. Bernays, *Grundlagen der Mathematik* I, §1, pp. 1–2.

[94]Fichte's work was the subject of Martial Gueroult's doctoral thesis.

latter allows a "*general* critique of pure reason."[95] Adopting the structural method that Gueroult applied to make explicit the internal coherence of a philosophical system (Descartes, Spinoza, Leibniz, Malebranche, Fichte) Vuillemin classifies philosophical systems without true consideration of any historical philosophical doctrine (Platonic, Aristotelian, Epicurean, etc.) and with the aim to highlight the rational a priori links between different effective systems, links which therefore have nothing to do with effective links and can be unearthed only by structural analysis. Granger qualified this combinatorial enterprise as "metasystematic."[96] I will not go into the details of this metasystematic approach, which reminds us of Leibniz.[97] It was the subject of the last part of Baptiste Mélès' thesis, which leads to a volume now published.[98] I just will comment briefly in the part 4 of this section devoted to Vuillemin's views.

2.4.3 Mathematical Reflexivity and Philosophical Reflexivity: What Is Critique?

The Philosophy of Algebra can be read as a long and patient effort to set out the possible meanings of the concept of critique, in mathematics and philosophy, and to add a new sense to existing meanings. At the same time the various figures of dogmatism are explained.

Critique is a "reflexive activity", attributed by philosophers to the ego (Descartes, Husserl) or to the thinking Subject (Kant). The question arises of what the subject of this attribution is when it comes to mathematical methods. Is it still the conscious subject, who investigates first the limits of his understanding? Or is it about the objective limits of this or that mathematical method?

---

[95] *La philosophie de l'Algèbre*, chapter 3, §25, pp. 218–221; Conclusion, § 60, p. 517 (emphasis added).

[96] Axiomatic Method and the idea of system in the work of Jules Vuillemin, in *Causality, Method and Modality, Essays in Honor of Jules Vuillemin* (Gordon G. Brittan ed.) Dordrecht/Boston/London, Kluwer Academic Publishers, 1991, cited by Jacques Bouveresse, Vuillemin between intuitionism and realism, in *Philosophie des mathématiques et théorie de la connaissance*, (R. Roshdi & P. Pellegrin eds.), Paris, Librairie Scientifique et Technique, 2005, p. 77.

[97] "Necessary truths, such as we find in pure mathematics and particularly in arithmetic and geometry, must have principles whose proof doesn't depend on instances (or, therefore, on the testimony of the senses), even though without the senses it would never occur to us to think of them" wrote Leibniz in *New Essays on Human Understanding*, Preface. The idea of a priori combinatorics returns many times to analyses by Vuillemin of the structural method, and of course it is common also among mathematicians, as, for example, in the fine text of Poincaré summarizing the work of Sophus Lie, cited in *La Philosophie de l'Algèbre*, p. 426.

[98] Paris, Vrin, October 2016. See also the article by Elisabeth Schwartz "History of mathematics and philosophy," and that of Stéphane Chauvier, "Philosophy of the classification of philosophical systems: criticism and decision-making", in *Philosophie des mathématiques et théorie de la connaissance. L'Œuvre de Jules Vuillemin*, R. Rashed & P. Pellegrin eds., Paris, Albert Blanchard, 2005, pp. 1–28 and 187–204 respectively.

The answer to this question depends basically on the relations one sets between mathematics and philosophy. We can advocate a separation of method between mathematics and philosophy, and consider mathematics as an autonomous positive science apt to provide us with exact methods for a mathematical treatment of philosophical or at least epistemological questions. We can also admit a separation of content between mathematics and philosophy, and consider that philosophy can simultaneously use structural methods and keep its own ability to address very general philosophical problems in a way that leads to round off the narrower views growing in the field of mathematical practice. It is in virtue of this alternative that Vuillemin distinguishes between intrinsic intuitionism, that of Kronecker and Brouwer, and extrinsic intuitionism, that of Kant, who demands "of concepts of understanding that they relate to necessarily sensory intuitions, and whose pure form thus relates to the externality of space and time."[99] In Vuillemin's opinion intrinsic intuitionism incorporates a metaphysical choice into mathematical practice, whereas extrinsic intuitionism holds that metaphysics is beyond the reach of human knowledge. Anyway, some issues call into question the analogy between axiomatics and critique, as it is used by both Hilbert and Vuillemin. Concerning Hilbert one will remember the analysis in the first part of this article. I now turn to Vuillemin and his *Philosophy of Algebra*.

3.1. It is clear that Vuillemin adopts Hilbert's idea that the axiomatic approach represents mathematics entering its critique phase. He bases this idea on the distinction between a posteriori and a priori and between genetic and structural methods, and he continually uses cross-referencing mathematical and philosophical texts in a way that often makes it difficult to follow his account. So Lagrange is paralleled with Fichte (Chapter I, § 13) to show how their shared a genetic method, where form is never separated from content, joins the rudiments of a thoughtful critical analysis that nevertheless has difficulty disengaging from its empirical origin. "Any genetic method is latent empiricism" says Vuillemin.[100] The study of the phase represented by the works of Gauss gives Vuillemin the opportunity to show how the prejudices of extrinsic intuitionism underlie the notion that geometric constructions are proof of existence in mathematics. Such a prejudice delayed full acceptance of complex numbers until their geometric representation by Gauss. Thank goodness for Abel, one might say! For it was he who was astonished that one would want to solve particular algebraic equations, without solving first the general question of the possibility or impossibility of some solution. In Vuillemin's opinion Abel's reasoning is comparable to Kant's critical method. And critical method is thought to be generally applicable in mathematics through the axiomatic method. However this opinion would be unquestionable only if one assumes Hilbert's view of the axiomatic method as the critical step in mathematics and if the term and concept of possibility had the same meaning in Abel's mathematical work and in Kant's transcendental philosophy. But I have shown above the philosophical difference between Kant's concept of possibility and the mathematical use of the term 'possibility'

[99] *La Philosophie de l'Algèbre*, p. 172.

[100] *La Philosophie de l'Algèbre*, p. 118.

Vuillemin adjusts Husserl's distinction between the Aristotelian "abstraction by generalisation" and an "abstraction by formalisation" to interpret Abel's perspective (p. 216) and Galois' theory (p. 288). The new kind of abstraction "is irreducible to the old one in that it requires that we first expose the general *conditions* of a problem and therefore the postulates on which the theory depends."[101] Notably with Galois, then with Dedekind and Kronecker, it appears that the concept of group or the concepts of field and domain of rationality "are a typical example of the idea of *critique* in algebra,"[102] which consists, according to Vuillemin, in using an a priori method, illustrated for example by the construction of Galois resolvent. In the conclusion of his book, Vuillemin reaffirms his belief in this filter of critique in mathematics, writing

> [In] material mathematics, [...] one can prove what is, but not the impossibility of what is not. Thus the long-held illusion under which algebraists remained concerning the solution of equations beyond the fourth degree could only be dispelled once, by a change of method, they sought to build not the solutions to equations they could solve, but the formal conditions rendering the solutions possible or impossible. Material mathematics remained necessarily dogmatic. Formal mathematics right away became critical.[...] It sought, indeed, [...] to define a priori the types of structures on which the solution to a problem depends, and consequently the *intrinsic* limits these structures entail...[103]

One can easily understand that the abstract axiomatic approach determines a priori the conditions for validity of *operations* defining a structure. But could one say inasmuch as it falls within the philosophical framework established by Kant's transcendental idealism? Let me try to address this question.

3.2. In footnote (1) of p. 262 of *The Philosophy of Algebra*, Vuillemin contrasts classical mathematics, which "defines an object independently of reflection", and modern mathematics, which "internalises reflection in its own methods". Vuillemin immediately adds a warning which seems to contradict somehow the aforesaid internalisation: "I will return later to the *confusions* insinuated here between the mathematical and the philosophical concept of operation, which consist in inferring untenable conclusions from the correct idea of *two levels of pure knowledge*" (emphasis added). The philosophical conception of operation is reflection,[104] it constitutes one of the two levels, the other being the mathematical and logical notion of operation. Does Vuillemin mean that reflection, qua "active operation of the soul", is internalised in mathematical methods?

It recently became a commonplace to stress the "reflective turn" of modern mathematics regardless of different levels of knowledge. Vuillemin's view is deeper and more subtle though complicated and limited to an inceptive state. According to it, the answer to the question posed above would demand a theory of operations of knowledge (conceiving, judging, reasoning) in relation to mathematical structures. This theory will be based on the difference between objective operators, such as con-

---

[101] Ibid., p. 216 (emphasis added).

[102] Ibid., p. 232 (Vuillemin's emphasis).

[103] Ibid., p. 471 (Vuillemin's emphasis).

[104] See for example p. 290, end of 1st paragraph.

junction, disjunction, negation or modal operators, and transcendental operations of the sort "I think that..." which, inter alia, do not have certain properties common, for example, to conjunction and disjunction such as symmetry and associativity. In any case, if we distinguish between mathematical or logical operations and operations of the mind, that is to say between objective and subjective operations, then it becomes hard to interpret the axiomatic method as the critical or Kantian stage of mathematics. Aware of this difficulty, Cavaillès and Granger used the Husserlian concept "thematisation" to describe the specific mathematical process which goes from an operation, let us say addition, to *the theory* of its possible properties: associativity, commutativity, invertibility, etc. and from the set of those properties to the concept of additive structure. Vuillemin does not discuss "thematisation", though this concept might have shed some light on his purpose. But he rightly reproaches post-Kantians for unduly associating the structural approach and idealism (p. 273), and, I would add, the objective method and the subjective act. However to break this very common but undue association, one must, I believe, explicitly and rigorously distinguish between *philosophical reflection*, which necessarily refers to an empirical or transcendental subject, and *mathematical reflexivity*, which refers to the superimposed layers of ideational objects. If one truly admits this distinction, the Kantian interpretation of the axiomatic method becomes highly questionable, unless one wishes to refer mathematical reflexivity and any other mathematical process as objective products to a subjective, empirical or transcendental, thinker or producer. Making this reference or not is a philosophical choice. In his last metamathematical writings Hilbert, for instance, moved away from his initial Kantian interpretation of the axiomatic method in order to advocate a more objectivist point of view.

But if one agrees with Vuillemin that the structural approach is not idealistic and not subjective, what philosophical sense could we give to the research of the mathematical general conditions of possibility for solving some problem? Certainly not, I think, a sense consistent with the Kantian *Critique*, which is inseparable from its transcendental setting. First of all, as I have observed in my analysis of Hilbert, in mathematics the conditions of possibility are neither universal nor immutable, but local and variable depending on the particular problem to solve, its formulation and its links with known results and conjectural hypotheses. The a priori and formal character of the conditions posed at the outset, that is to say of the axioms, is not external to the content they determine: they are *necessary and sufficient* conditions, constitutive of the content of a deductive theory rather than only regulative. Vuillemein is perfectly aware of this fact, and the mathematical material analysed by him illustrates these points abundantly. It follows that a non-Kantian conception of reason is required, since according to Vuillemin reason is "the faculty of thinking about structure" independently of the objects to which it is applicable and free from the constraints of intuition.[105] Actually, Vuillemin does not totally abandon Kant. Like many scientists and philosophers of the early 20th century who sought to adapt Kant to new developments in mathematics and physics (Hilbert, Poincaré, Brouwer, Cassirer, to name but a few), Vuillemin wanted only to redefine the extension of

[105] *La Philosophie de l'Algèbre*, p. 467.

critical philosophy "by relating it to the necessary plurality of the choices between axiomatic systems and not to the internal[106] limitation of knowledge required by the fact of experience". However, the relationship between axiomatic systems and critical philosophy remains problematic. For it rests on a shaky analogy, and that for at least three reasons: philosophy and mathematics are two heterogenous domains with different aims and ways, their operations are fundamentally dissimilar, and the vocation of philosophy, according to Vuillemin, is to seek legitimacy *by right* and to pose foundational questions that the positive sciences usually push aside.

The last trait engaged Vuillemin in an analysis of Husserl, who posed exactly the same question about the legitimacy of *every* judgement's "being claim", and who in his *Logical Investigations* studied the structures of meaning corresponding to different formal levels. But the *objective structures of intentionality* cede to the *constituent power of intentional consciousness* and to the *ultimate authority of categorial intuition*, which reestablishes the power of intuition and the conception of truth as adequacy between knowledge and Being. Vuillemin very justifiably concludes that phenomenology is dogmatism.[107] To respect the criticism in spirit, if not in letter, Vuillemin decided to overturn the interdiction which since Kant had weighed heavily on metaphysics and assumes that "all understanding—whatever it may be—is metaphysical through and through in that, at its core, it entails decisions and choices which do not in themselves belong to the internal jurisdiction of this understanding". Working on this assumption, the task of philosophy consists not of "ignoring metaphysical choices, but studying motives in relation to man's free will". If one cannot build a formal ontology founded on Kantian principles, which actually forbid any kind of ontology, it is worth examining *the hypothesis* that the "*general* critique of pure reason" is compatible with formal ontology. As a philosopher of human free will, Vuillemin shines a light on the metaphysical choices and decisions which preside, independently of all experience, he stresses, over every thought—including those in concepts and scientific theories—and every action. But given his intention to substitute "human cogito in a universe of gods with human doing in the world of man", Vuillemin claims he respects the spirit of Kantian philosophy, if not the letter, in his removal of formal ontology from theology without undermining its metaphysical foundations. Thus, "Critique should study the convenience of these general decisions with beings and values", that is to say with basic ontological and ethical assumptions.

2.4.4 From Abstract Algebra to Metaphysics

In contrast with classical algebra, which deals with theories, abstract algebra allows us to determine classes of theories, whose axioms systems may have multiple interpretations. For example, abstract field theory is instantiated by the field of real numbers,

[106] *La Philosophie de l'Algèbre*, p. 476 (here one would have to substitute "external" for "internal" in order to remain consistent with the analysis Vuillemin made previously in his book).

[107] Ibid., p. 493 and following pages.

the field of complex numbers, the finite fields **Z**/*n***Z** where *n* is a prime integer, the field of p-adic numbers—completion of the **Q** field of rational numbers supplied by a so-called ultrametric distance, etc. Analogically Vuillemin defines classes of philosophical systems based on sets of compatible structural properties. He writes:

> When we encounter the new and worn notion of *structure*, its use in the study of philosophical systems no doubt produces effects analogous to those suffered by pure reason at the hands of formal algebra. Distinguished both from a chronicle where thoughts are narrated as events and a philosophy of history where they constitute the materials of a theodicy, the technological history of systems of philosophy[108] abandons the intolerable idea of an absolute rule…[109]

The possibility of varying mathematical structural settings according to the problem to be resolved or to a preferred method of solving it is frequently realised. A familiar example is that of divisibility, treated differently by Dedekind and Kronecker. The first introduces the set theoretical concept of ideal, defining it by the conjunction of two very simple axioms, the second considers an algebraic number (the root of a polynomial with rational coefficients) as a "divisor."[110] Vuillemin states analogically that it is the fundamental plurality of *standards* that drives the various possible philosophical systems. Drawing up a rational distribution of philosophical systems shows that there is a choice at the root of every system. Now it will not be a question of saying which set of standards is the best but of laying out the range of standards with the purpose of providing the base for rational choice. Vuillemin gives a personal interpretation to the Kantian distinction between understanding and reason and replaces intuition by decision, that is to say seeing or thinking by doing.

> For the intuitions of *the understanding*, it [formal algebra] substitutes the order according to which *reason* combines operations, which do not in themselves have representative value and which, rather than ideas that one can see, embody the *decisions* which one can take and for which truth, without any substantive appropriateness, comes down to formal compatibility (emphasis added).

Moreover, Vuillemin's aim is to join together thinking and choosing, reflection and will. Hence the consequence for Fichte's *Theory of Science* [Wissenschaftslehre], which

---

[108]Martial Gueroult gave the name "History and technology of systems of philosophy" to the chair he occupied at the Collège de France from 1951 to 1962.

[109]*Leçon inaugurale* [Inaugural lecture] *au Collège de France*, 5 December 1962, p. 20 (emphasis added). Current mathematical usage of the qualifier 'formal' is now reserved for a theory written in logical language, that is, which specifies the primitive logical constants and the permitted rules of inference.

[110]The two axioms defining an ideal can be found in the second edition of the *Vorlesungen über Zahlentheorie von Dirichlet*, §§ 159–170 (1871), and in the article on algebraic theory (1877), partially reproduced in Dedekind's *La création des nombres*, Paris, Vrin, 2008, in particular p. 248. Kronecker's theory of divisors can be found in the Grundzüge einer arithmetischen Theorie des algebraischen Zahlen, *Journal für die reine und angewandte Mathematik*, vol. 92, 1882, 1–123. It is summarised in an accessible manner in the excellent article by Alain Michel, Après Jean Cavaillès, l'histoire des mathématiques, *Philosophia Scientiae*, vol. 3, cahier 1, 1998, 113–137; the two principal theorems of the theory are set out in note 22, p. 131.

> necessarily loses its absolute character and becomes permeable to the presence of different and relative interpretations that assign new content to the task of critique, in relating to the necessary plurality of *choices* between axiomatic systems, not to the internal[111] limitation that *the fact* of experience imposes on knowledge (emphasis added).

On the one hand, considered not only from the internal perspective of mathematics, but also and ultimately from the autonomous and overlooking standpoint of philosophy, the choice engages metaphysics. But on the other hand, annulling the limitations imposed by the Kantian principle of possible experience permits the reintroduction of metaphysics to the very heart of Critique. In Vuillemin's hands the structural point of view has a paradoxical consequence in giving the leading and ultimate role to metaphysics.

> In the new critical metaphysics, the being addresses itself[112] over and above all exterior affection, to the level of structures which define the relationship of reason with itself.[113]

This is a typically anti-Hilbertian and anti-positivist programme with a Hegelian flavour. Far from dismissing metaphysics and even philosophy itself in favour of science, which has, according to Hilbert, critical aspirations thanks to the axiomatic method and proof theory, Vuillemin uses the tools of science, principally the axiomatic method, to return philosophy its autonomy and its truly critical specificity with regards to theology and science as well, and to confer to metaphysics new credence in its contribution not as a thought of the One or of the Absolute, but as a thought of multiplicity, without it necessarily leading to ontological relativism. Finally, the true philosophical lesson to be learned from the mathematical concept of structure is an invitation to revive and invert the Platonic relation between the One and the Multiple.

It is true that Vuillemin believes that an updating of scientific methods, mathematics in particular, offers the opportunity to rethink the mission of philosophy. But as tight as this bond between science and philosophy may be, it precisely remains *a bond* between two visions whose intransigent duality prohibits one being absorbed into the other—as it happens in the opposite views of Hilbert and Husserl—and allows only external and retrospective enlightenment of science by philosophy, which simultaneously illuminates the philosophical systems chosen by Vuillemin as candidates for his analogical analysis. In Vuillemin's mind the analogy with science does not by any account mean that philosophy is or could be a kind of science, not even of a different type from that of the positive sciences as Husserl wished. Indeed, a choice, when it concerns not scientific axioms, which are always more or less local, confined in scope to a particular domain of a particular science, but general principles of interpretation of the constitution of the world as a whole, is a question of ethical decision, that is,

[111] Read 'external'.

[112] 'The being addresses itself' would not be understood without the contribution of Hegelian dialectics, finally rejected by Vuillemin for the finalist nature of its dynamic, mesmerised by divine Absolute. Vuillemin preserves certain souvenirs from his foray into Hegelian territory.

[113] *Leçon inaugurale au Collège de France*, p. 30. The last sentence belongs also to Hegel's legacy. "The relationship of reason with itself" is a consequence of the Hegelian reduction of instant sensory experience to an illusion.

of problems falling under *The Critique of Practical Reason*. Stéphane Chauvier[114] sheds light on what Vuillemin's reflections have in common with those of Cavaillès, which makes them both heirs of Pascal: the human being as man of science and man of action is basically forced to gamble. In Vuillemin's ultimate ontological view the gamble concerns the being as being, whereas Cavaillès' urgent scientific and ethical worries are about gambling on the future.[115]

2.5 Conclusion

In this paper, I have tried to analyse in some detail the *mutual* relationships between philosophy and mathematics in the modern era. The goal of the Lisbon Congress was to explain what philosophers thought about mathematics of their time and how they used mathematical innovations in the development of their own doctrines. As examples, I have thus examined the question of the "intrinsic possibility of pure knowledge in relation to the faculty of thought" that philosophers—in this case Husserl and Vuillemin—pose in thinking about the autonomy acquired by abstract mathematics from sensory intuition. Would it not be possible, indeed, to revive Kantian Critique through a theory of objective structures of reason whose origins or motivation are not to be found in sensory experience and whose keystone does not lie in formal subjectivity of pure apperception? This question was vital to French philosophers, Cavaillès then Vuillemin and Granger. This paper deals with Vuillemin point of view, while I have studied Cavaillès' and Granger's contributions in previous works.[116] Besides this, I have also wished to study here the inverse impact of philosophical doctrines not on the mathematical methods, concepts and techniques themselves, but on the ideas used by mathematicians to introduce them, while explaining their novelty and their advantages. Hence Richard Dedekind intended to break the Kantian link of pure thought to empirical experience, while Henri Poincaré, David Hilbert and L. E. J. Brouwer thought, each in his own way, that intuition supplies initial inspiration, a beginning or an origin, i.e. a fundamental material basis versus a metaphysical justification for elementary mathematical concepts. It is to the crossroads of lines of reasoning drawn by Kant, Husserl and Hilbert that I have dedicated this essay. The choice of this triad was motivated by my re-reading of *La Philosophie de l'Algèbre* by Jules Vuillemin, whose reflections interweave, in an original and sometimes confusing fashion, the results of abstract algebra with the philosophies of Descartes, Leibniz, Kant, Fichte and Husserl.

In summary one might say that in the classical era (Descartes, Spinoza, Leibniz, etc.), philosophy and science went hand-in-hand. In the modern era, by contrast,

---

[114]Article cited above in note 98.

[115]Cf. Hourya Benis Sinaceur, *Cavaillès*, Paris, Les Belles Lettres, 2013.

[116]Formes et concepts, *La connaissance philosophique, Essais sur l'œuvre de Gilles-Gaston Granger*, Joëlle Proust et Elisabeth Schwartz eds., Paris, PUF, 1995, 93–120; Style et contenus formels chez Gilles Gaston Granger, *La pensée de Gilles-Gaston Granger*, A. Soulez et Arley R. Moreno eds., Paris, Hermann, 2010, 161–206.

each protagonist, fascinated by the other, nevertheless battles to remain undisputed captain of the ship. On the one hand, the scientism of Hilbert's mathematical epistemology aimed to reduce, if not to eliminate the ambition of philosophy to submit mathematical practices and problems to its own principles and methods, be they transcendental or metaphysical. On the other, phenomenology has promoted the idea of a non-exact philosophical rigour and advocated a point of view encompassing positives sciences, ontology, and ethical values in connection with the dominant category of sense/meaning, while Jules Vuillemin assumed the inseparability of thought—-scientific or philosophical—from the metaphysics of free will. The battle has been, and still is in some circles, about the meaning and effective content of the idea of scientific philosophy and its putative links to metaphysics. A "disputatio" practiced from the beginnings of philosophy.

References

Abel, Niels Henrik. (1826). Beweis der Unmöglichkeit der algebraischen Gleichungen von höheren Grade als dem vierten allgemein aufzulösen. *Journal für die reine und angewandte Mathematik, 1,* 65–84.

Benis Sinaceur, H. (1994). *Jean Cavaillès. Philosophie mathématique*. Paris: Presses Universitaires de France.

Benis Sinaceur, H. (1995). Formes et concepts. In J. Proust & E. Schwartz (Eds.), *La connaissance philosophique, Essais sur l'œuvre de Gilles-Gaston Granger* (pp. 93–120). Paris: PUF.

Benis Sinaceur, H. (2010). Style et contenus formels chez Gilles Gaston Granger. In A. Soulez & A. R. Moreno (Eds.), *La pensée de Gilles-Gaston Granger*. Paris: Hermann.

Benis Sinaceur, H. (2013). *Cavaillès*. Paris: Les Belles Lettres.

Benoist, J. (1994). *Autour de Husserl*. Paris: Vrin.

Bouveresse, J. (2005). Vuillemin between intuitionism and realism. In R. Roshdi & P. Pellegrin (Eds.), *Philosophie des mathématiques et théorie de la connaissance: l'oeuvre de Jules Vuillemin*. Paris: Albert Blanchard.

Chauvier, S. (2005). La philosophie de la classification des systèmes philosophiques: criticisme et décisionnisme, in R. Rashed, & P. Pellegrin.

Dedekind, R. (1932). Über die Einführung neuer Funktionen in die Mathematik (The introduction of new functions in mathematics). In *Gesammelte mathematische Werke III* (pp. 428–438). Vieweg & Sohn, Braunschweig. In French *La création des nombres* (2008). Traduction, introduction et notes par Hourya Benis Sinaceur. Paris: Vrin.

Fries, J. F. (1822). *Die mathematische Naturphilosophie nach philosophischer Methode bearbeitet: ein Versuch*. Heidelberg: Mohr und Winter.

Gottlob F. (1967). *Kleine Schriften*. In I. Angelelli (Ed.), Darmstadt: Wissenschaftliche Buchgesellschaft (2nd ed., 1990). Hildesheim: Olms.

Granger, G. G. (1991). Axiomatic Method and the idea of system in the work of Jules Vuillemin. In G. G. Brittan (Ed.), *Causality, Method and Modality, Essays in Honor of Jules Vuillemin*. Dordrecht: Kluwer Academic Publishers.

Hegel, G. W. F. (2006). *Phénoménologie de l'esprit, introduction, traduction et notes par Bernard Bourgeois*. Paris: Vrin. Original: *Phänomenologie des Geistes* (1807). Bamberg und Würzburg: Joseph Antos Goebbardt.

Herzog, K. (1978). *"Kritische Mathematik"— ihre Ursprünge und moderne Fortbildung*. Dissertation, Düsseldorf.

Hessenberg, G. (1904). Über die kritische Mathematik. *Sitzungsberichte der Berliner mathematischen Gesellschaft, 3,* 21–28.

Hilbert, D. (1935). *Gesammelte Abhandlungen* (Vol. III). Berlin: Springer.
Hilbert, D. (1968). *Grundlagen der Geometrie*, 10. Auflage, Stuttgart, B.G. Teubner.
Hilbert, D. (1992). *Natur und mathematisches Erkennen*. Herausgegeben von David. E. Rowe, Basel, Birkhäuser Verlag.
Hilbert, D., & Bernays, P. (1968). *Grundlagen der Mathematik I*. Berlin: Springer (erste Auflage, 1934).
Husserl, E. (1891). Philosophie der Arithmetik. *Psychologische une logische Untersuchungen*. Halle: Pfeffer
Husserl, E. (1900). *Logische Untersuchungen*. Halle: Max Niemeyer Verlag. English trans. J. N. Findlay, *Logical Investigations* (reprint 2001), Routledge & Kegan Paul Ltd.
Husserl, E. (1906). Personliche Aufzeichnungen edited by W. Biemel. *Philosophy and Phenomenological Research, XVI*(1956), 293–302.
Husserl, E. (1907). Die Idee der Phänomenologie. *Fünf Vorlesungen*. Hua II.
Husserl, E. (1910). Philosophie als strenge Wissenschaft, *Logos. Zeitschrift für Philosophie und Kultur*, Bd. 1, Tübingen 1910/11, 289–341. Online http://www.gleichsatz.de/b-u-t/archiv/phenomeno/husserl_streng1.html.
Husserl, E. (1913). *Ideen zur einen reinen Phänomenologie und phänomenologischen Philosophie I*. Halle: Max Niemeyer Verlag.
Husserl, E. (1929). *Formale und transzendentale Logik. Versuch einer Kritik der logischen Vernunft*. Halle: Max Niemeyer Verlag.
Husserl, E. (1935). *Die Krisis der europäischen Wissenschaften und die transzendentale Phänomenologie*, French transalation Depraz N., 2012. Online: http://www.ac-grenoble.fr/PhiloSophie/wp-content/uploads/ebooks/husserl_depraz.pdf.
Fries, J. F. (1828). *Neue oder anthropologische Kritik der Vernunft, 3 Band*. Heidelberg: Christian Friedrich Winter.
Kant, I. (1781). *Prolegomena zu einer jeden künftigen Metaphysik, die als Wissenschaft wird auftreten können*. Kant Riga: Hartknocht. English trans. (1997). Cambridge: Cambridge University Press
Kant, I. (1951). *Dissertation de 1770*, traduction avec une introduction et des notes par P. Mouy, Paris, Vrin, translated into English with an introduction and discussion by W. J. Eckoff, Columbia College, 1894. https://archive.org/details/cu31924029022329.
Kant, I. (1998). *Critique of Pure Reason* (P. Guyer & A. W. Wood, Trans. and eds.). Cambridge: Cambridge University Press. Original: *Critik der reinen Vernunft* 1781, 1787. Riga: Hartknoch.
Kronecker, L. (1882). Grundzüge einer arithmetischen Theorie des algebraischen Zahlen. *Journal für die reine und angewandte Mathematik, 92*(1882), 1–123.
Largeault, J. (1992). *Intuitionisme et théorie de la démonstration*. Paris: Vrin.
Lauer, Q. (French trans.), Paris: Presses Universitaires de France (1955). English trans. in M. Brainard (2002). *The New Yearbook for Phenomenology and Phenomenological Philosophy II* (2002). Online: https://fr.scribd.com/doc/63651073/Husserl-Philosophy-as-a-Rigorous-Science-New-Translation.
Lejeune Dirichlet, P.G. (1894). *Vorlesungen über Zahlentheorie*. vierte Auflage, herausgegeben mit Zusätzen versehen von Dedekind, R. Braunschweig: Vieweg und Sohn.
Mancosu, P. (Ed.). (2008). *The Philosophy of Mathematical Practice*. Oxford: Oxford University Press.
Mancosu, P. (2015). *Infini, logique, géométrie*. Paris: Vrin.
Maronne, S. (2014). Pierre Samuel et Jules Vuillemin: mathématiques et philosophie. In T. Lambre (Ed.), *Des mathématiques en Auvergne: histoire, progrès, interactions* (pp. 151–173), t. 1. Clermont-Ferrand, Revue d'Auvergne.
Mélès, B. (2016). *Les classifications des systèmes philosophiques*. Paris: Vrin.
Michel, A. (1998). Après Jean Cavaillès, l'histoire des mathématiques. *Philosophia Scientiae, 3*(cahier 1), 113–137.
Nelson, L. (1906). Die kritische Methode und das Verhältnis der Psychologie zur Philosophie. Ein Kapitel aus der Methodenlehre. *Abhandlungen der Fries'schen Schule, Neue Folge, 1,*

1–88. Online: http://archive.org/stream/abhandlungenderf01gtuoft/abhandlungenderf01gtuoft_djvu.txt.

Nelson, L. (1973). *Geschichte und Kritik der Erkenntnistheorie, Complete Works* (Vol. II). Hamburg: Meiner.

Peckhaus, V. (1990). *Hilbertprogramm und Kritische Philosophie. Das Göttinger Modell interdisziplinärer Zusammenarbeit zwischen Mathematik und Philosophie*. Göttingen: Vandenhoeck und Ruprecht.

Poincaré, H. (1902). *La science et l'hypothèse*. Paris: Flamamrion.

Poinxaré, H. (1905). *La valeur de la science*. Paris: Flamamrion.

Pradelle, D. (2016). On the notion of sense in phenomenology: Noematic sense and ideal meaning. *Research in Phenomenology, 46,* 184–204.

Rashed, R., & Pellegrin, P. (Eds.). (2005). *Philosophie des mathématiques et théorie de la connaissance, L'Œuvre de Jules Vuillemin*. Paris: Albert Blanchard.

Schwartz, E. (2005). History of philosophy and mathematics, in R. Rasched, & P. Pellegrin, pp. 1–28

Varga, P. A. (2010). *Ein bisher unbekanntes Portrait von Edmund Husserl*, revised in September 2013. http://hiw.kuleuven.be/hua/Media/mitteilungsblatt/portrait.

Vuillemin, J. (1960). *Mathématique et métaphysique chez Descartes*. Paris: Presses universitaires de France.

Vuillemin, J. (1962). *La Philosophie de l'algèbre I. Recherches sur quelques concepts et méthodes de l'Algèbre moderne*. Paris: Presses Universitaires de France.

Vuillemin, J. (1962). *Leçon inaugurale au Collège de France*.

Vuillemin, J. (1985). On two cases of the application of the axiomatic approach to philosophy: Zeno of Elea's analysis of motion and Diodorus Cronus' analysis of freedom. *Fundamenta Scientiae, 6,* 209–219.

Chapter 3
Avicenna and Number Theory

Pascal Crozet

Abstract Among the four mathematical treatises that Avicenna takes care to place within his philosophical encyclopaedia (*al-Shifāʾ*), the one he devotes to arithmetic is undoubtedly the most singular. Contrary to the treatise on geometry, which differs little from its Euclidean model, the philosopher takes as his point of departure the treatise of Nicomachus of Gerase, but modifies its spirit to incorporate results from the many disciplines which were dealing with numbers: Euclidean Theory of numbers, Nicomachean Aritmāṭīqī, Indian reckoning, Ḥisāb, Algebra, etc. We would like to show how Avicenna, taking note of the changes in the mathematics of his time and led by a philosophical questioning about the nature of the disciplines, proposes here one of the very few texts that gives to Theory of numbers a synthetic image, gathering the themes of mathematical research for the next centuries.

3.1 Introduction

In 1981, to mark Avicenna's millennium, Roshdi Rashed gave a lecture, published some time later under the title "Mathématiques et philosophie chez Avicenne."[1] He started by noting that the links between mathematics and philosophy had always been strong since the beginning of Hellenistic philosophy, and that Avicenna's direct predecessors, al-Fārābī and al-Kindī, had composed several treatises devoted exclusively to mathematics. However, in both al-Fārābī and in al-Kindī, he remarked, these writings were somewhat separate from the philosophical presentation. Yet, this was not so for Avicenna, since he conceived his mathematical treatises as an integral part of his philosophical encyclopaedia, *Kitāb al-Shifāʾ*. To understand this fact, which had never before been highlighted, by historians of science nor by historians of phi-

[1] Roshdi Rashed, "Mathématiques et philosophie chez Avicenne", *Études sur Avicenne*, dir. Jean Jolivet and Roshdi Rashed, Éditions du CNRS (Paris, 1984), pp. 29–39.

P. Crozet (✉)
CNRS SPHERE, Université Paris 7, Paris, France
e-mail: crozet@univ-paris-diderot.fr

H. Tahiri (ed.), *The Philosophers and Mathematics*, Logic, Epistemology, and the Unity of Science 43, https://doi.org/10.1007/978-3-319-93733-5_3

losophy, Roshdi Rashed intended to refer to these mathematical writings, limiting himself, not without reason, to the sole treatise on arithmetic from the *Shifā*'.

From this point of view, numbers clearly offered a privileged field of study. This is perhaps due to the fact that compared to other domains, the Aristotelian conception of the classification of science, broadly adopted by Avicenna, seems to go against mathematical practice itself, since studies on numbers fell, at the time, under several related but distinct disciplines:

- Euclidean Number theory, as presented in books VII to IX of the *Elements*;
- What is commonly designated by the term *al-Arithmāṭīqī*, which refers to the neo-Pythagorean tradition in general and particularly to Nicomachus of Gerasa's book (translated in the 9th century by Thābit ibn Qurra);[2]
- Integer Diophantine analysis, present in the work of mathematicians of the 10th century like al-Khāzin;[3]
- *Ḥisāb* ("calculation"), a sometimes-eclectic combination of calculation procedures or problem solving and results on numbers;[4]
- Algebra;
- Indian reckoning, a reasoned set of calculation procedures related to the decimal place-value system.

Reverting to certain specificities of these disciplines, Roshdi Rashed showed in particular how Avicenna made algebra and Indian reckoning "secondary parts (...) of the <science> of numbers,"[5] thus developing, following al-Fārābī, but in a slightly different way, "a non-Aristotelian area within a classification whose bias remains Aristotelian."[6]

How then can we describe this science of numbers that Avicenna seems to have conceived? In order to answer this question, I would like to return to the *Shifā*'s treatise on arithmetic, thus following in Roshdi Rashed's footsteps: not to question his conclusions in any way, on the contrary, but merely to clarify a number of points he has not had the leisure to truly address or detail.

---

[2]Wilhelm Kutsch, *Ṯābit b. Qurra's Arabische Übersetzung der* Ἀριθμητικὴ Εἰσαγωγή *des Nikomachos von Gerasa zum Esrten Mal Herausgegeben*, Recherches publiées sous la direction de l'Institut de lettres orientales de Beyrouth, tome IX, Imprimerie Catholique (Beyrouth, 1959).

[3]See Roshdi Rashed, *The Development of Arabic Mathematics: Between Arithmetic and Algebra*, Boston Studies in Philosophy of Science, Kluwer (Boston, Dordrecht, 1994), in particular pp. 205–236 ("Diophantine Analysis in the Tenth Century: al-Khāzin").

[4]See Pascal Crozet, "Aritmetica", *Storia della Scienza*, volume 3: *La civiltà islamica* (dir. Roshdi Rashed) Istituto della Enciclopedia Italiana (Rome, 2002), pp. 498–506.

[5]Roshdi Rashed, "Mathématiques et philosophie chez Avicenne" (*op. cit.*), p. 37.

[6]Ibid., p. 34.

3.2 Euclid and Nicomachus

The mathematical part of the *Shifā'*, as we know, consists of four treatises: geometry, arithmetic, astronomy, and music; it therefore represents, in a fairly classical way, the *quadrivium* disciplines. The first treatise of the series, that on geometry, is an abridgment of all thirteen books of Euclid's *Elements* and the two books attributed to Hypsicles usually associated with them. Here, Avicenna scrupulously follows the sequencing of his model and although the text does not lack interest, it is therefore in no way an original composition. Due to his decision to conform to Euclid's work, Euclidean arithmetic from books VII to IX is presented as early on as that first treatise.

The second treatise, dedicated to arithmetic and of particular interest to us, appears to follow a similar pattern in its relation with a Hellenistic model. This is the representation given in the 19th century by the eminent historian Franz Woepcke:

> It is divided into four books. It is a kind of paraphrase of Nicomachean arithmetic, and the whole has little value as an original work. It is rather curious that I did not notice a single mention of Nicomachus by name throughout the course of the treatise, although Avicenna mentions Euclid's *Elements*, to which he refers, and the Pythagoreans.[7]

Woepcke is undoubtedly correct on several points. The reference to Nicomachus, though implied, is necessarily present from the first: the title (*al-Arithmāṭīqī*) refers directly to the neo-Pythagorean[8] tradition and, as we shall see, the general structure is the same. Secondly, it is true that compared to the work of Avicenna's predecessors and contemporaries, the treatise doesn't really offer new results. Finally, the comment on the *Elements* is particularly good since we will show shortly that the text is based, from the outset, on Euclidean arithmetic discussed in the previous treatise.

However, we believe, the treatise can't be reduced to a paraphrasing of Nicomachus's text. Even if the reference is obvious, we are dealing with an original composition, which offers new perspectives and incorporates new findings unknown within the strict framework of Nicomachean arithmetic.

Avicenna's choice to formally distinguish Euclidean and neo-Pythagorean traditions should therefore not suggest that these traditions are treated symmetrically in the *Shifā'*. Neither does it imply that the distinction goes any further than this formal framework, the intention remaining in both cases the study of integers. In his article, Roshdi Rashed reminds us that for 10th century scientists, the gap between the two traditions was reduced to "a distinction of methods and norms of rationality."[9] Thus, citing Ibn al-Haytham:

> Properties of numbers are shown in two ways: the first is by induction, since if we take the numbers one by one and if we distinguish between them, we find by distinguishing and

---

[7] Franz Woepcke, "*Mémoire sur la propagation des chiffres indiens*", *Journal asiatique*, 6e série, 1863, I-27–79, 234–290, 442–520, pp. 501–502.

[8] As Roshdi Rashed had previously noticed, the cover title given by the editor of this text (*al-Ḥisāb*) is particularly misleading: Ibn Sīnā, *al-Shifā'—al-Fann al-thānī fī al-riyāḍiyāt—al-Ḥisāb*, éd. *'Abd al-Ḥamīd Luṭfī Maẓhar, al-Hay'a al-miṣriyya al-'āmma li-l-kitāb* (Cairo, 1975).

[9] Roshdi Rashed, *The Development of Arabic Mathematics: Between Arithmetic and Algebra*, Kluwer Academic Publishers (Dordrecht, Boston, 1994), p. 246.

> considering all their properties, and to find the number in this way is called *al-arithmāṭiqī*. This is shown in a book on *al-arithmāṭiqī* [Nicomachus of Gerasa]. The other way of showing properties of numbers proceeds by proofs and deductions. All the properties of the number grasped by proofs are contained in these three books [of Euclid] or those which refer to them.[10]

Avicenna's introduction to his *al-Arithmāṭiqī* treatise leaves us in no doubt on how he conceived the relationship between this work and the Euclidean treatise. He starts by pointing out that the *Elements* (*al-Usṭuqsāt*) provides the science of numbers (*'ilm al-'adad*) with most of its foundations (*uṣūl*), adding that "it is possible to transpose to the number many geometric propositions dealing with multiplication, division and ratios."[11] Then, he mentions the quiddity (*māhiyya*) of a number, as if to indicate what constitutes the subject of his book, contenting himself to refer to the treatise on *Categories*. Finally, he reminds the reader that a fair share of important topics have been covered in the *Elements,* such as odd, even, prime and composite numbers, even-times even, even-times odd and even-times-even-times odd, but also perfect, deficient and abundant numbers. Therefore, he deems it not necessary to go over them again.[12]

Thus, it is as if the philosopher's intention was not so much to create a summary of Nicomachus's *Arithmetic*, as to complete the Euclidian's work in order to draw the boundaries for a science of numbers placed directly in its wake. If the topics addressed lead to a certain closeness to the neo-Pythagorean treatise (and, how could it be otherwise?), the project seems however much broader in its principle, and consequently capable of incorporating results from different sources. This is at least what we intend to show by taking up his work from its table of contents. The four books that make up the work are therein entitled:

1. Properties of a number;
2. States of a number in terms of its relationship to others;
3. States of a number in terms of its composition of units;
4. The ten proportions.

Although Nicomachus's treatise is composed of only two books, we have found the same general progression therein. Nonetheless, the similarities affect mostly the last three points (more precisely: relationships between numbers, figurate numbers and proportions).[13] As Avicenna's first book, the most voluminous, is structurally very different from what we find in its Hellenistic predecessor, we will concentrate on this book in particular. It is easy to distinguish three parts:[14]

[10]Ibid., p. 31.

[11]Ibn Sīnā, *al-Ḥisāb*, p. 17.

[12]Ibid. However, note that while perfect numbers are examined in the *Elements*, the notions (admittedly related) of abundant and deficient numbers are not included.

[13]The theme of relationships between numbers occupies the end of Book I and the beginning of Book II of Nicomachus's treatise, those of figurate numbers and proportions, the remainder of Book II (see for example Nicomaque de Gérase, *Introduction arithmétique*, ed. Janine Bertier, Vrin (Paris, 1978), or Wilhelm Kutsch, op. cit.).

[14]This division is ours and is not formally present in Avicenna's text.

1. a study of the natural order and the succession of integers;
2. a first partition of integers: odd and even;
3. a second partition of integers: perfect, deficient and abundant numbers.

3.3 The Succession of Integers

Avicenna's first proposition, introduced by him as "the first and most renowned" property of numbers, also coincides with Nicomachus's first strictly mathematical proposition:[15]

> Any number is half the <sum> of the two numbers adjacent to it, i.e. two numbers bordering its sides <at an equal distance> whether higher or lower. Example: five is half of six plus four, half of seven plus three, half of eight plus two and half of one plus nine.[16]

From there follows a number of outcomes, such as "the square of any number is equal to the product of its two close adjacents plus one", i.e.:

$$n^2 = (n + 1)(n - 1) + 1,$$

And generalizing:

$$n^2 = (n + a)(n - a) + a^2.$$

Next comes the notion of *distance* between two numbers, which does not exactly cover the difference, but introduces a perception so to speak, "spatialising" the succession of numbers which is later found in a different form. Thus:

> The distance of any number to its double is equal to its product by one if you don't include it, and that plus one if it is included within the ranks.[17]

Expressed in terms of distances and hence often to within one unit, a number of results are then developed, such as:

$$n^2 - n = n \cdot (n - 1)$$
$$n \cdot (n + 1) - n = n^2$$
$$n^3 - n = n \cdot (n + 1) \cdot (n - 1)$$
$$n^4 - n = \left(n^2 + (n + 1)\right) \cdot n \cdot (n - 1)$$

---

[15]This proposition corresponds to the first in Chapter VIII of Nicomachus's treatise (see ed. Bertier, p. 61), Avicenna does not incorporate the general considerations and definitions of chapters I to VII.

[16]Ibn Sīnā, *al-Ḥisāb*, p. 18.

[17]This involves counting the number of ranks that separate the two numbers. Thus, according to Avicenna's own example, the distance between 4 and its double 8 is either 4 or 5, depending on whether one counts rank 4, the ranks to keep being, in the first case 5, 6, 7 and 8, and 4, 5, 6, 7 and 8 in the second.

Avicenna then returns to the study of numbers adjacent to a given number n—type $(n-a)$ and $(n+a)$—while providing other equalities, such as:

$$2n^2+2=(n+1)^2+(n-1)^2,$$
$$2n^2+8=(n+2)^2+(n-2)^2,$$
$$2n^2+18=(n+3)^2+(n-3)^2,$$

Or:

$$2n^2+4=(n-1)(n-2)+(n+1)(n+2),$$
$$2n^2+12=(n-2)(n-3)+(n+2)(n+3).$$

In both these last results, and for those expressed in terms of distance, a series of identities can be found indicating a proximity to algebra which confirm Avicenna's use of this discipline's lexicon: for example, his use of the expression *māl māl* to indicate the fourth power (square square).[18]

This presentation on the succession of integers continues quite naturally with the question of arithmetical sequences. Avicenna begins by announcing his intention to deal with "properties of successive numbers in natural (*ṭabīʻiyya*) succession,"[19] in other words arithmetical sequences with common difference one, in order to then generalize to sequences with any first term and any common difference. He also gives the rules to determine the last term as well as those that provide the sum of the terms. Thus, for a sequence (u_n) with common difference d, we have:

$$u_n = u_1(n-1)\cdot d$$
$$\sum_{i=1}^{n} u_i = \frac{1}{2}(u_n+u_1)\cdot n$$

The results relating to the sum of various progressions were well known at the time and collections are often found in *ḥisāb* treatises.[20] What is particularly remarkable here, and much less common, is the comprehensive study of arithmetical sequences for themselves.

Also note the importance of pairs of adjacent numbers throughout this study. By considering the first n integers, Avicenna indicates that, if n is odd, numbers adjacent to the median term can be grouped two by two; the sum of any couple is, of course always the same, which naturally leads to the result giving the total sum (a quite similar reasoning gives the same result if n is even).

[18]Ibn Sīnā, *al-Ḥisāb*, p. 19 (see also Roshdi Rashed, *op. cit*., p. 32).

[19]Ibid., p. 21.

[20]In his *ḥisāb* treatise, a contemporary of Avicenna, al-Baghdādī, thus reveals the sums of the first n integers, the first n evens, the first n odds, the first n squares, the first n cubes, etc. See ʻAbd al-Qāhir ibn Ṭāhir al-Baghdādī, *al-Takmila fī al-ḥisāb*, ed. Aḥmad Salīm Saʻīdān, Maʻhad al-makhṭūṭāt al-ʻarabiyya (Kuwait, 1985), p. 179 et sq.

In this use we can see one of the reasons that led the author to introduce his first proposition as "the most renowned" property of numbers. But other purposes are also likely to be the cause of such a qualification. This particularly applies to the construction of magic squares, where pairs of numbers adjacent to the magical constant play a decisive role. An anonymous 11th century author placed this first proposition by Nicomachus and Avicenna among the preliminaries of his treatise on the subject. Furthermore the paragraph he devotes to it shows many similarities to Avicenna's text: we find the same terminology for adjacent numbers (*ḥāshiyatān*) that is absent from the translation of the Hellenistic text by Thābit, and the same notion of *distance* between two numbers (*bu'd*).[21]

Yet, the art of magic squares is not unrelated to the tradition of Nicomachus's *Arithmetic*. In addition to the fact that several passages in Iamblichus tend to suggest the presence of this theme among the neo-Pythagorean school,[22] we can't fail to notice that a predecessor of Avicenna, 'Alī ibn Aḥmad al-Anṭākī, had inserted a presentation on the construction of magic squares within his commentary on Nicomachus's treatise.[23] Moreover, prestigious mathematicians such as Thābit ibn Qurra and Abū al-Wafā' al-Buzjānī, also predecessors of Avicenna, had composed treatises dedicated to the very subject.

It is indeed the art of magic squares that evokes the way in which the philosopher began to exhibit a certain number of results relating to the succession of odd numbers and even-times odd numbers. Let us give an example:

9	7	5	3	1
19	17	15	13	11
29	27	25	23	21
39	37	35	33	31
49	47	45	43	41

> If we construct a square table from odd numbers in natural succession, properties relating to this layout will appear; it is the same if we construct a triangular table. Let's start with the square and let it be <the square of> five. We say that any cross on <this table> is either a

[21] See Jacques Sesiano, *Un traité médiéval sur les carrés magiques*, Presses polytechniques et universitaires romandes (Lausanne, 1996), pp. 206–207. To translate the proposition on adjacent numbers, Thābit wrote: *kullu 'adad musāwin li-niṣf al-'adadayni al-laḏayni 'an janbatayhi* (ed. Kutsch, *op. cit.*, p. 20).

[22] See Jamblique, *In Nicomachi Arithmeticam*, ed. Nicolas Vinel, Fabrizio Serra Editore (Pisa, Roma, 2014), pp. 26–31.

[23] Jacques Sesiano, *Les carrés magiques dans les pays islamiques*, Presses polytechniques et universitaires romandes (Lausanne, 2004), p. 11.

diagonal of the <entire> figure or not. The sums of the two diagonals are equal: in relation to the diagonal, the sum of each of the two diagonals of this figure is one hundred and twenty-five; as to what is not on the diagonal <of the entire figure>, it is like the cross coming from two lines, one of which is three, fifteen, and twenty-seven, and the second seven, fifteen and twenty-three, <the sum of> each diagonal is forty-five. We find that the sum of the extremities of a line of any cross is equal to the sum of the extremities of the other line. We find that the sum of boxes of any square <constructed> from these numbers according to their succession is equal to the square of the square of the number of boxes on a side. Thus, if one constructs a square whose side is two, then its numbers are one, three, five and seven. The sum of it all is sixteen, and this corresponds to the square of the square of two.[24]

3	1
7	5

So, we find here a different, "spatialised" approach to the study of the succession of numbers, which, although it has little relation to magic squares—the theme is never mentioned by Avicenna—does however, take up the general layout to visually express results that could otherwise be given quite differently.[25]

3.4 Odd and Even

The developments seen above (on odds and even-times odds) are in reality part of the second section of this chapter dedicated to what Avicenna presents as "the first of the two divisions of numbers", namely the distinction between odd and even.[26] The author reverts to notions explained in the *Elements* (quoted several times therein), and in particular to the numerous subdivisions of odd and even. He adds results relative to the succession of numbers in each of these subdivisions, once again, insisting beforehand on the importance of the role of pairs of adjacent numbers in determining the sums of arithmetical sequences. Several results are then given, like those quoted above, and the fact that the sum of a sequence of consecutive odd numbers is a square.

But properties of another nature are also explored, and other topics evoked. For example, while even-times even numbers are subject to fairly classic comments on their formation by geometric progression, and also play an important role in obtaining perfect and amicable numbers.[27] While the reference to perfect numbers

[24] Ibn Sīnā, *al-Ḥisāb*, p. 25; see p. 30 for a similar development with even-times odd numbers.

[25] The result at the end of our quote, for example, ensures that $\sum_{p=1}^{n^2} (2p-1) = n^4$.

[26] Ibid., p. 23.

[27] Ibid., p. 28.

is very succinct, since the question of their formation has been dealt with in the *Elements*, amicable numbers that are neither found nor mentioned in either Euclid or Nicomachus,[28] generate a much larger development. Besides the definition and the detailed example of the couple (224, 280), Avicenna states, if we except a superfluous condition, Thābit ibn Qurra's theorem, which provides a general rule for their construction.[29]

However, perfect and amicable numbers are defined by properties conferred on them by an elementary arithmetical function, the sum of an integer's aliquot parts, i.e. the sum of its proper divisors. By calling this σ_0, let's remind ourselves that:

- If $\sigma_0(n) < n$, n is deficient;
- If $\sigma_0(n) = n$, n is perfect;
- If $\sigma_0(n) > n$, n is abundant;
- If $\sigma_0(p) = q$ and $\sigma_0(q) = p$, p and q are amicable.

In his treatise on amicable numbers, and in order to demonstrate his theorem, Thābit ibn Qurra had already given some properties of this function. A sign of the interest in this field of study is that Avicenna also gives results relating to this function, although he goes no further than his predecessor (for a more systematic study, we have to wait for the work of Kamāl al-Dīn al-Fārīsī).[30] Thus, ranked among the properties of even numbers, we find the following statements:

$$2n = \sigma_0\big((2n-1)^2\big)\,\text{if}\,(2n-1)\,\text{prime},$$

and

$$2n = \sigma_0(2(2n-3))\,\text{if}\,(2n-3)\,\text{prime}.$$

3.5 Perfect, Deficient and Abundant Numbers

This leads us to the last section of this first chapter, devoted to what Avicenna considers as another division of numbers after the partition between even and odd, the one that distinguishes perfect, deficient and abundant numbers.[31] Notice from this introduction that by straying away from the letter of Nicomachus's treatise, which reserves these denominations exclusively for even numbers only,[32] Avicenna confers

[28]However, let us recall, that according to Iamblichus' testimony (*op. cit.*, p. 103), amicable numbers are related in one way or another to neo-Pythagorean tradition.

[29]Ibn Sīnā, *al-Ḥisāb*, p. 28. Also see Roshdi Rashed who corrects some mistakes from the Cairo edition (*op. cit.*, pp. 37–38).

[30]See Roshdi Rashed, "Amicable Numbers, Aliquot Parts and Figurate Numbers in the Thirteenth and Fourteenth Centuries", *The Development of Arabic Mathematics* (*op. cit.*), pp. 275–320.

[31]Ibn Sīnā, *al-Ḥisāb*, p. 32.

[32]Nicomaque, ed. Bertier, p. 74.

this new division with greater importance. In fact, by bringing together the characteristics and properties of these numbers, which are quite numerous and often absent from neo-Pythagorean treatises, Avicenna is demonstrating the vitality of this field of research at this time.

These results are all given without proofs; some of them were conjectures, of which only a few have been found to be false. Seeming to consider that all perfect numbers are Euclidean, he starts by affirming that all perfect numbers are always even.[33] He pursues with the assertion stated by Nicomachus which, although faulty was regularly referred to until the 16th century, whereby a perfect number can be found in each interval between 1, 10, 100, 1000, 10,000, and so on with the following powers of 10; al-Baghdādī, one of his contemporaries, pointed out the erroneous character of this assertion.[34] He then notes that the units digit of a perfect number is always six or eight—properties also announced by al-Baghdādī,[35] which is always true for Euclidean perfect numbers.

Next, come a set of properties on these Euclidean perfect numbers, like the fact that one obtains a square if one multiplies them by eight and adds one,[36] then a process to obtain an abundant number from a perfect one which reveals another property of the function σ_0:

p a perfect number and q a prime number so that p and q are prime to one another; then pq is abundant and:

$$\sigma_0(pq) = pq + 2p$$

This result provides the first step in the demonstration of a more general theorem, stipulating that all multiples of a perfect number are abundant.

Avicenna continues with a series of criteria for whether a number is deficient or abundant, these include for example: prime numbers are all deficient; all even-times even are deficient by a unit; all multiples of six are abundant.[37] He ends by affirming in substance that all odd numbers are deficient unless they are composed of four consecutive odd numbers, giving the case of 945 which, he assures, is the first abundant odd. This result, also given by al-Baghdādī, was for a long time attributed to Bachet de Méziriac.[38]

---

[33]Let us recall that proposition IX-36 of the *Elements* offers a sufficient condition for the formation of perfect numbers (if $(2^{n+1} - 1)$ is prime, then $2^n \cdot (2^{n+1} - 1)$ is perfect), and the question of whether or not there are odd perfect numbers is still not decided.

[34]See al-Baghdādī, *op. cit.*, p. 227, as well as Roshdi Rashed, "Amicable Numbers, Aliquot Parts and Figurate Numbers".

[35]Ibid. The property is also stated by Iamblichus, who adds an incorrect detail: the succession of six and eight is always alternate. The way al-Baghdādī in turn introduced his assertion suggests he had a demonstration.

[36]Indeed, $8 \cdot 2^n \cdot (2^{n+1} - 1) + 1 = (2^{n+2} - 1)^2$.

[37]It is therefore a special case of the result we expressed above, for six is a perfect number.

[38]The second quadruple of odd numbers gives 3465, which is also an abundant odd number. However, the general condition that Avicenna appears to impart (if one believes the Cairo edition) is

So ends the first chapter of Avicenna's treatise. One may observe how remote, both in mind and content, it is from Nicomachus's work.

3.6 Congruencies

The remaining three chapters, while shorter, are unquestionably far more related to the neo-Pythagorean treatise. However, it would be wrong to assume they do not contain results from other sources. Particularly the chapter on figurate numbers, which deals with a subject that experiences further developments over the following centuries.[39] However in this chapter, among the properties of squares and cubes, there are results on congruencies, which are particularly noteworthy. Quoting Avicenna:

> Know that <the number of> units of the square number will always be either one, four, five, six or nine. If it is one, the units on its side are nine or one; if it's four, then eight or two; if five, then five; if six, either six or four; and if it is nine, then three or seven.[40]

In other words:

$$x^2 \equiv 1 \pmod{10} \Leftrightarrow \begin{cases} x \equiv 1 \pmod{10} \\ x \equiv 9 \pmod{10} \end{cases}$$

$$x^2 \equiv 4 \pmod{10} \Leftrightarrow \begin{cases} x \equiv 8 \pmod{10} \\ x \equiv 2 \pmod{10} \end{cases}$$

$$x^2 \equiv 5 \pmod{10} \Leftrightarrow x \equiv 5 \pmod{10}$$

$$x^2 \equiv 6 \pmod{10} \Leftrightarrow \begin{cases} x \equiv 6 \pmod{10} \\ x \equiv 4 \pmod{10} \end{cases}$$

$$x^2 \equiv 9 \pmod{10} \Leftrightarrow \begin{cases} x \equiv 3 \pmod{10} \\ x \equiv 7 \pmod{10} \end{cases}$$

Then immediately afterwards, in the same style:

> Proof of the square in the Indian method: it must imperatively be one, four, seven or nine. For one, it's either one or eight; for four, either two or seven; for seven, four or five; and for nine, it's three, six or nine.[41]

---

neither necessary nor sufficient: the third quadruple gives 9009 as a deficient number, and there are also three other odd abundant numbers between 945 and 3465.

[39] See Roshdi Rashed, "Amicable Numbers, Aliquot Parts and Figurate Numbers".

[40] Ibn Sīnā, *al-Ḥisāb*, pp. 55–56.

[41] Ibid., p. 56. Similar results are given a little further for cubes (p. 60). It's this passage which sparked Woepcke's comment, quoted above; see also Roshdi Rashed, "Mathématiques et philosophie chez Avicenne", p. 38.

The "Indian method", in this case the famous divisibility rule for 9, is often exhibited in treatises on Indian reckoning and proves itself very useful for checking procedures where intermediate results are deleted. Thus, if we use the language of modern mathematics, Avicenna claims in substance that:

$$x^2 \equiv 1 \,(\mathrm{mod}\, 9) \Leftrightarrow \begin{cases} x \equiv 1 \,(\mathrm{mod}\, 9) \\ x \equiv 8 \,(\mathrm{mod}\, 9) \end{cases}$$

$$x^2 \equiv 4 \,(\mathrm{mod}\, 9) \Leftrightarrow \begin{cases} x \equiv 2 \,(\mathrm{mod}\, 9) \\ x \equiv 7 \,(\mathrm{mod}\, 9) \end{cases}$$

$$x^2 \equiv 7 \,(\mathrm{mod}\, 9) \Leftrightarrow \begin{cases} x \equiv 4 \,(\mathrm{mod}\, 9) \\ x \equiv 5 \,(\mathrm{mod}\, 9) \end{cases}$$

$$x^2 \equiv 0 \,(\mathrm{mod}\, 9) \Leftrightarrow \begin{cases} x \equiv 3 \,(\mathrm{mod}\, 9) \\ x \equiv 6 \,(\mathrm{mod}\, 9) \\ x \equiv 0 \,(\mathrm{mod}\, 9) \end{cases}$$

The consecutive nature of these results (modulo 10 then modulo 9) suggests the very idea of congruence, common to both sets, brought out here by Avicenna, in a way that perhaps had never been so clear.

3.7 Conclusion

In his article on Avicenna, Roshdi Rashed identifies the gap between the philosopher's views and the neo-Pythagorean tradition, in terms of both the methods used, and the standards of rationality. He noted in particular the exclusion of "all ontological and cosmological considerations bearing on the notion of a number" from his *al-Arithmāṭiqī*, leaving only "the philosophical aim common to all branches of philosophy—theoretical or practical—that is the perfection of the soul".[42] Avicenna wrote explicitly on the subject:

> It is usual, among those dealing with the art of numbers, to mention in this and similar places, developments foreign to this art and, more than that, foreign to the use of those who proceed by demonstration, and closer to the discourses of the orators and the poets. We must renounce this.[43]

Obviously, here we are far removed from Hellenistic arithmetic, and Avicenna is drawing an entirely new landscape. The intention of the text, which could be considered as being to complement the *Elements* by synthetically drawing new contours for the science of numbers—at least as he conceives it—seems to us in fact to be two-fold.

---

[42] *Op. cit.*, p. 32.

[43] Ibid. See also Ibn Sīnā, *al-Ḥisāb*, p. 60.

Firstly, to collect and classify certain results on numbers and that can have many different origins: arithmetical sequences, odd and even numbers, aliquots parts, amicable, perfect, abundant, deficient and figurate numbers, congruencies, proportions, etc. The origin could of course be Nicomachus's treatise, which to a certain degree provides the general framework, however, we have alternately come across many other disciplines being solicited, such as algebra, Indian reckoning, *ḥisāb*, and even the art of magic squares. Avicenna may well recall the recent results of his contemporaries and immediate predecessors, but he also focuses, as few had before him, on the succession of numbers and, the natural order of integers and arithmetic sequences, as well as providing advances on congruencies, and raising the partition between perfect, deficient and abundant numbers to the same rank as the distinction between odd and even.

Secondly, to exclude from number theory all that could be foreign to the knowledge of integers alone, whether it relates to the numbers being rational or irrational, or to the clarifications of calculation procedures or problem solving. This is how Avicenna explains his project at the end of the treatise:

> This is what we said in the science of *Arithmāṭiqī*. We left some cases whose mention in this place we considered foreign to the rule of this art. What remains in the science of *al-ḥisab* is suitable for the use and the determination <of numbers>. Finally, what is left in the practice is like algebra and *al-muqābala*, Indian addition and separation, and what is similar. But it is better to mention these among the secondary parts.[44]

All these disciplines—algebra, Indian reckoning, and *ḥisāb*—thus appear endowed with an instrumental character, which seems to oppose the concept of the philosopher. Note that there is neither any mention nor examples of integer Diophantine analysis. It is possible that Avicenna had not grasped all the issues involved in this still relatively new discipline. In a passage from the *Logic* section of the *Shifā'*, where he evokes the first case of Fermat's conjecture—on whether the sum of two cubes is a cube—he seems to think that the problem falls within *ḥisāb*[45], which would render it foreign to his topic.

The number theory outlined by Avicenna is of course not specific to him: features date as far back as Thābit ibn Qurra's memoir on amicable numbers, which tackles a neo-Pythagorean subject in a deliberately Euclidean style. Nevertheless, his treatise remains remarkable. Not so much by its content that one can often find elsewhere, nor by his methods (few of which are detailed), but by his globalizing aims and sharpness of the boundaries he draws. In this, this text differs from most mathematical writings of the time, whether pamphlets devoted to more circumscribe topics—like Thābit's memoir on amicable numbers—or much broader texts like al-Baghdādī's *al-Takmila fī al-ḥisāb*, where properties of integers coexist with procedures from Indian reckoning and the arithmetic of irrational numbers.

Noting the changes taking place within the mathematics of his time, and driven by an enquiring philosophical mind which leads him to question the nature of the

[44]Ibn Sīnā, *al-Ḥisāb*, p. 69, quoted by Roshdi Rashed, "Mathématiques et philosophie chez Avicenne", p. 33.

[45]See a translation of this passage in Roshdi Rashed, ibid.

various disciplines, Avicenna thus proposes one of the very few texts to give the from then on reunited number theory a synthetic image bringing together, with the exception of integer Diophantine analysis, themes for future research.

References

Al-Baghdādī (1985). *Al-Takmila fī al-ḥisāb*. ed. Aḥmad Salīm Sa'īdān, Ma'had al-makhṭūṭāt al-'arabiyya, Kuwait.

Crozet, P. (2002). "Aritmetica", Storia della Scienza (Vol. 3). In R. Rashed (Ed.), *La civiltà islamica* (pp. 498–506). Rome: Istituto della Enciclopedia Italiana.

Ibn Sīnā. (1975). *Al-Shifā' – al-Fann al-thānī fī al-riyāḍiyāt – al-Ḥisāb*, éd. 'Abd al-Ḥamīd Luṭfī Maẓhar, al-Hay'a al-miṣriyya al-'āmma li-l-kitāb, Cairo.

Kutsch, W. (1959). *Ṯābit b. Qurra's Arabische Übersetzung der Ἀριθμητικὴ Εἰσαγωή des Nikomachos von Gerasa zum Esrten Mal Herausgegeben*. Recherches publiées sous la direction de l'Institut de lettres orientales de Beyrouth, tome IX, Imprimerie Catholique, Beyrouth.

Nicomaque de Gérase. (1978). *Introduction arithmétique*. Vrin, Paris: Janine Bertier.

Rashed, R. (1984). Mathématiques et philosophie chez Avicenne. In Jean Jolivet & R. Rashed (Eds.), *Études sur Avicenne* (pp. 29–39). Paris: Editions du CNRS.

Rashed, R. (1994). *The development of Arabic mathematics: Between arithmetic and algebra*. Boston/Dordrecht: Boston Studies in Philosophy of Science, Kluwer Academic Publishers.

Sesiano, J. (1996). *Un traité médiéval sur les carrés magiques*. Lausanne: Presses polytechniques et universitaires romandes.

Sesiano, J. (2004). *Les carrés magiques dans les pays islamiques*. Lausanne: Presses polytechniques et universitaires romandes.

Woepcke, F. (1863). Mémoire sur la propagation des chiffres indiens. *Journal asiatique*, 6e série, I-27–79, 234–290, 442–520, pp. 501–502.

Chapter 4
Zigzag and Fregean Arithmetic

Fernando Ferreira

Abstract In Frege's logicism, numbers are logical objects in the sense that they are extensions of certain concepts. Frege's logical system is inconsistent, but Richard Heck showed that its restriction to predicative (second-order) quantification is consistent. This predicative fragment is, nevertheless, too weak to develop arithmetic. In this paper, I will consider an extension of Heck's system with impredicative quantifiers. In this extended system, both predicative and impredicative quantifiers co-exist but it is only permissible to take extensions of concepts formulated in the predicative fragment of the language. This system is consistent. Moreover, it *proves* the principle of reducibility applied to concepts true of only finitely many objects. With the aid of this form of reducibility, it is possible to develop arithmetic in a thoroughly Fregean way.

4.1 Introduction

One of the *dicta* of Frege was never to lose sight of the distinction between concept and object. When Frege wrote this maxim in p. x of Frege (2013), he was embarking on the project of reducing arithmetic to logic. The pursuit of this project eventually led him to introduce so-called value-ranges as regulated by his famous Basic Law V. In a sense, the introduction of value ranges entails that the world of concepts has a counterpart in the world of objects (by way of their extensions, which are a special kind of value-ranges). Even though the distinction between object and concept is by no mean obliterated by the introduction of extensions, the end result is that every concept has an object as a proxy and this situation proved to be fatal. For our purposes, I take Frege's system of the *Grundgesetze der Arithmetik* Frege (2013)

We would like to thank the financial support of FCT by way of grant PEst-OE/MAT/UI0209/2013 to the research center CMAF-CIO.

F. Ferreira (✉)
Departmento de Matemática, Faculdade de Ciências, Universidade de Lisboa,
Campo Grande, ed. C6, piso 2, 1749-016 Lisbon, Portugal
e-mail: fjferreira@fc.ul.pt

H. Tahiri (ed.), *The Philosophers and Mathematics*, Logic, Epistemology, and the Unity of Science 43, https://doi.org/10.1007/978-3-319-93733-5_4

as second-order logic (with unrestricted comprehension) together with an extension operator—attaching a first-order term $\hat{x}.A(x)$ to each formula $A(x)$—regulated by the scheme

$$\hat{x}.A(x) = \hat{x}.B(x) \leftrightarrow \forall x(A(x) \leftrightarrow B(x)).$$

The above scheme is our version of Frege's Basic Law V. Frege was aware that this law could be disputed. In page VII of the foreword to the first volume of the *Grundgesetze*, Frege writes that

> (...) as far as I can see, a dispute can arise only concerning my Basic Law of value-ranges (V), which perhaps has not yet been explicitly formulated by logicians although one thinks in accordance with it if, e.g., one speaks of extensions of concepts. I take it to be purely logical. At any rate, the place is hereby marked where there has to be a decision.

As it is well-known, Frege met the collapse of his system precisely at this point. Characteristically of him, Frege was forthright in his comments of Basic Law V and, after knowing of Russell's paradox, wrote—in a retrospective and melancholic mood—that he would have been glad to prescind from value-ranges had he known how to get by without them:

> Hardly anything more unwelcome can befall a scientific writer than to have one of the foundations of his edifice shaken after the work is finished.This is the position into which I was put by a letter from Mr. Bertrand Russell as the printing of this volume was nearing completion. The matter concerns my Basic Law V. I have never concealed from myself that it is not as obvious as the others (...). Indeed, I pointed this very weakness in the foreword to the first volume, p. VII. I would glady have dispensed with this foundation if I had known of some substitute for it. Even now, I do not see how arithmetic can be formulated scientifically, how the numbers can be apprehended as logical objects and brought under consideration, if it is not—at least conditionally—permissible to pass from concept to its extension. May I always speak of the extension of a concept, of a class? And if not, how are the exceptions to be recognised? (Cf. afterword of volume II of *Grundgesetze der Arithmetik*.)

Indeed, how are the exceptions to be recognized? Russell's paradox refutes the simple and elegant view that there are no exceptions. The logicist view of sets as "something obtained by dividing the totality of all existing things into two categories" led to the paradoxes and today has been effectively replaced by the iterative conception of sets.[1] At present, there is no satisfactory set theory based on the logicist view. However, even it proves impossible to develop a logicist set theory (as I suspect), perhaps it is possible to develop a logicist arithmetic. By this, I mean founding arithmetic on a *strict* logicist view according to which logical objects are extensions (as in Frege), not numbers (as in neologicism) nor constructed in terms of other kinds of abstractions.

In one of his first attempts to salvage logicism from the wreck of Frege's system, Bertrand Russell toyed with so-called zigzag theories. According to him "in a zigzag theory we start from the suggestion that the propositional functions determine classes when they are fairly simple, and only fail to do so when they are complicated and recondite" (Russell 1973, pp. 145–6). If a propositional function (a concept, in Frege's

[1] The citation is from Kurt Gödel in p. 475 of Gödel (2005).

terms) does not determine an extension then, given any class (set), there must be either an element of the class that does not fall under the concept or an element that falls under the concept but is not in the class. In the pictoresque terminology of Russell, propositional functions that do not have extensions must zigzag between classes. Russell never worked out a zigzag theory to his own satisfaction and eventually gave up extensions altogether and adopted a so-called "no-classes" theory.

This paper shows how to set up a zigzag theory which is sufficient to develop full second-order arithmetic (it is also sufficient to develop a nice theory of finite sets). The origins of this theory can be traced to an idea of Michael Dummett who blamed Russell's paradox not on the extension operator but on the impredicative character of Frege's system, viz. on the acceptance of the unrestricted comprehension principle (cf. pp. 218–9 of Dummett 1991). A few years later, Richard Heck proved in Heck (1996) that Dummett had a point: Frege's system is consistent provided that the comprehension principle is suitably restricted. Let us describe in some detail the theory that Heck proved consistent. This theory only differs from (our rendering of) Frege's system by restricting the comprehension scheme to formulas without second-order quantifications. So, the restricted scheme is

$$\exists F \forall x (A(x) \leftrightarrow Fx),$$

for formulas $A(x)$ in which second-order quantifications do not occur (and in which the variable F does not occur free). I denote this predicative Fregean theory by H. Heck showed that this theory is not trivial, in the sense that it is able to interpret Robinson's theory Q. However, as was suspected, H is rather weak since it cannot even interpret primitive recursive arithmetic (see Cruz-Filipe and Ferreira (2015) for a proof of this fact).

The theory H cannot be considered a zigzag theory. Even though it restricts the existence of extensions to predicative concepts, the restriction is accomplished by the drastic move of allowing only predicative concepts in the language. In fact, *all* concepts of the predicative theory have extensions. There is no zigzagging in a landscape in which every available concept has an extension.[2] Nevertheless, the toll for adopting this frugal landscape (obtained via a restriction of the comprehension scheme) is very high because H is proof-theoretically very weak. If we follow Russell's idea that concepts that have extensions must be fairly simple and if we equate recondite and complicated concepts with the impredicative ones, then the theory H falls short of these terms because it does not make room for impredicative concepts. What one needs is a theory in which impredicative concepts can be formed but in which only predicative concepts admit extensions. The business of the next section is to define such a theory, prove its consistency, and show how to develop arithmetic in it in a thoroughly Fregean way. In doing this, we isolate a certain weak form of finite reducibility. In Sect. 4.4, we *prove* the full form of finite reducibility: Every concept which is true of only finitely many elements is co-extensive with a predicative

[2]Later on, in Sect. 4.5, we will discuss a purported distinction between sets and extensions.

concept. This form of reducibility can be used to develop a workable theory of finite sets. The paper also includes two sections of commentary.

4.2 The Zigzag Theory

Let us consider a second-order language with a sort for first-order variables, written in lower case Latin letters $x, y, z, \ldots$, and two sorts for second-order variables: the predicative sort, given by capital Latin letters $F, G, H, \ldots$ and the impredicative sort given by Gothic letters $\mathfrak{F}, \mathfrak{G}, \mathfrak{H}, \ldots$. Formulas of this second-order language are defined as usual, with both kinds of second-order variables behaving syntactically like unary predicates. I also allow both kinds of second-order quantifiers $\forall F$, $\exists F$ and $\forall \mathfrak{F}$, $\exists \mathfrak{F}$. Typical of Fregean theories, an extension operator is included. This operator is, in our case, restricted: the expression $\hat{x}.A(x)$ is a well-formed (first-order) term only if impredicative variables do not occur (at all) in the formula $A(x)$. Note that the fragment of the language in which second-order impredicative variables do not occur is exactly the language of the predicative theory H. We are now ready to state the axioms of our theory PE (an acronym for predicative extensions). The theory PE is framed in classical logic and its proper axioms are those of H complemented with the following unrestricted (impredicative) comprehension scheme:

$$\exists \mathfrak{F} \forall x (A(x) \leftrightarrow \mathfrak{F}x),$$

for any formula $A(x)$ of the language (in which the variable $\mathfrak{F}$ does not occur free). Therefore, PE includes two forms of comprehension. The above impredicative form, and the (already discussed) predicative comprehension scheme that comes from the theory H:

$$\exists F \forall x (A(x) \leftrightarrow Fx),$$

for a formulas $A(x)$ in the language of H and in which second-order predicative quantifications do not occur (note that neither second-order bound impredicative variables nor second-order impredicative parameters are allowed). By impredicative comprehension, we have $\forall F \exists \mathfrak{F} \forall x (Fx \leftrightarrow \mathfrak{F}x)$, i.e., the Gothic variables have a wider range of values than the Latin variables. Basic Law V is as before, with the said restriction that terms of the form $\hat{x}.A(x)$ only make sense for formulas $A(x)$ in which second-order impredicative variables do not occur (neither free, nor bound).

The theory PE is defined in the spirit of the systems discussed by John Burgess in section 2.3d of his book Burgess (2005). The difference lies in the fact that the predicative system H over which impredicative variables of PE "float" is based on a variable-binding term-forming operator (the extension operator) and not, as in Burgess's above mentioned systems, on an extension symbol which applies to concept variables. This has the effect that—contrary to Burgess's systems—the Humean operator "number of" can be defined along Fregean lines (see the definition of card below).[3]

It is easy to see that the theory PE is consistent. Take Heck's model of the theory H. It consists of a first-order domain M (in Heck's model, this domain is actually the natural numbers), a second-order domain $S \subseteq \mathcal{P}(M)$ where the second-order variables range, and a carefully defined function that provides the interpretation of the extension operator. If we expand this structure by saying that the impredicative variables of PE range over the full power set $\mathcal{P}(M)$ of M, then it is clear that one obtains a model of PE. This simple construction proves the consistency of the theory PE.[4,5]

The blunt addition of the impredicative sort to the theory H, with apparently no interference on the predicative fragment, looks like adding a pointless idle running of language. This is not the case, however. The impredicative sort not only allows a new and important means of expression, but also increases the proof-theoretic power of the theory immensely because PE is able to interpret full second-order arithmetic. This should be compared with the theory H, which is not even able to interpret primitive recursive arithmetic. How can this be, given that (according to our conjecture in footnote 5) PE is conservative over H? The answer lies in the fact that, within the new theory, one is able to define properly (impredicatively) the concept of natural number and that, of course, arithmetic is developed for the objects falling under this concept.

Let $\mathrm{Eq}(F, G)$ abbreviate the formula which states that the (predicative) concepts F and G are equinumerous via a predicative bijection:

$$\exists R[\forall x(Fx \rightarrow \exists^1 y(Gy \wedge R\langle x, y\rangle)) \wedge \forall y(Gy \rightarrow \exists^1 x(Fx \wedge R\langle x, y\rangle))].$$

For the sake of simplicity, I formulated second-order logic with only unary concepts. We apparently need a binary predicate R in the above definition. It is nevertheless well known that R can be taken to be unary because there is a definable ordered pair operation, namely:

---

[3]In Boccuni (2010), Francesca Boccuni proposed a system with two kinds of second-order variables (plural variables and concept variables) and, like us, a variable-binding term-forming operator for getting extensions. Comprehension for plural variables is unrestricted, as with our impredicative variables. However, Boccuni's comprehension for concept variables differs from ours and cannot be understood as a predicative restriction in the traditional sense (because Boccuni's system admits comprehension for certain formulas with bound plural variables). Also, the extension operator of Boccuni is different. In particular, one cannot form the extensions of concepts given by formulas with bound concept variables, a feature which seems to prevent—as in Burgess—the definition of the operator "number of" in a Fregean manner. As a consequence, the development of arithmetic by Boccuni in Boccuni (2010) (and by Burgess in section 2.3d of Burgess 2005) is un-Fregean, being a mere Dedekindian development (i.e., based solely on the fact that a simply infinite system is present within a theory enjoying full impredicative comprehension).

[4]A similar argument appears in Wang (1950) while proving the relative consistency of a theory of classes put on top of Quine's theory *New Foundations*.

[5]We conjecture the stronger statement that PE is conservative over H. A model-theoretic proof of this fact should follow well known lines. The missing ingredient is a completeness result for theories with the extension operator.

$$\langle x, y \rangle := \hat{z}.(z = \hat{w}.(w = x \vee w = y) \vee z = \hat{w}.(w = x)).$$

The cardinality operator is defined in the Fregean way. The number of elements falling under the concept F is the extension formed by all the extensions of concepts equinumerous with F. Formally:

$$\text{card}(F) := \hat{z}.\exists H(\text{Eq}(H, F) \wedge z = \hat{w}.Hw).$$

As Heck observed in (1996), Hume's principle can be proved in H (and, hence, in PE), i.e.,

$$\mathsf{PE} \vdash \forall F \forall G(\text{card}(F) = \text{card}(G) \leftrightarrow \text{Eq}(F, G)).$$

The development of arithmetic now proceeds in a thoroughly Fregean manner. The number zero is defined by $0 := \hat{z}.(z = \hat{x}.(x \neq x))$. It is clear that the theory PE proves $\forall F(\text{card}(F) = 0 \leftrightarrow \neg\exists x Fx)$. The binary successor relation $S(x, y)$ is given by the following formula:

$$\exists F \exists G(x = \text{card}(F) \wedge y = \text{card}(G) \wedge \exists z(Gz \wedge \forall u(Fu \leftrightarrow Gu \wedge u \neq z))).$$

With these definitions, it is easy to prove in H (and, therefore, in PE) that 0 is not the successor of any object, that an object cannot be the successor of two different objects, and that two different objects cannot be the successor of the same object. For future reference, I list these properties:

(i) $\mathsf{PE} \vdash \forall x \neg S(x, 0)$
(ii) $\mathsf{PE} \vdash \forall x \forall y \forall z(S(x, z) \wedge S(y, z) \rightarrow x = y)$
(iii) $\mathsf{PE} \vdash \forall x \forall y \forall z(S(x, y) \wedge S(x, z) \rightarrow y = z)$

The definition of the *concept* of natural number can now be made. This concept is defined impredicatively, as it should be. First, I define the notion of *hereditarity* with respect to the successor relation:

$$\text{Her}(\mathfrak{F}) := \forall x \forall y(\mathfrak{F}x \wedge S(x, y) \rightarrow \mathfrak{F}y).$$

Note that I have defined "hereditarity" for impredicative variables. An *inductive* concept is a concept which is hereditary and true of 0. A *natural number* is defined as an object which falls under every inductive concept (Frege uses the terminology "finite number"). Formally:

$$\mathbb{N}(x) := \forall \mathfrak{F}(\mathfrak{F}0 \wedge \text{Her}(\mathfrak{F}) \rightarrow \mathfrak{F}x).$$

It is clear that PE proves $\mathbb{N}(0)$ and $\forall x \forall y(\mathbb{N}(x) \wedge S(x, y) \rightarrow \mathbb{N}(y))$. It is also the case that PE proves $\mathbb{N}(y) \wedge S(x, y) \rightarrow \mathbb{N}(x)$. To see this, assume that $\neg\mathbb{N}(x)$ and $S(x, y)$. By the first assumption, there is $\mathfrak{G}$ such that $\mathfrak{G}0$, $\text{Her}(\mathfrak{G})$ and $\neg\mathfrak{G}x$. By (impredicative) comprehension take $\mathfrak{H}$ such that $\forall z(\mathfrak{H}z \leftrightarrow \mathfrak{G}z \wedge z \neq y)$. By (i) we have $\mathfrak{H}0$, and using (ii) it is easy to argue that Her($\mathfrak{H}$). Clearly, $\neg\mathfrak{H}(y)$ and, therefore, $\neg\mathbb{N}(y)$.

Since full comprehension for the Gothic variables is available, it is immediate to show that PE proves the full scheme of induction:

$$A(0) \wedge \forall x \forall y(\mathbb{N}(x) \wedge A(x) \wedge S(x, y) \to A(y)) \to \forall x(\mathbb{N}(x) \to A(x)),$$

for *any* formula $A(x)$ of the language of PE.

If we show that every natural number has a successor, then the concept $\mathbb{N}$ defines a simply infinite structure in the sense of Dedekind. We prove this fact again in a Fregean manner, using the Fregean trick of showing that the successor of a natural number is the number of natural numbers less than or equal to that number. The "less than or equal relation" is defined impredicatively:

$$x \leq y := \forall \mathfrak{F}(\mathfrak{F}x \wedge \mathrm{Her}(\mathfrak{F}) \to \mathfrak{F}y),$$

i.e., x is less than or equal to y if y falls under any hereditary concept which is true of x. The following are straightforward:

(iv) PE $\vdash \forall x(x \leq x)$
(v) PE $\vdash \forall x \forall y(S(x, y) \to x \leq y)$
(vi) PE $\vdash \forall x \forall y \forall z(x \leq y \wedge y \leq z \to x \leq z)$

Given a natural number x, the concept of being less than or equal to x was defined impredicatively. However, as will shall see in the end of this section, it can be proved that this concept is co-extensive with a predicative concept. This is a form of *finite reducibility*.

Lemma 1 PE $\vdash \forall x(x \leq 0 \leftrightarrow x = 0)$.

Proof The right-to-left direction is a particular case of (iv) above. For the left-to-right direction, suppose that $x \leq 0$ but $x \neq 0$. Consider the concept given by the formula $u \neq 0$. By (i) this concept is (trivially) hereditary and, by assumption, it is true of x. By hypothesis, it is therefore true of 0. This is a contradiction. □

Lemma 2 PE $\vdash \forall x \forall y(S(x, y) \to \forall u(u \leq y \wedge u \neq y \to u \leq x))$.

Proof Let us suppose that $S(x, y)$, $u \leq y$ and $u \neq y$. To see that $u \leq x$, consider $\mathfrak{F}$ a hereditary concept such that $\mathfrak{F}u$. We must show that $\mathfrak{F}x$. Assume not. By impredicative comprehension, take a concept $\mathfrak{G}$ such that $\forall v(\mathfrak{G}v \leftrightarrow \mathfrak{F}v \wedge v \neq y)$. Let v and w be such that $\mathfrak{G}v$ and $S(v, w)$. In particular, $\mathfrak{F}v$. By the hereditarity of $\mathfrak{F}$, we also have $\mathfrak{F}w$. Given that $v \neq x$ (because one has both $\mathfrak{F}v$ and, by assumption, $\neg\mathfrak{F}x$), we conclude by (ii) that $w \neq y$. Hence $\mathfrak{G}w$. We have argued that $\mathfrak{G}$ is hereditary. Since $u \leq y$ and $\mathfrak{G}u$, we get $\mathfrak{G}y$. This is absurd. □

From the above lemma and (iv), (v) and (vi), we readily have

(vii) PE $\vdash \forall x \forall y(S(x, y) \to \forall u(u \leq y \leftrightarrow u \leq x \vee u = y))$

We are busy trying to show that $\mathbb{N}$, together with 0 and the successor, forms a simply infinite structure. The arguments so far do not use impredicativity in any essential way. Actually, the arguments so far do not use induction at all and the above statements are true in the full domain of objects, not only in $\mathbb{N}$. The next results of this section, however, use induction and are no longer true in the full domain of objects (they hold in $\mathbb{N}$).

Lemma 3 $\mathsf{PE} \vdash \forall x(\mathbb{N}(x) \rightarrow \neg S(x, x))$.

Proof By induction on x. By (i), $\neg S(0, 0)$. Let us now suppose $\mathbb{N}(x)$, $\neg S(x, x)$ and $S(x, y)$. If we assume that $S(y, y)$ then, by (ii), we would get $x = y$. This is absurd. □

Lemma 4 $\mathsf{PE} \vdash \forall x \forall y(\mathbb{N}(x) \wedge S(x, y) \rightarrow \forall u(u \leq x \rightarrow u \neq y))$.

Proof We show that $\forall y(S(x, y) \rightarrow \forall u(u \leq x \rightarrow u \neq y))$ for all natural numbers x. The proof is by induction on x. The base case $x = 0$ is a consequence of Lemma 1 and (i). To argue the induction step, assume $S(x, y)$ and $S(y, z)$ and, by induction hypothesis, $\forall u(u \leq x \rightarrow u \neq y)$. We must show that $\forall u(u \leq y \rightarrow u \neq z)$. Take u so that $u \leq y$. In order to reach a contradiction, assume $u = z$. Then $z \leq y$. By Lemma 3, $z \neq y$. Hence, by Lemma 2, $z \leq x$. By (v) and (vi), we get $y \leq x$. Using the induction hypothesis, we get the contradiction $y \neq y$. □

From the above lemma and and (v), (vi) and (vii), we get

(viii) $\mathsf{PE} \vdash \forall x \forall y(\mathbb{N}(x) \wedge S(x, y) \rightarrow \forall u(u \leq x \leftrightarrow u \leq y \wedge u \neq y))$

Proposition 1 $\mathsf{PE} \vdash \forall x(\mathbb{N}(x) \rightarrow \exists y S(x, y))$.

Proof We prove instead the stronger and more explicit sentence

$$\forall x(\mathbb{N}(x) \rightarrow \exists F(\forall u(Fu \leftrightarrow u \leq x) \wedge S(x, \mathrm{card}(F)))).$$

(This is a roundabout way of stating that the successor of the natural number x is the number of numbers less than or equal to x. Formulating the statement this way is not without advantages because it makes clear that a weak form of finite reducibility is in operation in the argument below.) The proof is by induction on x.

The case $x = 0$ follows from Lemma 1, predicative comprehension and the definitions of 0 and S. Suppose $\mathbb{N}(x)$ and $S(x, y)$. By induction hypothesis, take a predicative concept F such that $\forall u(Fu \leftrightarrow u \leq x)$ and $S(x, \mathrm{card}(F))$. By (iii), $y = \mathrm{card}(F)$. By predicative comprehension, there is G such that $\forall u(Gu \leftrightarrow Fu \vee u = y)$. Hence, by (vii), $\forall u(Gu \leftrightarrow u \leq y)$. Also, by (viii), $\forall u(Fu \leftrightarrow Gu \wedge u \neq y)$. Given that Gy holds, we get—by the definition of the successor relation—that $S(\mathrm{card}(F), \mathrm{card}(G))$. Therefore, $S(y, \mathrm{card}(G))$, as wanted. □

As I have commented, a weak form of finite reducibility is in operation in the above argument. Let us isolate it:

Theorem 1 (Weak finite reducibility) $\mathsf{PE} \vdash \forall y(\mathbb{N}(y) \rightarrow \exists F \forall x(Fx \leftrightarrow x \leq y))$.

4.3 First Commentary

We have rehearsed, in our setting, Frege's development of arithmetic. We defined a concept $\mathbb{N}$ which, together with the successor operation, gives rise to a simply infinite structure (in the sense of Dedekind). By the very way $\mathbb{N}$ is defined and by the availability of unrestricted comprehension, we get induction for every formula of the language. In effect (by unrestricted comprehension again), we are able to interpret full second-order arithmetic in PE.

We should distinguish between two different issues when setting up arithmetic. One issue concerns the definition of a simply infinite structure. Another issue concerns forming a substructure of the simply infinite structure that satisfies strong forms of induction. They are mixed together in the above development but, in general, this need not be so. In the presence of unrestricted comprehension (as it is the case with the theory PE), the second issue is easily solved in the usual Frege-Dedekind way (by considering the "smallest" simply infinite structure). However, with restricted kinds of comprehension, namely forms of predicative comprehension, the definition of natural number is severely crippled and the availability of induction is very limited [way below primitive recursive induction: see Burgess and Hazen (1998) and Cruz-Filipe and Ferreira (2015)].

Thirteen years ago, I tried to amend this kind of situation in order to obtain, within a predicative setting, the scheme of induction for all arithmetical predicates. The idea of Ferreira (2005) was simple enough. I worked within Heck's setting of ramified predicative arithmetic Heck (1996) (a provably consistent theory) and tried to adjoin an axiom of finite reducibility. Reducibility has a long story in logicism and was introduced by Russell and Whitehead for purely pragmatic reasons. According to Russell, "[the] axiom [of reducibility] has a purely pragmatic justification: it leads to the desired results and no others."[6] This is a justification that clearly departs from a logicist perspective and one that a logicist cannot rest contented with. In chapter XVII of Russell (1993), Russell writes that "I do not see any reason to believe that the axiom of reducibility is logically necessary, which is what would be meant by saying that it is true in all possible worlds." That notwithstanding, the restriction of the axiom of reducibility which states that a concept true of only finitely many objects is co-extensive with a predicative concept seems to be necessary in Russell's sense. In effect, if a concept is true of only finitely many objects $a_1, a_2, \ldots, a_n$ then the concept is co-extensive with the predicative concept given by the formula (with parameters) '$x = a_1 \vee x = a_2 \vee \cdots \vee x = a_n$.' I thought that a form of logicism was defensible by *postulating* the axiom of finite reducibility in Heck's ramified predicative theory and hoped that *that* would be enough to develop (in a Fregean way) first-order Peano arithmetic. This approach was tried in Ferreira (2005) and subjected to serious philosophical criticism in Burgess (2005) (cf. p. 113). The problem lies in *stating* the axiom of finite reducibility. How does one define finiteness within a predicative theory? One would hope, perhaps, to be satisfied at first with a "deficient"

[6]See p. xiv of the introduction to the second edition of volume I of *Principia Mathematica* Russell and Whitehead (1927).

definition of finiteness and then, after introducing the axiom of finite reducibility, show that the definition has, after all, the desired properties of the notion of finiteness and proves, in the end, to be right. Of course, this strategy is very delicate and unstable since the axiom of finite reducibility would be postulated concerning a *prima facie* inadequate definition of finiteness, which would only be proven right after the postulation. A (hopefully) virtuous circle, as this was classified in Ferreira (2005). With the benefit of hindsight, the development of Ferreira (2005) rests upon a presumption that goes beyond finite reducibility. It rest on the presumption that predicates of the form "there are finitely many w such that $A(x, w)$" are predicative for predicative formulas $A(x, w)$. This is a more stringent condition than that of finite reducibility and lacks an argument supporting it from the logicist viewpoint.[7] The present paper came about with the realization that if one defines finiteness impredicatively, then one can actually *prove* finite reducibility instead of "helping oneself to intuitions about finitude as axioms, not proved as theorems from logical axioms and a suitable definition of finitude" (cf. p. 113 of Burgess 2005).

Let us now turn to the first issue mentioned above, the one concerning the definition of a simply infinite structure. There is, in fact, a straightforward manner of obtaining a simply infinite structure within PE (even within H): the map $x \rightsquigarrow \hat{z}.(z = x)$ is clearly injective and not surjective (the object $\hat{z}.(z \neq z)$ is not in the image of the map). Of course, this is not the way that Frege developed arithmetic. It is an un-Fregean development. He would not have rested contented with any old simply infinite structure. He is no structuralist. For Frege, numbers are extensions formed by extensions of equinumerous concepts. For instance, the number 3 is the extension of all extensions with exactly three elements. This is an important feature of Frege's development of arithmetic. His development is not merely a technical exercise in modeling arithmetic. It obeys the constraint "that a philosophically satisfactory foundation for a mathematical theory must somehow intimately build in its possibilities of applications" (cf. p. 91 of Hale and Wright 2005).

An interesting fact is that it is possible to obtain a simply infinite structure in H in terms of Frege's own definition of number. The development of a simply infinite structure given in Sect. 4.2 up to statement (vii) can essentially be formalized in H. For the remaining bit, it is enough to prove that every object of the form $\mathrm{card}(F)$, for some F, has a successor. Heck showed in Heck (1996) how this can be done (in an un-Fregean way) by a simple ordered pair trick.[8] An even more interesting fact, actually a rather striking one, is that this remaining bit can also be obtained in a (roughly) Fregean way within a predicative setting. This can be accomplished within Heck's ramified predicative theory described in Heck (1996). In the ramified setting there are several rounds of second-order variables enjoying acceptable forms of

[7] The condition was made explicit in Ferreira (2005). My present view is that the arguments in Ferreira (2005) for accepting this condition are not founded on a logicist point of view (nor on finite reducibility).

[8] The trick can be easily described. The predicate F can be put in bijection with the predicate H under which fall the ordered pairs of the form $\langle \hat{x}.(x \neq x), z \rangle$, with z such that Fz. Now, clearly, $\hat{x}.(x \neq x)$ does not fall under H. The successor of $\mathrm{card}(F)$ is $\mathrm{card}(G)$, where w falls under G if, and only if, $Hw \vee w = \hat{x}.(x \neq x)$.

comprehension for the predicativist. The zeroth round (corresponding to the second-order variables of H), the first round, the second round, etc. If in the development of arithmetic in Sect. 4.2 all the definitions are given in terms of zeroth round variables *except for* the definition of the concept of natural number (which should be defined with first round quantifications), then Lemmas 3 and 4, (viii), the proposition and the theorem can be proved when the predicative variables are rendered by zeroth round variables.[9] This was first shown by Heck in Heck (2011) (see, specially, Sect. 4.5) and the reader should consult his paper for details.

4.4 Finiteness and Reducibility

Some technical choices which were made in the development of arithmetic in Sect. 4.2 may be questioned on philosophical grounds. Take, for instance, the definition of equinumerousity. By definition, $\mathrm{Eq}(F, G)$ holds whenever there is a *predicative* bijection between the objects falling under F and the objects falling under G. Is this definition faithful to the meaning of equinumerousity? Why shouldn't an impredicative bijection count as a witness of equinumerousity? The problem is compounded because our definition of equinumerousity was not really a matter of choice. The reader can easily check that the statement of equinumerousity with an impredicative bijection blocks the definition of Frege's cardinality operator card. That notwithstanding, we will see below that the issues just raised do not really arise as long as we are dealing only with concepts that are true of only finitely many elements.

In §158 of *Grundgesetze der Arithmetik*, Frege uses the locution "the cardinal number of a concept is finite" as a way of expressing that there are only finitely many elements falling under the given concept. This manner of speaking about finitude is not directly available in PE for impredicative concepts because the cardinality operator is only defined for predicative concepts. Fortunately, Frege also gives in the *Grundgesetze* a characterization of finitude in pure second-order (impredicative) logic—one that does not need value-ranges (extensions). This characterization provides a Fregean handle for approaching the questions of the above paragraph. Before directing our attention to this issue, we need to establish some easy facts about the finitude of predicative concepts.

Proposition 2 *The theory* PE *proves*

$$\forall F, G \forall z (\forall x (Gx \leftrightarrow Fx \vee x = z) \to (\mathbb{N}(\mathrm{card}(F)) \leftrightarrow \mathbb{N}(\mathrm{card}(G)))).$$

Proof Suppose that $\forall x (Gx \leftrightarrow Fx \vee x = z)$. If Fz, then $\mathrm{card}(G) = \mathrm{card}(F)$ and we are done. Otherwise, by the definition of the successor relation, $S(\mathrm{card}(F), \mathrm{card}(G))$.

[9] However, the form of finite reducibility stated in the theorem of Sect. 4.2 undergoes a subtle change of meaning. It becames much weaker because the "less than or equal relation" is defined with zeroth round quantifiers—one round less than the variables used in the definition of the concept of natural number.

The equivalence $\mathbb{N}(\text{card}(F)) \leftrightarrow \mathbb{N}(\text{card}(G))$ is now clear by the properties discussed immediately after the definition of natural number in Sect. 4.2. □

The above proposition entails that the union and the cartesian product of two finite (predicative) concepts are still finite concepts. The first statement is formally

$$\forall F, G, H\, (\mathbb{N}(\text{card}(F)) \wedge \mathbb{N}(\text{card}(G)) \wedge \forall x(Hx \leftrightarrow Fx \vee Gx) \rightarrow \mathbb{N}(\text{card}(H)))$$

and it is easily proved by induction on card(G) using the above proposition. The second statement is the universal closure of

$$\mathbb{N}(\text{card}(F)) \wedge \mathbb{N}(\text{card}(G)) \wedge \forall x(Hx \leftrightarrow \exists u, v(x = \langle u, v\rangle \wedge Fu \wedge Gv)) \rightarrow \mathbb{N}(\text{card}(H))$$

and it is proved by induction on card(G) using the first statement.

We are now ready to discuss finitude in the context of impredicative concepts. In the sequel, we describe (an adaptation of) Frege's second-order characterization of finitude and show that "finite" concepts—according to this characterization—are co-extensive with predicative concepts of finite cardinality (finite reducibility). We also argue that impredicative bijections between two "finite" concepts are co-extensive with predicative ones. In fact, a robust theory of finite sets can be developed in PE.

Frege's characterization of finitude in terms of pure second-order logic is discussed between §158 and §179 of the volume I of *Grundgesetze der Arithmetik*. We will not follow Frege's treatment in a strict manner (see Heck's paper Heck (1998) for a discussion of why Frege sometimes did not choose the simplest way), but opt for a streamlined analysis adapted to our present purposes. The two theorems of this section can be seen as versions of Theorems 327 and 348 of Frege's *Grundgesetze* adapted to the setting of PE.

In Sect. 4.2, we defined the notion of hereditarity with respect to the successor relation. We need that notion with respect to an arbitrary relation $\mathfrak{R}$:

$$\text{Her}_{\mathfrak{R}}(\mathfrak{F}) := \forall x \forall y(\mathfrak{F}x \wedge \mathfrak{R}\langle x, y\rangle \rightarrow \mathfrak{F}y),$$

as well as the associated *ancestral* relation $\mathfrak{R}^*$ of $\mathfrak{R}$:

$$\mathfrak{R}^*\langle x, y\rangle := \forall \mathfrak{F}\, (\mathfrak{F}x \wedge \text{Her}_{\mathfrak{R}}(\mathfrak{F}) \rightarrow \mathfrak{F}y).$$

It is clear that $\mathfrak{R}\langle x, y\rangle \rightarrow \mathfrak{R}^*\langle x, y\rangle$ and that the relation given by $\mathfrak{R}^*$ is reflexive and transitive. We will need to rely on some facts about ancestral relations. These are known facts of pure second-order (impredicative) logic. As a matter of fact, ancestral relations made their appearance in Frege's first book (1967), where they were studied.

We lay the needed facts in the form of three lemmas and, for the sake of completeness, we provide their proofs in the appendix of this paper.

Lemma 5 PE $\vdash \mathfrak{R}^*\langle x, y\rangle \wedge \forall z(\mathfrak{R}\langle x, z\rangle \rightarrow z = x) \rightarrow y = x$.

Lemma 6 *The theory* PE *proves the universal closure of the conditional formula whose antecedent is*

$$\forall u \forall v\,(\mathfrak{R}\langle u, v\rangle \wedge \mathfrak{R}^*\langle x, u\rangle \wedge \mathfrak{R}^*\langle u, y\rangle \wedge \mathfrak{R}^*\langle x, v\rangle \wedge \mathfrak{R}^*\langle v, y\rangle \to \mathfrak{Q}\langle u, v\rangle),$$

and whose consequent is $\mathfrak{R}^*\langle x, y\rangle \to \mathfrak{Q}^*\langle x, y\rangle$.

In particular, it proves $\forall u \forall v(\mathfrak{R}\langle u, v\rangle \to \mathfrak{Q}\langle u, v\rangle) \wedge \mathfrak{R}^*\langle x, y\rangle \to \mathfrak{Q}^*\langle x, y\rangle$.

In the antecedent of the main result above, there are more conjunts than needed. We inserted them in order to emphasize that u and v are "between" x and y.

We say that $\mathfrak{R}$ is *functional* if $\forall x \forall y \forall z\,(\mathfrak{R}\langle x, y\rangle \wedge \mathfrak{R}\langle x, z\rangle \to y = z)$, and we write Func$(\mathfrak{R})$.

Lemma 7 *The following formulae are provable in* PE*:*

1. $\mathrm{Func}(\mathfrak{R}) \wedge x \neq y \wedge \mathfrak{R}^*\langle x, y\rangle \wedge \mathfrak{R}\langle x, z\rangle \to \mathfrak{R}^*\langle z, y\rangle$.
2. $\mathrm{Func}(\mathfrak{R}) \wedge \mathfrak{R}^*\langle z, x\rangle \wedge \mathfrak{R}^*\langle z, y\rangle \to \mathfrak{R}^*\langle y, x\rangle \vee \mathfrak{R}^*\langle x, y\rangle$.
3. $\mathrm{Func}(\mathfrak{R}) \wedge x \neq y \wedge \mathfrak{R}^*\langle x, y\rangle \wedge \mathfrak{R}^*\langle y, x\rangle \wedge \mathfrak{R}^*\langle x, z\rangle \to \mathfrak{R}^*\langle z, x\rangle$.

Frege's characterization of finiteness is based on the following definition:

$$\mathrm{Btw}(\mathfrak{R}, a, b, x) := \mathrm{Func}(\mathfrak{R}) \wedge \mathfrak{R}^*\langle a, x\rangle \wedge \mathfrak{R}^*\langle x, b\rangle \wedge \neg\exists z(\mathfrak{R}\langle b, z\rangle \wedge \mathfrak{R}^*\langle z, b\rangle).$$

In the words of Frege, according to the translation Frege (2013), "x belongs to the $\mathfrak{R}$-series starting with a and ending with b" (see §158 of the *Grundgesetze*). We also say that x lies between a and b in the $\mathfrak{R}$-series.

Definition We say that $\mathfrak{F}$ is *Fregean finite*, and write Fin$(\mathfrak{F})$, just in case

$$\exists \mathfrak{R} \exists a \exists b \forall x (\mathfrak{F}x \leftrightarrow \mathrm{Btw}(\mathfrak{R}, a, b, x))$$

or, else, $\mathfrak{F}$ is an empty concept.

We also use the notation Fin(F) for predicative variables F. We could have permitted this case in the above definition. Equivalently, we can see it as abbreviating $\exists \mathfrak{F}(\forall x(\mathfrak{F}x \leftrightarrow Fx) \wedge \mathrm{Fin}(\mathfrak{F}))$.

Lemma 8 PE $\vdash \mathrm{Btw}(\mathfrak{R}, a, b, x) \to \neg\exists z(\mathfrak{R}\langle x, z\rangle \wedge \mathfrak{R}^*\langle z, x\rangle)$.

Proof Suppose that Btw$(\mathfrak{R}, a, b, x)$ and assume that there is z such that $\mathfrak{R}\langle x, z\rangle$ and $\mathfrak{R}^*\langle z, x\rangle$. By definition, $x \neq b$. Since $\mathfrak{R}^*\langle x, b\rangle$, by (3) of Lemma 7, either $x = z$ or $\mathfrak{R}^*\langle b, x\rangle$. If $x = z$, by the functionality of $\mathfrak{R}$ and Lemma 5, we get $x = b$ which is impossible. If $\mathfrak{R}^*\langle b, x\rangle$, let y be such that $\mathfrak{R}\langle b, y\rangle$: this y exists by Lemma 5. By (1) of Lemma 7 and the fact that $x \neq b$, we get $\mathfrak{R}^*\langle y, b\rangle$. This contradicts the last clause of the definition of Btw$(\mathfrak{R}, a, b, x)$. □

Lemma 9 *The theory* PE *proves the universal closure of the conditional formula whose antecedent is* $\mathrm{Btw}(\mathfrak{R}, a, b, x) \wedge \mathrm{Btw}(\mathfrak{R}, a, b, y) \wedge \mathfrak{R}\langle x, y\rangle$ *and whose consequent is*

$$\forall w(\mathfrak{R}^*\langle a, w\rangle \wedge \mathfrak{R}^*\langle w, y\rangle \rightarrow \mathfrak{R}^*\langle w, x\rangle \vee w = y).$$

Proof Suppose that $\mathrm{Btw}(\mathfrak{R}, a, b, x)$, $\mathrm{Btw}(\mathfrak{R}, a, b, y)$, $\mathfrak{R}\langle x, y\rangle$, $\mathfrak{R}^*\langle a, w\rangle$ and $\mathfrak{R}^*\langle w, y\rangle$. By (2) of Lemma 7, either $\mathfrak{R}^*\langle w, x\rangle$ or $\mathfrak{R}^*\langle x, w\rangle$. In the first case, we are done. Assume $\mathfrak{R}^*\langle x, w\rangle$. By (1) of Lemma 7, either $x = w$ or $\mathfrak{R}^*\langle y, w\rangle$. In the former case, $\mathfrak{R}^*\langle w, x\rangle$ and we are done. The following situation remains to be studied: $\mathfrak{R}^*\langle x, w\rangle$ and $\mathfrak{R}^*\langle y, w\rangle$. If $y = w$, we are done. Otherwise $y \neq w$. In this case, by Lemma 5, there is z with $\mathfrak{R}\langle y, z\rangle$. Now, by (1) of Lemma 7, $\mathfrak{R}^*\langle z, w\rangle$ and, as a consequence, $\mathfrak{R}^*\langle z, y\rangle$. This is impossible by the previous lemma. □

Our version of Theorem 327 of the *Grundgesetze der Arithmetik* is the result below. Its formulation necessarily incorporates the principle of finite reducibility:

Theorem 2 $\mathsf{PE} \vdash \forall \mathfrak{F}\,(\mathrm{Fin}(\mathfrak{F}) \rightarrow \exists F(\forall x(Fx \leftrightarrow \mathfrak{F}x) \wedge \mathbb{N}(\mathrm{card}(F))))$.

Proof Let $\mathfrak{F}$ be such that $\mathrm{Fin}(\mathfrak{F})$. If $\mathfrak{F}$ is an empty concept, the conclusion is clear. Otherwise, take $\mathfrak{R}$, a and b such that $\forall x(\mathfrak{F}x \leftrightarrow \mathrm{Btw}(\mathfrak{R}, a, b, x))$. Using Lemma 6, we may suppose without loss of generality that

$$\forall u \forall v(\mathfrak{R}\langle u, v\rangle \rightarrow \mathfrak{R}^*\langle a, u\rangle \wedge \mathfrak{R}^*\langle u, b\rangle \wedge \mathfrak{R}^*\langle a, v\rangle \wedge \mathfrak{R}^*\langle v, b\rangle).$$

By impredicative comprehension, let $\mathfrak{G}$ be such that for all w, $\mathfrak{G}w$ if, and only if,

$$\exists G(\mathbb{N}(\mathrm{card}(G)) \wedge \forall x(Gx \leftrightarrow \mathfrak{R}^*\langle a, x\rangle \wedge \mathfrak{R}^*\langle x, w\rangle))).$$

We first claim that $\mathrm{Her}_{\mathfrak{R}}(\mathfrak{G})$. To see this, suppose that $\mathfrak{G}u$ and $\mathfrak{R}\langle u, v\rangle$. Hence, there is G such that $\mathbb{N}(\mathrm{card}(G))$ and $\forall x(Gx \leftrightarrow \mathfrak{R}^*\langle a, x\rangle \wedge \mathfrak{R}^*\langle x, u\rangle)$. By the previous lemma, $\forall x(\mathfrak{R}^*\langle a, x\rangle \wedge \mathfrak{R}^*\langle x, v\rangle \rightarrow \mathfrak{R}^*\langle x, u\rangle \vee x = v)$. Therefore,

$$\forall x(\mathfrak{R}^*\langle a, x\rangle \wedge \mathfrak{R}^*\langle x, v\rangle \leftrightarrow Gx \vee x = v).$$

By predicative comprehension (and Proposition 2), it follows that $\mathfrak{G}v$.

We also claim that $\mathfrak{G}a$ holds. This easily follows from

$$\forall x(\mathfrak{R}^*\langle a, x\rangle \wedge \mathfrak{R}^*\langle x, a\rangle \rightarrow x = a).$$

To see this, assume that $\mathfrak{R}^*\langle a, x\rangle$, $\mathfrak{R}^*\langle x, a\rangle$ but $x \neq a$. By Lemma 5, take z such that $\mathfrak{R}\langle a, z\rangle$. By (1) of Lemma 7, we get $\mathfrak{R}^*\langle z, x\rangle$ and, so, $\mathfrak{R}^*\langle z, a\rangle$. This contradicts Lemma 8.

Since $\mathfrak{R}^*\langle a, b\rangle$, we conclude $\mathfrak{G}b$. This is what we want. □

Theorem 348 of Frege's *Grundgesetze* says that "if the cardinal number of a concept is finite, then the objects that fall under it can be ordered into a simple series running from a specific object to a specific object" (cf. §172 of Frege 2013). In our setting, this is stated as $\forall F(\mathbb{N}(\mathrm{card}(F)) \rightarrow \mathrm{Fin}(F))$. For convenience, we prove a stronger property.

Theorem 3 $\mathsf{PE} \vdash \forall F \forall \mathfrak{H}(\mathbb{N}(\mathrm{card}(F)) \wedge \forall x(\mathfrak{H}x \rightarrow Fx) \rightarrow \mathrm{Fin}(\mathfrak{H}))$.

Proof By induction on $\mathrm{card}(F)$. If $\mathrm{card}(F) = 0$, $\mathfrak{H}$ is empty and there is nothing to prove. Suppose that $\mathbb{N}(n)$, $S(n, m)$, $\mathrm{card}(F) = m$ and $\forall x(\mathfrak{H}x \rightarrow Fx)$. Take c such that $\mathfrak{H}c$ (otherwise, $\mathfrak{H}$ is empty and there is nothing to prove). If G is such that $\forall x(Gx \leftrightarrow Fx \wedge x \neq c)$, clearly $\mathrm{card}(G) = n$,. By induction hypothesis, $\mathrm{Fin}(\mathfrak{L})$, where $\mathfrak{L}$ is such that $\forall x(\mathfrak{L}x \leftrightarrow \mathfrak{H}x \wedge x \neq c)$. If $\mathfrak{L}$ is empty then $\mathfrak{H}$ is true of only the element c and the Fregean finiteness of $\mathfrak{H}$ easily follows. Otherwise, take $\mathfrak{R}$, a and b so that $\forall x(\mathfrak{L}x \leftrightarrow \mathrm{Btw}(\mathfrak{R}, a, b, x))$. The idea is clear now: We want to tack c onto the end of the series given by $\mathfrak{R}$, a and b.

As a preliminary step, by Lemma 6, we may suppose that

$$\forall u \forall v(\mathfrak{R}\langle u, v\rangle \rightarrow \mathfrak{R}^*\langle a, u\rangle \wedge \mathfrak{R}^*\langle u, b\rangle \wedge \mathfrak{R}^*\langle a, v\rangle \wedge \mathfrak{R}^*\langle v, b\rangle).$$

By impredicative comprehension, let $\mathfrak{Q}(x)$ be defined as

$$\exists u \exists v[x = \langle u, v\rangle \wedge (\mathfrak{R}\langle u, v\rangle \vee (u = b \wedge v = c))].$$

We have $\mathrm{Func}(\mathfrak{Q})$ and, since $\neg\exists z \mathfrak{Q}\langle c, z\rangle$, *a fortiori* $\neg\exists z(\mathfrak{Q}\langle c, z\rangle \wedge \mathfrak{Q}^*\langle z, c\rangle)$. We argue that $\forall x(\mathfrak{H}x \leftrightarrow \mathrm{Btw}(\mathfrak{Q}, a, c, x))$, and this shows that $\mathfrak{H}$ is Fregean finite. So, we must argue

$$\forall x\, ((\mathrm{Btw}(\mathfrak{R}, a, b, x) \vee x = c) \leftrightarrow \mathrm{Btw}(\mathfrak{Q}, a, c, x)).$$

If $\mathrm{Btw}(\mathfrak{R}, a, b, x)$ then, by Lemma 6 (its particular case), $\mathrm{Btw}(\mathfrak{Q}, a, b, x)$. It readily follows that $\mathrm{Btw}(\mathfrak{Q}, a, c, x)$. Of course, $\mathfrak{Q}^*\langle a, b\rangle$. Therefore $\mathfrak{Q}^*\langle a, c\rangle$, and we get $\mathrm{Btw}(\mathfrak{Q}, a, c, c)$. Conversely, assume that $\mathrm{Btw}(\mathfrak{Q}, a, c, x)$. We may suppose that $x \neq c$. We have $\mathfrak{Q}^*\langle a, x\rangle$. By Lemma 6, $\mathfrak{R}^*\langle a, x\rangle$. On the other hand, by (2) of Lemma 7, $\mathfrak{Q}^*\langle x, b\rangle$ or $\mathfrak{Q}^*\langle b, x\rangle$. In the first case, by Lemma 6, we get $\mathfrak{R}^*\langle x, b\rangle$, and we are done. In the second case, we are in a situation where both $\mathfrak{Q}^*\langle b, x\rangle$ and $\mathfrak{Q}\langle b, c\rangle$ hold. By (1) of Lemma 7, either $b = x$ or $\mathfrak{Q}^*\langle c, x\rangle$. In the first case, $\mathfrak{R}^*\langle x, b\rangle$ and we are done. The second case is impossible by Lemma 5. □

Theorems 2 and 3 permit the development in PE of a very robust theory of finiteness. As an illustration, we show that an impredicate bijection between two predicative concepts of finite cardinality is co-extensive with a predicative concept. So, let us consider $\mathfrak{R}$ a bijection between concepts F and G of finite cardinality. Of course, $\mathfrak{R}$ is a sub-relation of the cartesian product between F and G. The latter has finite cardinality, as we have observed at the beginning of the present section. Hence, by Theorem 3, $\mathrm{Fin}(\mathfrak{R})$. The conclusion follows from an application of Theorem 2.

4.5 Second Commentary

Frege's system of the *Grundgesetze* is a theory of extensions (in fact, a theory of value-ranges). Extensions are, of course, extensions *of* concepts and they satisfy the form of extensionality given by Frege's Basic Law V. Sets, for Frege, are extensions (of concepts). They are not autonomous from concepts, nor is their basic relation: membership. In Frege's system of the *Grundgesetze*, membership is a defined notion given by

$$x \in y :\equiv \exists F(y = \hat{z}.Fz \wedge Fx).$$

This defined notion is fully operational in Frege's *Grundgesetze* in the sense that the *law of concretion* is derivable:

$$\forall x(x \in \hat{z}.A(z) \leftrightarrow A(x)),$$

for every formula of the language A.[10] The problem, of course, is that this law leads to Russell's paradox. This situation was analyzed in detail by Heck in (1996) who observed that, in the predicative setting H, the law of concretion does not hold. The derivation from right to left is blocked by lack of concept-comprehension.

It is very instructive to discuss Heck's theory H because it brings to light some interesting phenomena. As with Frege's (inconsistent) theory, in (the consistent) H the extension operator applies to every formula of the language. Also, extensions satisfy the form of extensionality given by Frege's Basic Law V.

Lemma 10 *The theory* H *proves the following form of the law of concretion:*

$$\forall F \forall x(x \in \hat{z}.Fz \leftrightarrow Fx).$$

Proof Let F be given. Suppose that $x \in \hat{z}.Fz$. By the definition of membership, there is G such that $\hat{z}.Fz = \hat{z}.Gz \wedge Gx$. By Basic Law V, we conclude that Fx. Conversely, given Fx it is clear (by the definition of membership) that $x \in \hat{z}.Fz$. □

The following fact is illuminating:

Proposition 3 *The theory* H *proves* $\exists y \forall x(x \in y \leftrightarrow A(x)) \leftrightarrow \exists F \forall x(Fx \leftrightarrow A(x))$, *for every formula* A *of the language (in which the variables* y *and* F *do not occur free).*

Proof Suppose that there is y such that $\forall x(x \in y \leftrightarrow A(x))$. If there is no element x such that $A(x)$, then clearly $\exists F \forall x(Fx \leftrightarrow A(x))$: just take the predicative concept associated with the formula '$x \neq x$'. Suppose that there is w such that $A(w)$. Hence

---

[10] Frege works with value-ranges, of which extensions are a particular case. The analogue of membership for value-ranges is Frege's application operator $\frown$ as defined in §34 of Frege (2013). The analogue of the law of concretion for value-ranges is discussed by Frege in the same Section and formally proved in §55 of the *Grundgesetze*. The designation 'law of concretion' comes from Quine (see p. 16 of Quine 1963).

$w \in y$ and, in particular, there is F such that $y = \hat{z}.Fz$. We claim that, for this F, one has $\forall x(Fx \leftrightarrow A(x))$. By the above Lemma, $\forall x(x \in y \leftrightarrow Fx)$. The claim follows.

Let us now assume that there is F such that $\forall x(Fx \leftrightarrow A(x))$. Take y as $\hat{z}.Fz$. Applying the above Lemma again, we immediately get $\forall x(x \in y \leftrightarrow A(x))$. □

The above proposition says that set-comprehension and concept-comprehension go hand in hand. The right notion of set in H, one abiding by the law of concretion, is to say that y is a *set* if $\exists F(y = \hat{z}.Fz)$. In Heck's predicative setting, one must distinguish concepts which are in the range of second-order variables from (more generally) concepts expressed by a formula of the language. On purely philosophical grounds, the predicativist can accept every concept expressed by the language of H. However, he is at technical odds to make comprehension available for the latter concepts (but not to obtain their extensions, since every concept expressed by the language of H has an extension in H). One must not comprehend these concepts with the same round of second-order variables (i.e., within the language of H), on pain of falling into impredicative comprehension and, indeed, on pain of contradiction. But, of course, these concepts can be comprehended in an enlargement of the language with a *new* round of concept variables. In short, when the predicativist brand of logicism is formalized in a language, there are concepts which have extensions and which are fit for being sets (by enlarging the language) but which are not sets according to the language. We may dub these extensions which are not sets as 'protosets.' Interestingly, the existence of these protosets allows the theory H (and the theory PE) to explore some features of the (technically) missing corresponding sets. For instance, protosets are instrumental in defining Frege's cardinality operator card (in H) and in developing numbers in a Fregean manner (in PE).

As discussed in the last paragraph, protosets in H may be gathered as sets (via concept-comprehension) with a new round of second-order variables. Of course, with this move, new protosets will emerge in the extended language and a further round of second-order variables is necessary to gather them as sets. Etcetera. Etcetera. One is led into theories of ramification and into envisaging ramified systems in which the extension operator applies to every formula of the language and in which Basic Law V is unrestricted (as Heck does in 1996). If there is a last round of second-order variables, a most general membership relation can be defined, but the law of concretion fails. If the ramified theory has no last round of variables, membership relations necessarily ramify. Even if one takes the position that no formal theory is able to express all the predicative concepts (presumably because of the inexistence of a natural ordinal bound for ramification), in this informal theory there is no overall membership relation, but only local ones for fixed rounds. Be that as it may, we do not get a decent theory of sets for all extensions (because either there is no global membership relation or the law of concretion fails for extensions).

How does the theory PE fare with respect to the above issues? Well, PE is essentially a round of impredicative variables on top of H. The membership operation $x \in y$ has to be defined with predicative variables because the alternative '$\exists \mathfrak{F}(y = \hat{z}.\mathfrak{F}z \wedge \mathfrak{F}x)$' is not a well-formed formula (we remind the reader that the extension operator does not apply to formulas in which impredicative variables occur). As with H, the theory PE has extensions which are not sets. However, the situation is quite different for *finite* extensions. Let us say that an extension $\hat{x}.A(x)$ is finite if $\exists \mathfrak{F}(\mathrm{Fin}(\mathfrak{F}) \wedge \forall x(\mathfrak{F}x \leftrightarrow A(x)))$. This is a well-defined notion for extensions (thanks to Basic Law V). Now, the principle of finite reducibility (Theorem 2) entails that finite extensions are sets. In other words, as long as we are only dealing with finite extensions, no new round of variables is necessary to gather finite extensions as sets. There are no finite protosets in PE, and this permits the development of a robust theory of finite sets.

The main goal of this paper was to convince the reader that arithmetic can be developed in a strict logicist manner by working on a (consistent) subsystem of Frege's original (inconsistent) theory. The restriction is simple and memorable: one is only allowed to take extensions of predicative concepts. Indeed, on this view of extensions, we believe that we have shown that Frege's own logicist program, when restricted to arithmetic, is successful.

Acknowledgements I would like to thank the invitation of *Argument Clinic* (an association of students of Philosophy at the University of Lisbon) for inviting me to give a presentation at the conference *Principia Mathematica (1913–2013)*, held at the University of Lisbon in February 6–7, 2014. I came with the main idea of this paper while preparing a talk for this conference. Afterwards, I also had the chance to speak about the issues of this paper at the meeting "2014: Abstractionism/Neologicism" (University of Storrs, Connecticut, U.S.A.), at "Journée sur les Arithmétiques Faibles 33" (University of Gothenburg, Sweden) and at the conference "The Philosophers and Mathematics" (University of Lisbon, Portugal). I want to thank the organizers of these meetings (Marcus Rossberg, Ali Enayat and Hassan Tahiri, respectively), as well as the participants for their remarks (specially to Marco Panza, for detailed comments and questions). My final thanks are to an anonymous referee. In a preliminary version of this paper, Section 4.4 used a definition of finiteness due to Paul Stäckel in 1907 (for a reference see Parsons 1987). The referee complained about the artificiality of using this definition. We reformulated the section using a characterization of finiteness given by Frege in volume I of the *Grundgesetze der Arithmetik* (1893). I believe that this change made the paper more natural and tighter.

Appendix

In order to prove Lemma 5, let x and y be given such that $\mathfrak{R}^*\langle x, y\rangle$. Suppose that $\forall z(\mathfrak{R}\langle x, z\rangle \to z = x)$. Take $\mathfrak{F}$ such that $\forall w(\mathfrak{F}w \leftrightarrow w = x)$. Clearly, $\mathrm{Her}_{\mathfrak{R}}(\mathfrak{F})$ and $\mathfrak{F}x$. By $\mathfrak{R}^*\langle x, y\rangle$, we get $\mathfrak{F}y$, i.e., $y = x$.

To see Lemma 6, assume the antecedent condition and $\mathfrak{R}^*\langle x, y\rangle$. By impredicative comprehension, let $\mathfrak{F}$ be such that $\forall w(\mathfrak{F}w \leftrightarrow \mathfrak{R}^*\langle x, w\rangle \wedge (\mathfrak{R}^*\langle w, y\rangle \to \mathfrak{Q}^*\langle x, w\rangle)))$. $\mathfrak{F}x$ is immediate. We claim that $\mathrm{Her}_{\mathfrak{R}}(\mathfrak{F})$. Suppose $\mathfrak{F}u$ and $\mathfrak{R}\langle u, v\rangle$. We get $\mathfrak{R}^*\langle x, u\rangle$ and, therefore, $\mathfrak{R}^*\langle x, v\rangle$. Now, suppose that $\mathfrak{R}^*\langle v, y\rangle$. We infer $\mathfrak{R}^*\langle u, y\rangle$ and, by $\mathfrak{F}u$, we also get $\mathfrak{Q}^*\langle x, u\rangle$. By the antecedent condition, it is clear that we have $\mathfrak{Q}\langle u, v\rangle$. Hence, $\mathfrak{Q}^*\langle x, v\rangle$. We have shown $\mathfrak{F}v$ and, therefore, proved $\mathrm{Her}_{\mathfrak{R}}(\mathfrak{F})$. Given that we have $\mathfrak{R}^*\langle x, y\rangle$, we may conclude $\mathfrak{F}y$ and, hence, $\mathfrak{Q}^*\langle x, y\rangle$.

Let us now prove Lemma 7.

For (1), assume Func($\mathfrak{R}$), $x \neq y$, $\mathfrak{R}^*\langle x, y\rangle$ and $\mathfrak{R}\langle x, z\rangle$. By impredicative comprehension, take $\mathfrak{F}$ such that $\forall w(\mathfrak{F}w \leftrightarrow (x = w \vee \mathfrak{R}^*\langle z, w\rangle))$. We claim that $\text{Her}_{\mathfrak{R}}(\mathfrak{F})$. To see this, take u and v such that $\mathfrak{F}u$ and $\mathfrak{R}\langle u, v\rangle$. Then either $x = u$ or $\mathfrak{R}^*\langle z, u\rangle$. The latter case obviously entails $\mathfrak{R}^*\langle z, v\rangle$ and, therefore, $\mathfrak{F}v$. If $x = u$, then we have both $\mathfrak{R}\langle x, z\rangle$ and $\mathfrak{R}\langle x, v\rangle$. By the functionality of $\mathfrak{R}$, $v = z$. Hence $\mathfrak{R}^*\langle z, v\rangle$, as wished. Now that we have established $\text{Her}_{\mathfrak{R}}(\mathfrak{F})$, note that $\mathfrak{F}x$ and (by assumption) $\mathfrak{R}^*\langle x, y\rangle$. Therefore, $\mathfrak{F}y$ and we are done.

To see (2), assume Func($\mathfrak{R}$), $\mathfrak{R}^*\langle z, x\rangle$ and $\mathfrak{R}^*\langle z, y\rangle$. By impredicative comprehension, take $\mathfrak{F}$ such that $\forall w(\mathfrak{F}w \leftrightarrow \mathfrak{R}^*\langle y, w\rangle \vee \mathfrak{R}^*\langle w, y\rangle)$. We claim that $\text{Her}_{\mathfrak{R}}(\mathfrak{F})$. To see this, take u and v such that $\mathfrak{F}u$ and $\mathfrak{R}\langle u, v\rangle$. Either $\mathfrak{R}^*\langle y, u\rangle$ or $\mathfrak{R}^*\langle u, y\rangle$. The former case entails $\mathfrak{R}^*\langle y, v\rangle$. In the latter case, by (1) above, either $u = y$ or $\mathfrak{R}^*\langle v, y\rangle$. Note that $u = y$ together with $\mathfrak{R}\langle u, v\rangle$ entails $\mathfrak{R}^*\langle y, v\rangle$. In both cases, $\mathfrak{R}^*\langle y, v\rangle \vee \mathfrak{R}^*\langle v, y\rangle$. We have showed $\mathfrak{F}v$ and, hence, proved $\text{Her}_{\mathfrak{R}}(\mathfrak{F})$. Clearly $\mathfrak{F}z$ and, by assumption, $\mathfrak{R}^*\langle z, x\rangle$. Hence $\mathfrak{F}x$, and we are done.

Finally, assume Func($\mathfrak{R}$), $x \neq y$, $\mathfrak{R}^*\langle x, y\rangle$, $\mathfrak{R}^*\langle y, x\rangle$ and $\mathfrak{R}^*\langle x, z\rangle$. By (2), either $\mathfrak{R}^*\langle z, y\rangle$ or $\mathfrak{R}^*\langle y, z\rangle$. In the former case, we get our conclusion $\mathfrak{R}^*\langle z, x\rangle$. We study the case $\mathfrak{R}^*\langle y, z\rangle$. By comprehension, take $\mathfrak{F}$ such that $\forall w(\mathfrak{F}w \leftrightarrow \mathfrak{R}^*\langle w, x\rangle)$. We claim that $\text{Her}_{\mathfrak{R}}(\mathfrak{F})$. To see this, take u and v such that $\mathfrak{F}u$ and $\mathfrak{R}\langle u, v\rangle$. By (1), either $u = x$ or $\mathfrak{R}^*\langle v, x\rangle$ (i.e., $\mathfrak{F}v$). We only need to study the case $u = x$. We have $\mathfrak{R}\langle x, v\rangle$ and, since $\mathfrak{R}^*\langle x, y\rangle$ and $x \neq y$, we get $\mathfrak{R}^*\langle v, y\rangle$ by another application of (1). Given that $\mathfrak{R}^*\langle y, x\rangle$, by transitivity, we conclude $\mathfrak{R}^*\langle v, x\rangle$ (i.e., $\mathfrak{F}v$). We have just proved $\text{Her}_{\mathfrak{R}}(\mathfrak{F})$. Since $\mathfrak{R}^*\langle x, z\rangle$ and $\mathfrak{F}x$, we conclude $\mathfrak{F}z$, i.e., $\mathfrak{R}^*\langle z, x\rangle$.

References

Boccuni, F. (2010). Plural Grundgezetze. *Studia Logica*, *96*(2), 315–330.

Burgess, J. (2005). *Fixing Frege*. Princeton: Princeton University Press.

Burgess, J., & Hazen, A. (1998). Predicative logic and formal arithmetic. *Notre Dame Journal of Formal Logic*, *39*, 1–17.

Cruz-Filipe, L., & Ferreira, F. (2015). The finitistic consistency of Heck's predicative Fregean system. *Notre Dame Journal of Formal Logic*, *56*(1), 61–79.

Dummett, M. (1991). *Frege. Philosophy of Mathematics*. Cambridge, MA: Harvard University Press.

Ferreira, F. (2005). Amending Frege's Grundgesetze der Arithmetik. *Synthese*, *147*, 3–19.

Frege, G. (1967). Begriffsschrift, a formula language, modeled upon that of arithmetic, for pure thought. In J. van Heijenoort (ed.), *From Frege to Gödel* (pp. 5–82). Harvard: Harvard University Press. A translation of Frege's *Begriffsschrift, eine der arithmetischen nachgebildete Formelsprache des reinen Denkens*, which appeared in German in 1879. Translated by J. van Heijenoort.

Frege, G. (1980). *The Foundations of Arithmetic*. Evanston: Northwestern University Press. A translation of Frege's *Die Grundlagen der Arithmetik*, which appeared in German in 1884. Translated by J. L. Austin.

Frege, G. (2013). *Basic Laws of Arithmetic*. Oxford: Oxford University Press. A translation of Frege's two volumes of the *Grundgesetze der Arithmetik*, which appeared in German in 1893 and 1903. Translated and edited by P. A. Ebert and M. Rossberg with a foreword of Crispin Wright.

Gödel, K. (2005). What is Cantor's continuum problem. In P. Benacerraf and H. Putnam (eds.), *Philosophy of Mathematics (selected readings)* (pp. 470–485). Cambridge: Cambridge University Press (This is a revised and expanded version of a paper first published in 1947).

Hale, B., & Wright, C. (2005). Logicism in the twenty-first century. In S. Shapiro (Ed.), *The Oxford Handbook of Philosophy of Mathematics and Logic* (pp. 166–202). Oxford: Oxford University Press.

Heck, R. (1996). The consistency of predicative fragments of Frege's Grundgesetze der Arithmetik. *History and Philosophy of Logic*, *17*, 209–220.

Heck, R. (1998). The finite and the infinite in Frege's Grundgesetze der Arithmetik. In M. Schirn (Ed.), *The Philosophy of Mathematics Today* (pp. 429–466). Oxford: Clarendon Press.

Heck, R. (2011). Ramified Frege arithmetic. *Journal of Philosophical Logic*, *40*(6), 715–735.

Quine, W. O. (1963). *Set Theory and its Logic*. Harvard: Harvard University Press.

Parsons, C. (1987). Developing arithmetic in set theory without infinity: Some historical remarks. *History and Philosophy of Logic*, *8*, 201–213.

Russell, B. (1973). On some difficulties in the theory of transfinite numbers and order types. In D. Lackey (Ed.), *Essays in Analysis* (pp. 135–164). Sydney: George Allen and Unwin Ltd. (This paper was first published in 1906.)

Russell, B. (1993). *Introduction to Mathematical Philosophy*. New York: Dover Publications. (First published in 1919.)

Russell, B., & Whitehead, A. N. (1927). *Principia Mathematica* (2nd ed.). Cambridge: Cambridge University Press.

Wang, H. (1950). A formal system of logic. *The Journal of Symbolic Logic*, *15*, 25–32.

Chapter 5
The Foundations of Geometry by Peano's School and Some Epistemological Considerations

Paolo Freguglia

Abstract The aim of this paper is to individualize some contributions by Peano and his school (in particular Burali-Forti and Pieri) to the foundations of synthetic geometry. In particular, we propose: some remarks regarding Peano's axiomatic foundation approach to geometry (1889) and some considerations on the fundamental role of geometrical calculus (Grassmann-Peano system) (1888).

5.1 Introduction

It is evident for those who dwell upon the history of mathematics during the second half of the 19th century, and in particular between the 19th and 20th century, that the research into the foundations of geometry, during that period, had led to remarkable developments: from the discovery and analysis of non-Euclidean geometry and relative Euclidean models, to the search for and the finding of suitable and adequate axiomatization for elementary and projective geometry.[1] Moreover, we can also take into consideration, in another mathematical context, the proposals linked to Grassmann-Peano's geometrical calculus, and, before that, Bellavitis's equipollences calculus. In fact, Peano's geometrical studies were also influenced by the reading of H. G. Grassmann *Ausdehnungslehre* (1844). From which Peano and his disciples derive an interesting reconstruction of projective geometry, according to an approach which is based on an adequate geometrical calculus. Even this approach has a significant fundamental implication. The study of the foundations of geome-

[1] See i.e. Henkin et al. (1959), Rosenfeld (1988)

This paper is a revision and an enlargement of the article: Freguglia, P., "Geometric calculus and Geometry Foundations", in *Giuseppe Peano between Mathematics and Logics* (F. Skof ed.), Springer Verlag, Milano, Dordrecht, London, New York, 2011.

P. Freguglia (✉)
Department of Information Engineering, Computer Science and Mathematics (DISIM), University of L'Aquila, L'Aquila, Italy
e-mail: pfreguglia@gmail.com

H. Tahiri (ed.), *The Philosophers and Mathematics*, Logic, Epistemology, and the Unity of Science 43, https://doi.org/10.1007/978-3-319-93733-5_5

try represents a crucial topic in order to provide philosophical and epistemological reflections. Consequently, the philosophy of geometry must give the essential basis of geometry, that is, the (algebraic or axiomatic) starting points, from which one can show important propositions. We can find a lot of critical literature about this research.[2].

Firstly, we find two important well-known works about the hypotheses of geometry: B. Riemann, *Ueber die Hypothesen welche der Geometrie zu Grunde liegen* (1854) and H. Helmholtz, "Ueber die thtsachlichen Grundlagen der Geometrie" (1866). The aim of these works was to establish crucial ideas, which increased awareness of elementary geometrical bases, on which consistent geometrical theories could be developed (with their theorems, problems etc.). This fact has an evident epistemological meaning. Furthermore, other mathematicians, logicians and philosophers took an interest in the foundations of geometry. Above all, we must remember G. Frege and B. Russell: Frege and Hilbert corresponded on these topics; Russell wrote *The Principles of Mathematics* (1903), where the sixth part is devoted to analyzing the notion of space through the different geometrical approaches (projective geometry, metric geometry). One of the main questions that has been tackled by these authors is this: in what way can formal logic be used in a rigorous way to explain geometrical theorems? Or, is it possible to individualize a kind of logical nature of geometrical objects?

However, Pasch, who had influenced Peano (1889) and Hilbert (1899) by means of his work entitled *Vorlesungen über neuere Geometrie* (1882), where he had stated his thought that the foundations of geometry were autonomous and unrelated to Arithmetic, in a letter to Frege dated 11 February 1894 changed his opinion, and wrote:

> [...] today from a general point of view, I disagree in part with the convictions which I have held until now. The rigorous foundations of geometry must be preceded by those of Arithmetic. [...].

Likewise, Pasch in *Vorlesungen über neuere Geometrie* asserted:

> [...] If we want geometry to be truly deductive, the proof procedure must be completely unrelated with the meaning of geometrical concepts, in the same way that [this procedure] must be unrelated with the figures [...].

But, for instance, Veronese said: "[...] the geometrical treatment must be elementary and based on constructive procedure and on spatial intuition."[3].

During this period, the main studies on this subject are: M. Pasch, *Vorlesungen über neuere Geometrie* (1882), G. Peano, "I principi di geometria logicamente esposti" (1889), M. Pieri, "I principi della geometria di posizione composti in sistema logico deduttivo" (1897–1898), G. Veronese, *Fondamenti di geometria a più dimensioni e a più specie di unità rettilinee* (1891), D. Hilbert, *Grundlagen der Geometrie* (1899), F. Schur, *Grundlagen der Geometrie* (1909), E. V. Huntington,

---

[2]See *infra*, for instance, Toretti (1978) and Magnani (2001)

[3]See Veronese G. (1891) p. XVI

"A Set Postulates for Abstract Geometry, Expressed in Terms of the Simple Relation of Inclusion" (1913) and H. P. Manning, *Geometry of Four Dimensions* (1914).

Peano's programme of research on this topic is surely different in comparison to Frege's or Hilbert's, and also Pasch's. Peano's studies consisted of carrying out a very rigorous analysis of the various theories of the mathematical corpus, by means of the formalism of his mathematical language of logic. His research disregarded the established idea of the theoretical unity of all mathematical knowledge around basic concepts of a logical nature. According to Peano, every mathematical theory is autonomous, that is to say that it has its statute because it has its own specific origins. The difference between the birth and the first developments of geometry and the arithmetic unitary epistemological vision of mathematics was felt, in particular, in the late XIX and early XX century.

5.2 The Axiomatic Foundations of Geometry

Let us begin our account of Peano's and his school's ideas by quoting Wylie Jr:

> The abstract logical plan which Euclid conceived and so ably illustrated in his "Elements" we now refer to as *axiomatic method*, and any particular instance of it we call an *axiomatic system*. The impact of the axiomatic method upon mathematics, and upon other sciences as well, has been so profound that not only the scholar who must be prepared to use it in own work but also the intelligent layman who would achieve some understanding of the nature of scientific thought must be familiar with it.[4]

Peano's works which concern the study of Geometry axioms are: *I principi di geometria logicamente esposti* (Fratelli Bocca, Torino, 1889) and «Sui fondamenti della geometria» [Rivista di Matematica IV (1894)]. Historically, these works were published between Pasch's *Vorlesungen über neuere Geometrie* (1882) and Hilbert's *Grundlagen der Geometrie* (1899). In the first work, Peano's explicit aim is to propose as theorems the fundamental propositions of *position geometry*, i.e. the propositions about the *order* and the *belonging* properties. Peano's goal is to establish the smallest possible set of geometrical concepts as *primitive*, i.e. not reducible to preceding concepts (from which we can define other concepts) and the smallest possible set of axioms (from which we can show other propositions).[5] Peano proposes "point" and "segment" as primitive concepts. When he writes about "points" and "segments" he is semantically considering these concepts according to the Euclidean proto-physical idea,[6] even if the concept of segment is expressed through the ternary relation $c \in ab$,[7] which means "c is between (the points) a and b". Therefore **1** indicates the set

[4] See Wylie C.R. (1964) p. 2.

[5] We have taken into account Beth E. W and Tarski A. (1956), pp. 462–467), Robinson (1959) pp. 68–85, Royden (1959) pp. 86–96, Scott (1956) pp. 456–461, Tarski A. and Lindenbaum (1926) pp. 111–113.

[6] That is, the geometrical concepts derive from immediate physical observations.

[7] Typographically we find in the Peano's text the symbol ε instead of the modern $\in$.

of points and does not denote any generic set (or class). By means of these primitive concepts it is possible to express the fundamental ternary constant predicate: $\beta(a, c, b)$ or $c \in ab$, which means "c stays between (the points) a and b". See Alfred Tarski's *What is elementary geometry?* (1969).

Peano denotes: **1** as the set or the class of (all) points, – as the negation, $\supset$ as "if...than...", $\cup$ as "or", ε as "to belong", ab as a segment (of extremes a and b), $=_x$ as $\forall x$, $-=_x \Lambda$, that is, it is not true for every x, because the symbol Λ means "false" or "empty set", as $\exists x$ and various kinds of punctuation as brackets.

If a and b are points, we can define the following two sets of points which we call *radii:*

$a'b =: \mathbf{1}.\ [x\ \varepsilon]\ (b\ \varepsilon\ ax)$, that is, $a'b$ is a class of points **1** which determines the segment ax, when x varies and where $b \in ax$

$$ab' =: \mathbf{1}.[x\ \varepsilon]\ (a\ \varepsilon\ xb)$$

i.e. $a'b$ is the set of points x so that b is inside the segment ax, while ab' is the set of points x so that a is inside the segment xb. The axioms, introduced by Peano are taken from Pasch's *Vorlesungen,* even if he presents some interesting additions and modifications.

Peano's preliminary axioms about the segments:

0.1. $a, b\ \varepsilon\ \mathbf{1}\ .\supset .\ ab\ \varepsilon\ \mathrm{K}\mathbf{1}$, that is ab is a class or set of points, where K denotes "class of". Therefore, if a and b are points, then the segment ab is a set of points.

Besides "[...] The symbol = between two points denotes their identity", hence:

0.2. $a,b,c,d\ \varepsilon\ \mathbf{1}\ .\ a=b\ .\ c=d : \supset .\ ac=bd$ (the three dots have the meaning of the connective "and") i.e. if $a, b, c, d,$ are points and a is equal to b and c is equal to d, then the segment ac is equal to the segment bd.

These axioms seem to contradict with what was mentioned earlier, i.e. segment as a primitive notion. In which case the notion of segment should be introduced directly and independently of the other primitive notions. But Peano's aim is only to describe the segments by means of points. In fact a segment is a particular set of points, and the previous axioms do not characterize the segments. As Peano expresses, by **K1,** that a segment is a class or set of points and he does not indicate a segment by **1**, that should mean "a point".

Actually. Peano also utilizes the constant predicate "=" when he writes $ac=bd$. Afterwards Tarski (1969) in order to express the same thing, introduces the *predicate* of arity equal to 4, so: $\delta(a, c, b, d)$, which means "a is distant from c as b from d."

Peano's ***straight line*** axioms[8]:

1. $\mathbf{1}-=\Lambda$, that is $\mathbf{1} \neq \emptyset$
2. $a\ \varepsilon\ \mathbf{1}.\supset.\because x\ \varepsilon\ \mathbf{1}.\ x -=a: -=_x \Lambda$, that is, if a is a point, then a point x exists and it is different from a. In other words, 'the set of x which are **1** is not empty and is not a singleton'.

[8]See also Vailati (1892) pp. 71–75, (1895) pp. 75–78, 183–185.

3. $a \,\varepsilon\, \mathbf{1}.\supset: aa = \Lambda$, that is, if a is a point, the segment aa does not exist. aa should be a point, but, according to Peano, a segment can be defined by two distinct points, and a point cannot be understood as a degenerate segment.
4. $a, b \,\varepsilon\, \mathbf{1}.\ a -= b: \supset.\ ab -=_x \Lambda$, i.e. if a and b are points and a is different from b then points x, that is a segment ab contains other points in addition to the boundaries.
5. $a, b \,\varepsilon\, \mathbf{1}.\ \supset.\ ab = ba$, i.e. the segments ab are not oriented.
6. $a, b \,\varepsilon\, \mathbf{1}.\ \supset.\ a - \varepsilon\, ab$, i.e. the extremes do not belong to the respective segment (this proposition can depend on the axiom 3)
7. $a, b \,\varepsilon\, \mathbf{1}.\ a -= b: \supset.\ a'b -= \Lambda$, i.e. if a and b are points and a is different from b, then the "radium" $a'b$ exists.
8. $a, b, c, d \,\varepsilon\, \mathbf{1}.\ c \,\varepsilon\, ad\ .\ b \,\varepsilon\, ac: \supset.\ b \,\varepsilon\, ad$, i.e.: if a, b, c, d are points and c belongs to the segment ad and b belongs to the segment ac, then b belongs to the segment ad.
9. $a, d \,\varepsilon\, \mathbf{1}.\ b, c \,\varepsilon\, ad: \supset: b = c.\ \cup.\ b \,\varepsilon\, ac.\ \cup.\ b \,\varepsilon\, cd$, that is, if a, b, c, d are points and b, c belong to the segment ad then or b coincides with c, or b belongs to the segment ac, or b belongs to the segment cd.
10. $a, b \,\varepsilon\, \mathbf{1}.\ c, d \,\varepsilon\, a'b: \supset: c = d.\ \cup.\ d \,\varepsilon\, bc.\ \cup.\ c \,\varepsilon\, bd$, that is, if a, b, c, d are points and c, d belong to the "radium" $a'b$ then or c coincides with d, or d belongs to the segment bc or c belongs to the segment bd.
11. $a, b, c, d \,\varepsilon\, \mathbf{1}.\ b \,\varepsilon\, ac.\ c \,\varepsilon\, bd: \supset.\ c \,\varepsilon\, ad$, i.e., Let a, b, c, d be points, if b belongs to the segment ac and c belongs to the segment bd, then c belongs to the segment ad.

    Moreover, Peano adds other axioms respectively to obtain the plane (2D) geometry and the solid (3D) geometry. So the *plane geometry* is obtained by adding the following axiom:
12. $r \,\varepsilon\, \mathbf{2}.\supset: x \,\varepsilon\, \mathbf{1}.\ x - \varepsilon\, r.\ -=_x \Lambda$

    where **2** is the class of straight lines. A straight line can be obtained and defined as two opposite and collinear radii, i.e.

    if r is a straight line, then a point x exists which does not belong to r.

    Peano gives other three axioms for the plane. Finally, Peano obtains the *3D space geometry* by adding the following axiom:
16. $h \,\varepsilon\, \mathbf{3}.\supset: x \,\varepsilon\, \mathbf{1}.\ x - \varepsilon\, h.\ -=_x \Lambda$

    where **3** is the set of planes, and another axiom. The plane can be seen as a sheaf of straight lines, i.e.

    if h is a plane, then a point x exists which does not belong to h.

With regards to Peano's axioms 2, 12 and 16, it is important to notice a considerable analogy with Veronese's related axioms and geometrical statement, even if Veronese is referring to greater than 3 dimensions spaces. The logical language proposed by Peano can be compared to a language of first order predicates logic. Peano does not always consider metalogic properties, except the independence. Peano's geometrical-epistemological point of view is not unlike the Euclidean position (in particular when compared to Hilbert). But, the introduction of a symbolic and logical apparatus and the axiomatic enrichment enables him to establish a rigorous analysis and presentation of the basis of geometry. In this case, the comparison with Pasch's *Vorlesungen* is very important.

Now, let us look at the controversy[9] between Peano and Giuseppe Veronese (1854–1917). During this period, Veronese was the Italian mathematician who presented an alternative approach to the foundations of Geometry which we can see in his treatise *Fondamenti di geometria* (1891) (see also non-Archimedean quantities approach). He begins as follows[10]:

> Empirical remark: When we consider outside bodies, which appear to us by means of our senses, in particular touch and sight, we associate these bodies with the object that contains them. We call this *outer environment* or *intuitive space* where each body occupies a certain *place* [...]

To determinate a nD Euclidean space S^n, Veronese introduces the following (methodological) principle:

- if S^{n-1} is a $(n-1)$D space, then a point x exists outside this. By means of Peano's symbolism we can write:

$$s \in (\mathbf{n{-}1}). \supset : x \in \mathbf{1}. x - \in s - = {}_x\Lambda$$

Hence:

- According to Veronese, the hyperstar of straight lines which pass by the point x and intersect the hyperspace S^{n-1} constitutes S^n.

This generating space principle has an intuitive validity if we use, according to Veronese, projective and descriptive geometry: from a S^n, through consecutive projections we arrive at S^2 (or S^3) and the geometrical properties of the plane S^2 or of the space S^3 are intuitively verifiable. This principle – as Veronese explicitly says – is based on the following logic or epistemological law:

> (*) Given a determined object A, if we do not establish that A is the set of all possible things to be considered by us, then we can think of another thing which does not belong to A, which is outside it and independent of A.[11]

Peano observes that proposition (*) is equivalent to:
Given a set A, if A does not contain all objects, then A does not contain all objects.[12]

Peano's remark is understandable, but Veronese's epistemological point of view is different from Peano's foundational ideas. Peano's aim is to propose an optimal and rigorous system of axioms, while Veronese wants to present a large number of theorems and properties and to study the *general space* which is a container with

---

[9]See Palladino D. § 5 *Appendix* of Borga et al. (1985), pp. 244–250.

[10]See Veronese G. (1891) Parte Prima, Libro I, Capitolo I, § 1, p. 209. See also Cantù P. (1999), *infra*.

[11]Ibid, from p. 1.

[12]See Peano G. "Recensione al volume *Fondamenti di geometria a più dimensioni e a più specie di unità rettilinee, esposti in forma elementare*, Padova, 1891, pp. XLVIII-630", Rivista di Matematica, 2 (1892), pp. 143–144. Actually in this review Peano blasts the Veronese's treatise.

a very high number of dimensions. Besides, Peano's point of reference is Pasch while Veronese's point of reference is Riemann, so their respective approaches are necessarily different.

5.3 The Fundamental Aspects of Geometrical Calculus

Is it possible to find another criterion for the foundations of geometry? That is, is it possible to have another epistemological point of view which is different from the axiomatic method? The answer could be affirmative if we are thinking of a foundation based on an algebraic structure or calculus. This is the case of Grassmann–Peano calculus.

Now we would like to present a brief analysis of the theoretical development which led to the realization of a vector calculus, and also to the possibility to construct the proofs (by means of this calculus) of geometrical theorems, in particular of projective geometry. Peano and his school wrote papers and books which were based on the reading of H. Grassmann's work. It should be remembered that Hermann Günther Grassmann (1809–1877) published *Ausdehnungslehre* in 1844. This is a work full of philosophical reflections, written in a language that had little to do with the mathematical mentality. In Germany this work had no impact on the mathematical world. In Italy Giusto Bellavitis read it and began an exchange of letters with Grassmann. Likewise, Luigi Cremona also appreciated this work. In 1862, Grassmann published a second edition in which he devoted considerable space to geometrical interpretations and applications, but this edition was no more successful than the first. Grassmann's fundamental idea was advanced in (1862).[13] So when Giuseppe Peano (1858–1932) held the post of lecturer of "Geometrical applications of infinitesimal calculus" at the University of Turin from the academic year 1885–86 to that of 1888–89, he was well aware of the problems regarding geometric calculus. Therefore, when in 1887 he published his lectures in a book entitled *Geometrical applications of infinitesimal calculus*, he had in mind Bellavitis, Möbius, Hamilton and Grassmann. In particular, in his first treatise on this subject, he gave importance to Bellavitis' manner of expression, in part thanks to the influence of his colleague Genocchi,[14] who was linked to Bellavitis by friendship and respect. However, it was in 1888 that Peano published his fundamental work on these topics: *Geometric Calculus according to H. Grassmann's Ausdehnungslehre preceded by operations of deductive logic*, a work which is also crucial for the history of logic. Here, he shows that he is decidedly convinced by Grassmann's approach, taking into account the 1844 edition of *Ausdehnungslehre*. Cesare Burali Forti (1861–1931) was Peano's disciple who studied geometric calculus the most; but Filiberto Castellano

[13]See Grassmann H. (1862) p. 415.

[14]The Genocchi-Bellavitis letters are analyzed in Canepa, G., Freguglia, P., "Alcuni aspetti della corrispondenza Giusto Bellavitis-Angelo Genocchi", *Angelo Genocchi e i suoi interlocutori scientifici* (edited by Conte, A. and Giacardi, L.), Deputazione subalpina di storia patria, Torino, 1991.

(1860–1919), Tommaso Boggio (1877 1963) and Mario Pieri (1860–1904) also took an interest in the subject.

According to Peano and Burali-Forti, the Cartesian co-ordinates method constitutes a "numerical intermediation", that is the representation by means of co-ordinates of geometrical entities, in order to study the same geometrical objects and their properties, while geometric calculus has the advantage of being absolute and concise and offers an immediate and direct approach to the study of geometrical problems. However, this calculus does not exclude the use of co-ordinates. In the volume published in 1888, Peano presents Grassmann's ideas in an original way: as we have already said, he gives a Euclidean interpretation to the fundamental Grassmannian notions, not beyond three dimensions. Firstly, he introduces the notion of *geometrical formation* as finite sum:

$$\sum_{i=1}^{n} m_i \tau_i^q$$

where m_i are real numbers and τ_i^q are q—hedrons, with $1 \leq q \leq 4$, which are Peano's interpretations of Grassmann's unities. These unities are analogous to the versors of a linear combination. That is, we have the following possibilities, according to Peano, for the system of unities:

1-hedron	point
2-hedron	segment
3-hedron	triangle
4-hedron	tetrahedron

With Peano, we call the geometrical formations which have as system of unities: points (and only points) of the first kind (or degree), points of the second kind if the unities are segments, points of the third kind if the unities are triangles and finally points of the fourth kind if the unities are tetrahedrons. We will generally denote a geometrical formation by 'F_q', where 'q' expresses the kind of formation under consideration. Even if the possibility of geometrical formations with $q \geq 5$ is not contemplated by Peano, it should be an easy and natural generalization. Veronese, who was a contemporary of Peano, should have done this but he did not. But Peano maintains only a traditional vision of geometry. Between two geometrical formations we can establish the operation of algebraic addition, which complies with the rules of the algebra of polynomials. However, conceptually the more important operation is the *alternate product*,[15] which is introduced by Peano (and by Burali-Forti[16]) as follows:

[15]See Peano, G. (1888) pp. 110–111.

[16]See Burali-Forti, C. (1926) pp. 5–6.

If we have two geometrical formations in 3D, F_r and F_s, the alternated product is a product which complies with the rules of the algebra of polynomials, but without changing the order of the letters which denote points. If $r+s \leq 4$ the product is called *progressive*[17] and expresses the geometrical operation of projection. If $r+s > 4$ the product is called *regressive*[18] and represents the geometrical operation of section. In the plane case, 2D, in the definition we must respectively replace $r+s \leq 3$ and $r+s > 3$. The alternate product is not commutative.

For instance, a segment is represented by the product AB, if A and B are two points which determine the segment. But a segment can be represented by the formation of the first kind B – A that it mean that AB is equivalent to B – A. We also have: AB = –BA and AA = 0. Some particular progressive products, equalized to 0, have interesting geometrical interpretations:

ABCD = 0: points A, B, C, D are coplanar
ABC = 0: points A, B, C are in a straight line
AB = 0: points A and B coincide

By means of their geometric calculus, Peano and Burali-Forti are able to prove theorems of projective geometry. Even Bellavitis[19] proposed the application of his equipollence calculus to elementary geometry and to projective geometry. Peano does not utilize figures, while we in this essay will utilize them in order to have greater clarity. He considers the figures which we find in traditional treatises of synthetic (elementary and) projective geometry as heuristic representations. Indeed, a figure can psychologically influence the actual proof of a theorem and the solution of a problem, for instance, when we are not able to draw a general figure. Instead, Peano and Burali-Forti only take into consideration chains of expressions-identities of our calculus and subsequently they interpret the last expression geometrically. Hence, according to our mathematicians, a geometrical theorem is an interpretation, a model of an identity which concludes a sequence of geometrical calculus. Now, for instance, we will present Peano's rendering of Menelaus-Ptolemy, Desargues and Pascal theorems,[20] three fundamental theorems of projective geometry. To simplify, we will make the content of these theorems clear by means of figures, but the reader should be able to realise that these figures are superfluous for the proofs.

Let us begin with the following lemma:

[17]See Peano, G. (1888) p. 30.
[18]Ibid. from p. 107.
[19]See Bellavitis (1854) from p. 13.
[20]See Peano G. (1888) p. 47

Lemma *If C is a point which belongs to the straight line AB, then we can write:*

$$C = \frac{CB}{AB}A + \frac{AC}{AB}B \tag{3.1}$$

Proof If C is a point which belongs to the straight line AB (A $\neq$ B), then we can write

$$\mathrm{C} = x\mathrm{A} + y\mathrm{B}, \text{ with } x \text{ and } y \text{ real numbers}$$

multiplying the previous expression by A, we obtain:

$$\mathrm{AC} = x\mathrm{AA} + y\mathrm{AB}$$

Because AA = 0 then y = AC/AB. Likewise for x.

The Menelaus-Ptolemy theorem is, then, rendered as follows:

Theorem 3.1 *If the points* $AB \,.\, B'C' = C'; BC \,.\, B'C' = A'; AC \,.\, B'C' = B'$ *are collinear then* $AC' \cdot BA' \cdot CB' = BC' \cdot CA' \cdot AB'$

that is:

$$\frac{AC' \cdot BA' \cdot CB'}{BC' \cdot CA' \cdot AB'} = 1 \tag{3.2}$$

Proof: In virtue of the previous Lemma, insofar as $C' \in AB$, $B' \in AC$ *and* $A' \in BC$ *we have:*

$$A' = \left(\frac{A'C}{BC}\right)B + \left(\frac{BA'}{BC}\right)C$$
$$B' = \left(\frac{B'A}{CA}\right)C + \left(\frac{CB'}{CA}\right)A$$
$$C' = \left(\frac{C'B}{AB}\right)A + \left(\frac{AC'}{AB}\right)B$$

Now if we form the progressive product $A'B'C'$ *we have:*

$$A'B'C' = \left[\left(\frac{A'C}{BC}\right)\left(\frac{B'A}{CA}\right)\left(\frac{C'B}{AB}\right) + \left(\frac{B'A}{BC}\right)\left(\frac{CB'}{CA}\right)\left(\frac{AC'}{AB}\right)\right]ABC \tag{3.3}$$

Hence Peano says: "if ABC is different from zero, a necessary and sufficient condition so that the three points A′, B′, C′ are in a straight line (that is when $A'B'C' = 0$*), is to equalize to zero the coefficient [in the square brackets of (3.3)] of ABC". This proposition is equivalent to saying that, if the points A′, B′, C′, which lie respectively on the sides BC, CA, AB of a triangle ABC, are in a straight line,*

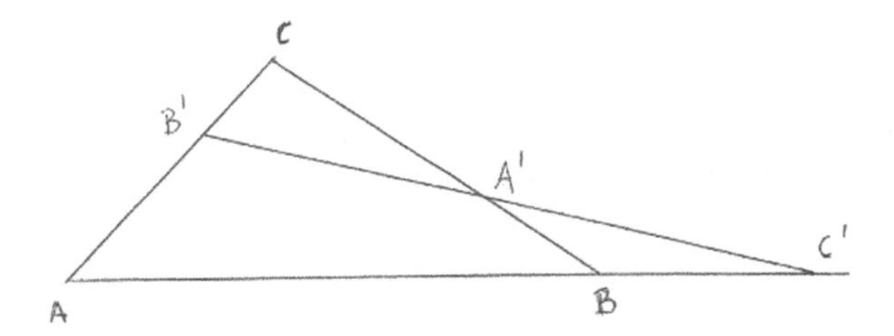

Fig. 5.1 Representation of Menelaus-Ptolemy theorem

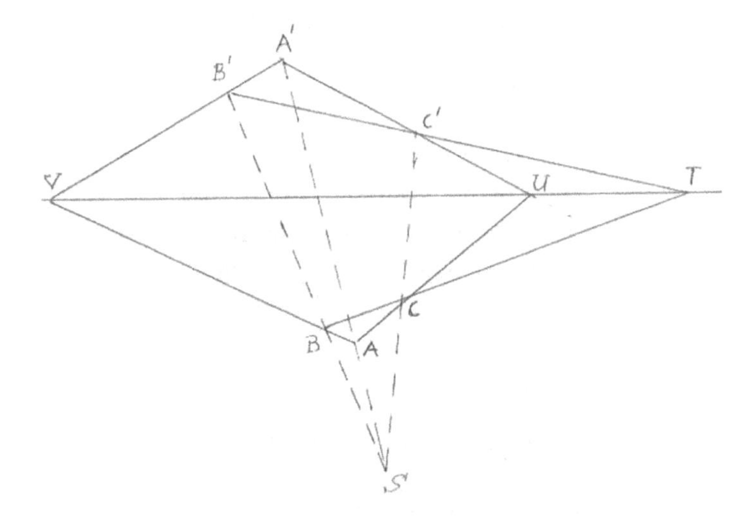

Fig. 5.2 Representation of homological triangles (plane case) Desargues theorem

then (3.3) is true and vice versa. In (3.2) the symbol « · » denotes both the progressive product (geometrical operation of section) and the arithmetical product (Fig. 5.1).

Let us now pass to Peano's renderings of Desargues' classic theorem[21] (plane case of homological triangles) and Pascal's *hexagon* theorem.

Theorem 3.2 *The points T, U, V, intersections respectively of the pairs of sides BC, B′C′; CA, C′A′; AB, A′B′ are collinear if only if the straight lines AA′, BB′, CC′ intersect at the point S.*

Proof Independently of the drawing (Fig. 5.2), we obtain the following steps.

We can start from the identity (see previous Lemma [where (ABD), (ABC) as coeff.]:

$$\mathrm{AB.CD} = \mathrm{(ABD).C} - \mathrm{(ABC).D} \tag{3.4}$$

[21] Ibid., p. 92.

which says that the product AB.CD between the segments AB and CD is regressive and hence it expresses an intersection on the plane, i.e. a point, which is represented by the right-hand side of (3.4). We shall denote the points with capital letters and the segments with small letters. Thus, by writing ‘p’ instead of ‘CD’, (3.4) becomes:

$$\mathrm{AB.p} = \mathrm{Bp.A} - \mathrm{Ap.B} \tag{3.5}$$

Let us now take into consideration the following expression:

$$(\mathrm{BC.a})(\mathrm{CA.b})(\mathrm{AB.c}) \tag{3.6}$$

Applying (3.5)–(3.6) we obtain consecutively:

$$\begin{aligned}
&(\mathrm{Ba.C} - \mathrm{Ca.B})(\mathrm{Cb.A} - \mathrm{Ab.C})(\mathrm{Ac.B} - \mathrm{Bc.A}) \\
&= (\mathrm{Ba.Cb.CA} + \mathrm{Ca.Ab.BC})(\mathrm{Ac.B} - \mathrm{Bc.A}) \\
&= \mathrm{Ba.Cb.Ac.CAB} - \mathrm{Ca.Ab.Bc.ABC} \\
&= \mathrm{Ba.Cb.Ac.ABC} - \mathrm{Ca.Ab.Bc.ABC} \\
&= (\mathrm{Ba.Cb.Ac} - \mathrm{Ca.Ab.Bc})\mathrm{ABC}
\end{aligned}$$

Hence we have the identity:

$$(\mathrm{BC.a})(\mathrm{CA.b})(\mathrm{AB.c}) = (\mathrm{Ba.Cb.Ac} - \mathrm{Ca.Ab.Bc})\mathrm{ABC} \tag{3.7}$$

Dually, i.e. by putting capital letters in the place of the small letters and vice versa, we obtain the true expression:

$$(\mathrm{bc.A})(\mathrm{ca.B})(\mathrm{ab.C}) = (\mathrm{bA.cB.aC} - \mathrm{cA.aB.bC})\mathrm{abc} \tag{3.8}$$

Multiplying (3.7) by abc and (3.8) by ABC and adding member to member, we obtain:

$$\mathrm{abc}(\mathrm{BC.a})(\mathrm{CA.b})(\mathrm{AB.c}) + \mathrm{ABC}(\mathrm{bc.A})(\mathrm{ca.B})(\mathrm{ab.C}) = 0 \tag{3.9}$$

If in (3.9) we put $\mathrm{a} = \mathrm{B'C'}$, $\mathrm{b} = \mathrm{C'A'}$, $\mathrm{c} = \mathrm{A'B'}$ we will have:

$$\mathrm{A'B'C'}\big(\mathrm{BC.B'C'}\big)\big(\mathrm{CA.C'A'}\big)(\mathrm{AB.A'B'} + \mathrm{ABC}\big(\mathrm{A'A.B'B.C'C}\big) = 0 \tag{3.10}$$

Assuming that (the triangles) ABC and $\mathrm{A'B'C'}$ are different from zero, (3.10) leads us to:

$$\big(\mathrm{BC.B'C'}\big)\big(\mathrm{CA.C'A'}\big)\big(\mathrm{AB.A'B'}\big) = 0$$

and

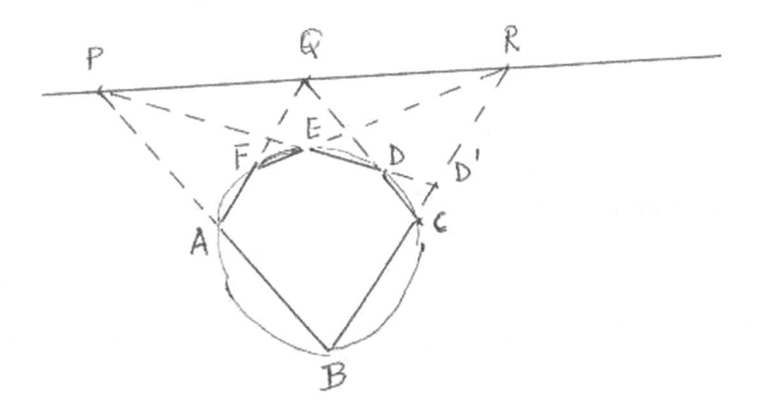

Fig. 5.3 Representation of Pascal's hexagon theorem

$$(AA'.BB'.CC') = 0$$

In conclusion and interpreting this, we have (see Fig. 5.2):
(BC.B′C′)(CA.C′A′)(AB.A′B′) = TUV = 0, i.e. "*the points* T, U, V, *where the corresponding sides* BC, B′C′; CA, C′A′; AB, A′B′ *respectively intersect, are collinear*", if (AA′.BB′.CC′) = abc = 0, i.e. "*the straight lines* AA′, BB′, CC′ *intersect at the point* S"

While Bellavitis' calculus of equipollence is more directly connected with elementary (plane) synthetic geometry, Peano uses his calculus in a more compact way. For instance, Pascal's hexagon theorem[22] is only presented through the following equality (see Fig. 5.3):

$$(AB.DE)(BC.EX)(CD.AX) = 0 \tag{3.11}$$

and in the interpretation of (3.11) also the proof of the theorem remains. In fact (3.11) is a second degree monomial for X (because the number of times that X appears is two). Because X is an unknown variable and A, B, C, D, E, X are points, then (3.11) is a second degree equation, i.e. a conic equation which passes through the previous points. Besides, (3.11) is identically equal to zero if X = A, or X = B, or X = C, or X = D, or X = E. Therefore, when X = F the six different points A, B, C, D, E, F belonging to the conic, determine a hexagon inscribed in these same conic. We can observe that: (AB.DE) = P, (BC.EF) = R and (CD.AF) = Q. In virtue of (3.11) we have: PQR = 0, i.e. the points P, Q, R are aligned. Hence:

Theorem 3.3 *If a hexagon ABCDEF is inscribed in a conic then the three intersections P, Q, R of opposite sides belong to the same straight line, i.e. P, Q, R are aligned, and vice versa.*

[22] Ibid. p. 95.

If we write (3.4) in lower case, we obtain, by means of duality, Brianchon's theorem. Hence Peano and his disciple propose a very synthetic geometric analysis of previous projective fundamental theorems.

5.4 Some Considerations

Certainly, from a foundational point of view, the Grassmann-Peano calculus shows its relevance. But the goal of the geometrical calculus, or the vector calculus or the homographies, goes beyond a "philosophical" justification of the bases of geometry. In fact, the applications to mathematical physics and to transformation geometry are very important and, notwithstanding the matrix calculus, Peano's geometrical calculus is a general theory which is of intrinsic theoretical and mathematical interest. It is the opinion of a lot of historians that Peano's philosophical attitude is similar to a position which combines logical rigour and Euclidean vision of geometry. In fact, the introduction of a symbolic and logical apparatus and the axiomatic enrichment, establishes a rigorous presentation of classical geometry bases.

From a historical and epistemological point of view, perhaps the geometrical calculus approach can seem more interesting. But the development of studies on the foundations of geometry in the 20th century has preferred the axiomatic approach (Hilbert, Huntington, Tarski, ...). In our opinion, the protophysical role of Euclidean geometry in Peano's works is essential and decisive even if Peano distinguishes his *position geometry* from Euclidean geometry. In his "Sui fondamenti della geometria" the congruence theory is well determined and regulated. Classical geometry constitutes the crucial model for the study of the foundations of geometry. Even Hilbert, deep down, takes Euclid into account.[23] Hence, we can observe equivalent theories for the foundation of geometry, and in this way we have a "theoretical pluralism" regarding the choice of primitive elements and fundamental axioms.

References

Bellavitis, G. (1854). Sposizione del metodo delle equipollenze. *Memorie della Società Italiana delle Scienze, XXV*.

Borga, M., Freguglia, P., & Palladino, D. (1985). *I contributi fondazionali della scuola di Peano*. In F. Angeli (Ed.), Milano.

Bottazzini, U. (2000). I geometri italiani e i *Grundlagen der Geometrie* di Hilbert, (Atti XVI Cong.UMI), Monograf, Bologna.

Cantù, P. (1999). *Giuseppe Veronese e i fondamenti della geometria*. Milano: Unicopli.

Freguglia, P. (2011). Geometric calculus and Geometry Foundations. In F. Skof (Ed.), *Giuseppe Peano between Mathematics and Logics*. Milano, Dordrecht, London, New York: Springer.

Henkin, P., Suppes, P., & Tarski, A. (Eds.). (1959). *The axiomatic method*. Amsterdam: North-Holland.

[23] See Bottazzini (2000), *infra*.

Magnani, L. (2001). *Philosophy and geometry. Theoretical and historical issues*. Dordrecht: Kluwer.
Robinson, R. M. (1959). Binary relations as primitive notions in elementary geometry. In P. Henkin, P. Suppes, & A. Tarski (Eds.), *The axiomatic method*. Amsterdam: North-Holland.
Rosenfeld, B. A. (1988). *A history of non-euclidean geometry. Evolution of the Concept of a Geometric Space*. New York: Springer.
Royden, H. L. (1959). Remarks on primitive notions for elementary Euclidean and non-Euclidean plane geometry. In P. Henkin, P. Suppes, & A. Tarski (Eds.), *The axiomatic method*. Amsterdam: North-Holland.
Scott, D. (1956). A symmetric primitive notion for Euclidean geometry. *Indagationes Mathematicae, 18*.
Tarski, A., & Lindenbaum, A. (1926). Sur l'indépendance des notions primitives dans les systems mathématiques. In *Annales de la Société Polonaise de Mathématiques* (Vol. 5).
Toretti, R. (1978). *Philosophy of geometry from Riemann to Poincaré*. Dordrecht: Reidel.
Vailati, G. (1892). Sui principi fondamentali della geometria della retta. *Rivista di Matematica, 2*, 71–75.
Veronese, G. (1891). *Fondamenti di Geometria a più dimensioni e a più specie di unità rettilinee*. Padova: Tipografia del Seminario.
Wylie, C. R., Jr. (1964). *Foundations of geometry*. New York: McGraw-Hill.

Supplementary

Beth, E. W. & Tarski, A. (1953). Equilaterality as the only primitive notion of euclidean geometry. In *Indagationes Mathematices*, 15.
Borga, M., Freguglia, P., & Palladino, D. (1983). Il problema dei fondamenti della matematica nella scuola di Peano. *Epistemologia, VI*, 45.
Bottazzini, U. (1883). Sul Calcolo geometrico di Peano, (Atti Conv. Intern. di Storia della Logica), CLUEB, Bologna.
Burali Forti, C. (1986). *Geometria analitico proiettiva*. Torino: Ed. Petrini.
Burali-Forti, C., & Boggio, T. (1924). *Espaces courbes. Critique de la relativité*. Torino: Sten Editrice.
Burali-Forti, C., & Marcolongo, R. (1909). *Elementi di calcolo vettoriale con numerose applicazioni alla geometria, alla meccanica e alla fisica-matematica*, Nicola Zanichelli, Bologna, (2d ed., 1921).
Burali-Forti, C., & Marcolongo, R. (1912–13). *Analyse vectorielle générale* (I et II vol.). Pavia: Mattei & C. Editeurs.
Cartan, E. (1907)."Nombres Complexes, exposé d'après l'article allemand de E. Study. *Encyclopédie des Sciences Mathématiques* (Vol. I). Paris: Gauthier-Villars.
Freguglia, P. (1982). Il contributo di G. Peano agli studi sui fondamenti della geometria. *Atti del convegno 'La storia delle matematiche in Italia'*.
Freguglia, P. (1992). *Dalle equipollenze ai sistemi lineari. Il contributo italiano al calcolo geometrico*. Urbino: QuattroVenti ed.
Freguglia, P. (1993). Il contributo di Giuseppe Peano e della sua scuola al calcolo geometrico (Atti Conv. 'Peano e i fondamenti della matematica'), Mucchi, Modena.
Freguglia, P. (2000). Gli studi sui fondamenti della geometria nella seconda metà dell'ottocento con particolare riferimento alla situazione italiana. *Le Matematiche, LV*, 161–205.
Freguglia, P. (2006). *Geometria e numeri. Storia, teoria elementare ed applicazioni del calcolo geometrico*. Torino: Bollati Boringhieri.
Freguglia, P. (2010). Hermann Grassmann's work and the Peano School. In H. J. Petsche et al. (Eds.), *Hermann Grassmann. From past to Future: Grassmann's Work in Context*. Basel: Birkhauser Verlag.

Freguglia, P., & Bocci, C. (2008). Dall'eredità grassmanniana alla teoria delle omografie nella scuola di Peano. *La Matematica nella Società e nella Cultura (Journal of Unione Matematica Italiana), 1,* 131–164.

Grassmann, H. G. (1894–1911). *Gesammelte mathematische und physikalische Werke* [*G. W.*] ed. by F. Engel, 3 voll., Leipzig: Teubner. *Die lineale Ausdehnungslehre, eine neuer Zweig der Mathematik*, 1844, in *G.W.* (Teubner), Leipzig, 1894, vol. I_1; *Die Ausdehnungslehre von 1862* (1862), in *G.W.* (Teubner), Leipzig, 1894, vol. I_2 (1896).

Lolli, G. (2004). *Da Euclide a Goedel*. Bologna: Società editrice Il Mulino.

Lützen, J. (Ed.). (2001). *Around Caspar Wessel and the geometric representation of complex numbers*. Copenhagen: C. A. Reitzels Forlag.

Marcolongo, R. (1905). *Meccanica razionale* (2 vol.). Milano: Manuali Hoepli.

Marcolongo, R. (1921). *Relatività*. Messina: Casa Editrice Giuseppe Principato.

Peano, G. (1887). *Applicazioni geometriche del calcolo infinitesimale*. Torino: Fratelli Bocca Editori.

Peano, G. (1988). *Calcolo geometrico secondo l'*Ausdehnungslehre *di H.Grassmann preceduto dalle operazioni della logica deduttiva*. Torino: Fratelli Bocca Editori.

Schubring, G. (Ed.). (1996). *Hermann Günther Grassmann (1809–1877): visionary mathematician, scientist and neohumanist scholar*. Dordrecht: Kluwer Academic Publ.

Vailati, G. (1895a). Sulle relazioni di posizione tra i punti di una linea chiusa. *Rivista di Matematica, 5,* 26–29.

Vailati, G. (1895b). Sulle proprietà caratteristiche delle varietà a una dimensione. *Rivista di matematica, 5,* 183–185.

Chapter 6
Μηδεὶς ἀγεωμέτρητος εἰσίτω

Reinhard Kahle

Abstract This paper provides a discussion to which extent the Mathematician David Hilbert could or should be considered as a Philosopher, too. In the first part, we discuss some aspects of the relation of Mathematicians and Philosophers. In the second part we give an analysis of David Hilbert as Philosopher.

6.1 Philosophers as Mathematicians and Mathematicians as Philosophers

> Let us note first that it is as rare to see a mathematician in possession of a strong philosophical culture as to see a philosopher who has an extensive knowledge of mathematics; the opinions of mathematicians on topics in philosophy, even when these questions are concerned with their field, are most often opinions received at second or third hand, coming from doubtful sources. But, precisely because of this, it is these average opinions which interest the mathematical historian, at least as much as the original views of thinkers such as Descartes or Leibniz (to mention two who were also mathematicians of the first rank), Plato (who at least kept up with the mathematics of his time), Aristotle or Kant (of whom the same could not be said).
>
> Bourbaki (1994, p. 11).

In the following, we aim to isolate some criteria which characterize Philosophers as Mathematicians or Mathematicians as Philosophers.

Work partially supported by the Portuguese Science Foundation, FCT, through the projects *The Notion of Mathematical Proof*, PTDC/MHC-FIL/5363/2012, *Hilbert's 24th Problem*, PTDC/MHC-FIL/2583/2014, and the *Centro de Matemática e Aplicações*, UID/MAT/00297/2013.

The title phrase was recorded at the entrance of the Platonic Academy. In a free English translation it reads: "Let none but geometers (i.e., mathematicians) enter here." The earliest references to this phrase is from the sixth century and can be found, in slight variations, in works of John Philoponus (1897, p. 117.27) and Elias (1900, p. 118, 18–19). It is in line with Diogenes Laertius (1959, VI.10; p. 384) who reports that Xenocrate, the third leader of Platon's Academy, classified Mathematics as part of the handles of philosophy.

R. Kahle (✉)
CMA and DM, FCT, Universidade Nova de Lisboa, 2829-516 Caparica, Portugal
e-mail: kahle@mat.uc.pt

H. Tahiri (ed.), *The Philosophers and Mathematics*, Logic, Epistemology, and the Unity of Science 43, https://doi.org/10.1007/978-3-319-93733-5_6

Philosophical Tradition and Mathematical State-of-the-Art

To bridge Philosophy and Mathematics, it is, of course, important to show respect for the other area. For Philosophers this would mean they need to have an up-to-date knowledge of the Mathematics of their time; for Mathematicians that they would need to be at home at the philosophical debates, both historically and of their times.

This criteria can be verified for Philosophers like Plato (of course), but also Avicenna[1] and Carnap, just to mention three names. It fails, on the other hand, for instance for Kant who—despite the important role Mathematics plays in his Philosophy—was not abreast with the Mathematics of his times (Volk 1925) and, in particular, for Hegel.[2]

Likewise, Gauß did not show particular respect for Philosophy[3]; and also for Euclid we do not have sources which would show that he was appreciating Philosophy. On the other hand, Descartes, Pascal, and Leibniz are examples of Mathematicians which were at the very same time Philosophers and, of course, were not only at home at the philosophical debates but shaped them. Finally, we may mention Cantor who was staying much more in the philosophical tradition—and even theological (see Tapp 2005)—than one might expect at first glance.

Philosophers as Mathematicians

To be a Mathematician (as Philosopher or anybody else), one is expected to contribute to the body of mathematical knowledge. As a matter of fact, this is hardly the case for anybody coming from Philosophy. It could, for instance, be claimed neither for Plato nor for Wittgenstein, even if both influenced Philosophy of Mathematics—more or less, respectively.

There are, of course, examples where scientists advanced in both disciplines "in parallel"—like Descartes, Pascal, and Leibniz, already mentioned, but also Bolzano[4] and Ramsey. In these cases, it is rather debatable whether it makes sense to give "preference" to one of the areas to be the one they started from to reach also the other.

---

[1] Crozet (2018), Rashed (2008, 2018).

[2] Hegel's dissertation of 1801 finishes with a discussion—although not a "proof"—that there should be exactly eight planets in the solar system. His work was characterized as "*Monumentum Insaniae Saec. XIX*" by Ernest II, Duke of Saxe-Gotha-Altenburg and discussed as such by Gauß in the correspondence with his friend Schumacher in 1842 (Peters 1862, letters 763–765).

[3]
> Just look around at the modern philosophers, at Schelling, Hegel, Nees von Esenbeck and consorts—don't their definitions make your hair stand on end? Read in the history of ancient philosophy what the men of the day, Plato and others (I except Aristotle), gave as explanations. And even in Kant matters are often not much better; his distinction between analytic and synthetic propositions seems to me to be either a triviality or false.

Gauß in a letter to Schumacher on 1 November 1844; translation from (Ewald 1959, p. 293); German original in (Peters 1862, letter 944).

[4] "Bolzano, der große Gegner Kants, ist seit LEIBNITZ[sic!] der erste philosophische Mathematiker und mathematische Philosoph." ("Bolzano, the great opponent of Kant, is the first philosophical Mathematician and mathematical Philosopher since Leibniz.") (Korselt 1903, p. 405).

Russell and Husserl were Mathematicians by education. When they turned later into Philosophers, their initial work on Mathematics should not be considered as that of a Philosopher. The additional question is whether the mathematical community recognized this education. This might not have been the case for Husserl; but it was the case for Russell. Yet when he became one of the world's most famous Philosophers of the 20th century, he became increasingly disconnected from Mathematics. Only at the time he was publishing his opus magnum, the *Principia Mathematica* written together with Whitehead, he was (still) considered as a Mathematician. Now, a closer inspection shows that despite the tremendous *logical* work expounded in the *Principia*, there is essentially no new *mathematical* inside which could be attributed to Russell, and, from a modern perspective, one could actually question his status as Mathematician.

Mathematicians as Philosophers

How a Mathematician can be considered as a Philosopher is a rather intricate question.

One could take the position that everybody is doing Philosophy, and becomes a Philosopher, as soon as (s)he is reflecting in one or another way on his or her scientific activity. In this view, nearly every Scientist would turn into a Philosopher—even against his/her explicit will, as it would probably be the case for Bourbaki.

On the other hand, one could reserve the title "Philosopher" to those who enter in a predefined philosophical career and expose his/her reflections in terms of scholarly philosophical standards. With this criterion one would probably find only a handful of distinguished Mathematicians who could be considered Philosophers (like Descartes, Pascal, and Leibniz—but not even Frege).

We would like to steer a middle course, or better: two middle courses.

First, we may qualify a Mathematician as a Philosopher if (s)he is building his/her mathematical work on a clear philosophical conviction. And this conviction should have priority in the sense that, in case of problematic mathematical results, Mathematics is rethought but not the philosophical foundation. Pythagoras falls into this category, but the prime example is probably Brouwer (van Dalen 2013).

Secondly, a Mathematician can be considered as Philosopher, if there is an echo from Philosophy proper in the specific contribution of the Mathematician; i.e., the philosophical discussion is taking up ideas from the Mathematician. And the Mathematician takes part in this discussion. Here, Frege will be the best example. Cantor, however, who tried to take part in philosophical discussions, was, at the end of the day, rather neglected by the philosophers.

The Working Mathematician

So far, we have only mentioned leading Mathematicians and their relation to Philosophy. For the average Mathematician, it is probably the case that (s)he reflects even less on philosophical issues. In a reply to Whitehead concerning Swift's description of the Mathematicians in Gulliver's voyage to Lapute, Philip Jourdain (1915, p. 638) writes:

> ...Swift, like everybody else, could not doubt the usefulness, importance, and correctness of the mathematician's work, but shared, with the philosopher, a doubt of the mathematician's being able to state his principles clearly and reasonably, just as we may doubt the existence of a knowledge of thermodynamics in a man who drives a railway engine.

This might be true for the working Mathematician in the role of the engine driver. But the comparison is flawed if the Philosopher should take over the role of the Physicist studying thermodynamics: such a Physicist might not be able to drive a locomotive, in the same way as a Philosopher does not do Mathematics. But, as a rule, the theoretical study of the Physicist contributes to the development of better engines. This is, unfortunately, rarely the case with contributions of Philosophers to Mathematics.[5]

6.2 David Hilbert as Philosopher

> Don't raise great hopes from Hilbert's philosophical "results". I'm already disappointed by what he produced of it so far. [...] That he feels the need for more, is quite fine, but as soon as he leaves the purely mathematical, he is simply silly.
>
> Nelson to Hessenberg, June 1905[6]

Hilbert as Mathematician

David Hilbert was one of the greatest Mathematicians of all time. While sharing the title with Henri Poincaré of leading Mathematician around the turn of the 19th to 20th century, he occupied this position unchallenged after Poincaré's death in 1912 up to the 1930s. In this time, he also promoted a foundational programme in Mathematics, which has a strong philosophical rationale. But it is clear that the authority of Hilbert was based on his mathematical work and the influence he had on the mathematical community.

In his mathematical work, two results are of importance from a philosophical perspective.

In his solution of *Gordan's Problem* Hilbert provided a *non-constructive* existence proof; i.e., he was proving an existential statement without providing a method to find a witness for it. This was quite contrary to the understanding of existential statements by that time, and the German Mathematician Gordan, who raised the problem said[7]:

[5]Von Plato (2016) pointed to such a rare case, when Kant triggered with his discussion of the equation $7 + 5 = 12$ the development of modern recursive foundations of Arithmetic through the obscure figures of Johann Schultz and Michael Ohm; via Grassmann, Hankel, Schröder, Dedekind it finally reaches Peano, Skolem, and Bernays.

[6]The translation is ours; German original in (Peckhaus 1990, p. 166): "Von Hilberts philosophischen 'Resultaten' mach' Dir nun lieber keine großen Hoffnungen. Ich bin bereits durch das, was er bisher davon produziert hat, recht enttäuscht. [...] Daß er das Bedürfnis nach mehr fühlt, ist ja auch sehr schön, aber so wie er das rein Mathematische verläßt, wird er einfach albern."

[7]The anecdote, including Hilbert's relation to Kronecker, is told by Reid (1970, Ch. V); the citation of Gordan is on page 34. McLarty (2012) gives a detailed account on the story.

"That's not Mathematics. That's Theology." The use of non-constructive methods was also contrary to the philosophical standpoint of Kronecker, one of the leading Mathematicians at that time. Kronecker was already involved in a philosophical debate with Dedekind on the use of abstract methods in Mathematics, and was also famous as an opponent of Cantor's set theory. Hilbert was clearly aware of these debates, and Reid (1970, p. 31) reports that he was very much impressed by a personal discussion with Kronecker, giving him 4 pages in his notebook, while any other mathematician only had at most one page.

The second contribution was his new axiomatization of Euclidean Geometry (Hilbert 1899). This work has, in the first place, a mathematical objective: finding a new axiomatization of Euclidean Geometry which is more appropriate with respect to modern developments (including the discovery of Non-Euclidean Geometry); he highlights, in particular, Desargues's Theorem and Pascal's (Pappus's) Theorem, and the way they are derivable in the new axiomatization.[8] Conceptionally it also builds on prior work by Pasch.

Both issues, non-constructive proofs and axiomatizations, started to merge when Hilbert raised the question of consistency of axiom systems. As Geometry can be reduced to a theory of the real numbers, the consistency problem is stated for arithmetic (including analysis)—quite prominently—in the second of his famous 23 problems given at the ICM in Paris in 1900. It was Hilbert's hope that consistency proofs would not only secure mathematical reasoning, but also provide proper meaning to non-constructive features of proofs.

Hilbert's Foundational Programme

Hilbert's Foundational Programme is often caricatured as a naive formalist endeavor to prove the consistency of formal systems by finitistic means. While, of course, consistency was its driving force, and formalistic elements occupied at some stage a prominent place, the full history of the programme is much more subtle and involve various philosophical viewpoints.

Hilbert aimed to secure usual mathematical reasoning. Being aware of the failure of Frege's *Grundgesetze*, it was Logic which attracted Hilbert's attention first. In the 1910s he submitted Whitehead and Russell's *Principia* to a thorough evaluation (see Kahle 2013). By this time, one finds evidence that could allow to count Hilbert as a Logicist. However, in 1920 at latest, Hilbert came to the conviction that Russellian Logicism fails because of the problematic status of the reducibility axiom.

By that time, Hilbert's own approach, his *Beweistheorie*, already took shape. It was first sketched in his contribution to the ICM 1904 in Heidelberg, but put aside for a while, probably because of criticism by Poincaré (see Kahle 2014). Apparently, it was Brouwer who gave him the idea to overcome this criticism by distinguishing between induction on the object and on the meta level (as we would call it now).[9] With this tool at hand, he proposed the search for consistency proofs in a weak meta theory for a formalized strong object theory. While there were some promising first

[8] A detailed account to the development is given in (Toepell 1999).

[9] See (Brouwer 1927, Footnote 1).

results by Ackermann and von Neumann, Gödel's incompleteness theorems show the unfeasibility of this programme in its original form.

This negative result, however, did not stop Hilbert. In contrast to Frege with respect to his logicistic programme, he did not give up hope, but simply proposed a change of the underlying philosophical framework.[10] And, as soon as 1936, Gentzen was able to establish a consistency proof for Arithmetic which is in line with the new foundational standpoint. In fact, proof theory continues to pursue this *revised Hilbert programme* to this day (Kahle 2015).

Was Hilbert a Philosopher?

We opened this section with the disappointing report from Nelson to Hessenberg about Hilbert's "silliness" in philosophical questions. Could such a judgement be justified in view of Hilbert's importance in the Philosophy of Mathematics? We think that, indeed, it can be.

Based on textual evidence, there are two motivations for Hilbert's foundational interests: the set-theoretical paradoxes and his opposition to the *Ignorabimus* of Bois-Reymond (see, e.g., Tapp 2013).

It is important to note that, as much as the paradoxes are concerned, Hilbert's interest was mathematically motivated, not philosophically motivated. He was quite explicit that the paradoxes should be resolved by mathematical means and he aimed to "eliminate once and for all the questions regarding the foundations of mathematics" (Hilbert 1967, p. 464). At best, we have here a "meta-philosophical" position, which, in fact, tries to "overcome Philosophy" by mathematical means.[11]

When it comes to Hilbert's epistemological optimism, he exposes, however, a genuine philosophical position. Although Hilbert tried to use Mathematics (and Meta-Mathematics) to underpin this optimism, a proper defense of it would need to be provided on philosophical grounds.[12]

Let us now compare Hilbert with the criteria which we have worked out in the previous section.

The first observation is that his texts lack a proper respect for the philosophical tradition. Of course, he might have read Plato and/or Aristotle (we have no evidence for or against it, but is might well have been part of his high school education); he

---

[10]"Besonders interessiert hat mich der neue meta-mathematische Standpunkt, den Sie jetzt einnehmen und der durch die Gödelsche Arbeit veranlaßt worden ist." ("I was particularly interested in the new meta-mathematical standpoint which you now adopt and which was provoked by Gödel's work.") Ackermann in a letter to Hilbert, August 23rd, 1933 (Ackermann 1933, p. 1f).

[11]Gentzen expressed it in these words: (Gentzen 1938, p. 237 in the english translation)

> A foremost characteristic of *Hilbert's* point of view seems to me to be the endeavour to withdraw the problem of the foundations of mathematics from *philosophy* and to tackle it as far as in any way possible with methods proper to mathematics.

[12]This was, in fact, one of the central points of origin of Brouwer's criticism (see Brouwer 1927, Third insight).

read some work of Kant, as Reid reports (1970, p. 194): "[...] he had smilingly commented to a young relative that a lot of what Kant had said was 'pure nonsense' ".[13] And Maurice Janet, who was as a student in Göttingen in 1912, recorded Hilbert's contempt of Philosophy by the following quote[14]: "[Hilbert] has uttered the thought that he would be happy if all libraries in the world burned down, 'the mathematicians alone could reconstruct the mathematics, the philosophers would be quite embarrassed.' " In his writings there are essentially no scholarly references to Philosophy. One even has to be careful with his later publications as they are, in a good part, influenced by Paul Bernays, who, indeed, had a firm philosophical background.[15]

Even in the discussions on the Foundations of Mathematics with Brouwer, he was not taking proper notice of his opponent's work. As Smoryński (1994) argues, Hilbert was rather responding to the ghost of Kronecker or to Weyl than to Brouwer. And even more, his arguments are not carefully pondered philosophical reasons but sometimes just polemic statements that could hardly stand up to scientific standards. An extreme case is reported by Reid (1970, p. 184): " 'With your methods,' [Hilbert] said to Brouwer, 'most of the results of modern mathematics would have to be abandoned, and to me the important thing is not to get fewer results but to get more results.' "

Thus, even for his contributions to the Philosophy of Mathematics—which are, of course, tremendous—one cannot claim that Hilbert was properly involved in it on philosophical ground.

Insofar as a philosophical conviction is concerned, let us assume, for the sake of the argument, that Hilbert could be judged as a Finitist and Formalist. But these positions could, on no account, be considered as a primary philosophical standpoint which "goes ahead" of the mathematical results. It is known that for a certain period preceding Formalism, Hilbert had a quite strong interest in Logicism. But even at that time he was able to astonish philosophers by asking for a *reform of logic*: "To eliminate the contradiction in set theory, [Hilbert] wants to reform (not set theory but) logic."[16] His attitude towards *inhaltliche Mathematik* (contentual Mathematics) can well be described in terms of Platonism.[17] We are, in fact, confronted with a mixup

---

[13]For the more positive evaluation of the influence of Kant on Hilbert, see Sinaceur (2018) in this volume.

[14]French original from Janet's notebook: "[Hilbert] a émis l'idée qu'il serait heureux que toutes les bibliothéques du monde brûlassent, ≪les mathématiciens seuls pourraient reconstruire les mathématiques, les philosophes seraient bien embarrassés≫." (Mazliak 2013, p. 55).

[15]This is explicit, for instance, in (Hilbert 1967, p. 479): "I would like to note further that P. Bernays has again been my faithful collaborator. He has not only constantly aided me by giving advice but also contributed ideas of his own and new points of view, so that I would like to call this our common work." It is also known that the opus magnum of Hilbert and Bernays, the *Grundlagen der Mathematik* (Hilbert and Bernays 1934; 1939), was essentially entirely written by Bernays.

[16]"Um den Widerspruch in der Mengenlehre zu beseitigen, will er [Hilbert] (nicht etwa die Mengenlehre sondern) die Logik reformieren." Nelson in a letter to Hessenberg, June 1905, cited in (Peckhaus 1990, p. 166).

[17]See the discussion of this point by Bernays (1935).

of philosophical positions. And, after Gödel's results,[18] Hilbert had no problem at all with abandoning the original Finitism and promoting transfinite methods in Meta-Mathematics.[19]

As another example let us mention the apparent centrality of *consistency* in Hilbert's foundational programme. Kreisel (2011, p. 43) reports the following anecdote: "According to Bernays [...] Hilbert was asked [...] if his claims for the ideal of consistency should be taken literally. In his (then) usual style, he laughed and quipped that the claims serve only to attract the attention of mathematicians to the potential of proof theory."

Thus, in conclusion, Hilbert hardly fulfills any of the criteria we have described for a Mathematician being a Philosopher. But having said this, there is something which distinguishes him among other Mathematicians (and Philosophers of Mathematics): he had a deep *philosophical vision*, namely to solve philosophical questions in Mathematics by mathematical means. One may even call it a *Meta-Philosophy*. And it is due to this vision—and due to the success of this vision—that there can be no doubt that Hilbert would have been more than welcome to enter Plato's Academy.

References

Ackermann, W. (1933). Letter to David Hilbert, August 23rd, 1933, Niedersächsische Staats- und Universitätsbibliothek Göttingen, Cod. Ms. D. Hilbert 1.

Bernays, P. (1935). Hilberts Untersuchungen über die Grundlagen der Arithmetik. In *David Hilbert: Gesammelte Abhandlungen* (Hilbert 1935) (Vol. III, pp. 196–216). Berlin: Springer.

Bernays, P. (1954). Zur Beurteilung der Situation in der beweistheoretischen Forschung. *Revue Internationale de Philosophie*, *27–28*(1–2), 1–5.

Bernays, P. (1935). Über den Platonismus in der Mathematik. In *Abhandlungen zur Philosophie der Mathematik* (pp. 62–78). Darmstadt: Wissenschaftliche Buchgesellschaft, 1976 (German translation of a talk given in 1934 and published in French in 1935).

Bourbaki, N. (1994). *Elements of the History of Mathematics*. Berlin: Springer.

Brouwer, L. E. J. (1927). Intuitionistische Betrachtungen über den Formalismus. *Koninklijke Akademie van wetenschappen te Amsterdam, Proceedings of the section of sciences*, *31*, 374–379. Translation in Part in *Intuitionistic reflections on formalism* (van Heijenoort 1967) pp. 490–492.

Crozet, P. (2018). Avicenna and number theory. In H. Tahiri (Ed.), *The Philosophers and Mathematics* (pp. 67–80). Berlin: Springer.

Diogenes Laertius (1959). *Lives of Eminent Philosophers*. Volume I. Cambridge Mass.: Harvard University Press.

Eliae (1900). In Aristotelis Categorias. Commentarium. In A. Busse (Ed.), *Commentaria in Aristotelem graeca* (Vol. XVIII, Part I, pp. 1-255). Berlin: Georg Reimer.

Ewald, W. (Ed.). (1996). *From Kant to Hilbert* (Vol. 1). Oxford: Clarendon Press.

Gentzen, G. (1938). Die gegenwärtige Lage in der mathematischen Grundlagenforschung. *Forschungen zur Logik und zur Grundlegung der exakten Wissenschaften, Neue Folge*, *4*, 5–18. also in *Deutsche Mathematik*, *3*, 255–268, 1939. English translation in (Gentzen 1969 #7).

---

[18] According to Bernays (1935, p. 215) the first steps towards such a shift took already place before Gödel's result became known.

[19] For the necessary "philosophical switch" see, for instance, Bernays (1954, p. 4), and the letter of Ackermann cited in Footnote 10 above.

Gentzen, G. (1969). *Collected Works*. Amsterdam: North-Holland.
Hilbert, D., & Bernays, P. (1934). *Grundlagen der Mathematik I, volume 40 of Die Grundlehren der mathematischen Wissenschaften in Einzeldarstellungen*. Berlin: Springer. (2nd ed. 1968). Partly translated into English in the bilingual edition (Hilbert and Bernays 2011).
Hilbert, D., & Bernays, P. (1939). *Grundlagen der Mathematik II, volume 50 of Die Grundlehren der mathematischen Wissenschaften in Einzeldarstellungen*. Berlin: Springer (2nd ed. 1970).
Hilbert, D., & Bernays, P. (2011). *Grundlagen der Mathematik I/Foundations of Mathematics I*. Washington, DC: College Publications (Bilingual edition of Prefaces and §§1–2 of Hilbert and Paul Bernays 1934).
Hilbert, D. (1899). Grundlagen der Geometrie. *Festschrift zur Feier der Enthüllung des Gauss-Weber-Denkmals in Göttingen, herausgegeben vom Fest-Comitee* (pp. 1–92). Leipzig: Teubner.
Hilbert, D. (1935). *Gesammelte Abhandlungen, vol. III: Analysis, Grundlagen der Mathematik, Physik, Verschiedenes, Lebensgeschichte*. Berlin: Springer (2nd ed. 1970).
Hilbert, D. (1967). The foundations of mathematics. In J. Heijenoort (Ed.), *From Frege to Gödel: A Source Book in Mathematical Logic, 1879–1931* (pp. 464–479) (van Heijenoort 1967). Cambridge, MA: Harvard University Press.
Jourdain, P. E. B. (1915). Mathematicians and philosophers. *The Monist, 25*(4), 633–638.
Kahle, R. (2013). David Hilbert and the Principia Mathematica. In N. Griffin & B. Linsky (Eds.), *The Palgrave Centenary Companion to Principia Mathematica* (pp. 21–34). London: Palgrave Macmillan.
Kahle, R. (2014). Poincaré in Göttingen. In M. de Paz & R. DiSalle (Eds.), *Poincaré, Philosopher of Science, volume 79 of The Western Ontario Series in Philosophy of Science* (pp. 83–99). Berlin: Springer.
Kahle, R. (2015). Gentzen's consistency proof in context. In R. Kahle & M. Rathjen (Eds.), *Gentzen's Centenary* (pp. 3–24). Berlin: Springer.
Korselt, A. (1903). Über die Grundlagen der Geometrie. *Jahresbericht der Deutschen Mathematiker-Vereinigung, 12*, 402–407.
Kreisel, G. (2011). Logical hygiene, foundations, and abstractions: Diversity among aspects and options. In M. Baaz, et al. (Eds.), *Kurt Gödel and the Foundations of Mathematics* (pp. 27–53). Cambridge, MA: Cambridge University Press.
Mazliak, L. (Ed.) (2013). *La voyage de Maurice Janet á Göttingen*. Les Éditions Materiologiques.
McLarty, C. (2012). Hilbert on theology and its discontents. In A. Doxiadis & B. Mazur (Eds.), *Circles Disturbed* (pp. 105–129). Princeton: Princeton University Press.
Peckhaus, V. (1990). *Hilbertprogramm und Kritische Philosophie, volume 7 of Studien zur Wissenschafts-, Sozial- und Bildungsgeschichte der Mathematik*. Vandenhoeck & Ruprecht, Göttingen.
Peters, C. A. F. (Ed) (1862). *Briefwechsel zwischen C. F. Gauss und H. C. Schumacher*, vol. 4. Gustav Esch, Altona.
Philoponus, J. (1897). Ioannis Philoponi in Aristotelis De anima libros commentaria. In M. Heyduck (Ed.), *Commentaria in Aristotelem graeca* (Vol. XV). Berlin: Georg Reimer.
von Plato, J. (2016). In search of the roots of formal computation. In F. Gadducci, M. Tavosanis (Eds.), *History and Philosophy of Computing. HaPoC 2015, volume 487 of IFIP Advances in Information and Communication Technology* (pp. 300–320). Berlin: Springer.
Rashed, R. (2008). The philosophy of mathematics. In S. Rahman, T. Street, & H. Tahiri (Eds.), *The Unity of Science in the Arabic Tradition, volume 11 of Logic, Epistemology, and the Unity of Scienced* (pp. 153–182). Berlin: Springer.
Rashed, R. (2018). Avicenna: Mathematics and philosophy. In H. Tahiri (Ed.), *The Philosophers and Mathematics* (pp. 67–80). Berlin: Springer.
Reid, C. (1970). *Hilbert*. Berlin: Springer.
Sinaceur, H. B. (2018). Scientific Philosophy and Philosophical Science. In H. Tahiri (Ed.), *The Philosophers and Mathematics* (pp. 25–66). Berlin: Springer.
Smoryński, C. (1994). Review of Brouwer's Intutionism by W. P. van Stigt. *The American Mathematical Monthly, 101*(8), 799–802.

Tapp, C. (2005). *Kardinalität und Kardinäle, volume 53 of Boethius*. Stuttgart: Franz Steiner.

Tapp, C. (2013). *An den Grenzen des Endlichen. Mathematik im Kontext*. Berlin: Springer.

Toepell, M. (1999). Zur Entstehung und Weiterentwicklung von David Hilberts "Grundlagen der Geometrie". In *David Hilbert: Grundlagen der Geometrie* (14th ed., pp. 283–324). Stuttgart: Teubner.

van Dalen, D. (2013). *L.E.J. Brouwer*. Berlin: Springer.

van Heijenoort, J. (Ed.). (1967). *From Frege to Gödel: A Source Book in Mathematical Logic, 1879–1931*. Cambridge, MA: Harvard University Press.

Volk, O. (1925). Kant und die Mathematik. In *Mathematik und Erkenntnis* (pp. 74–77). Königshausen & Neumann (German translation of a paper originally written in Lithuanian and published in *Kosmos*, vol. 6, pp. 320–323

Chapter 7
Enthymemathical Proofs and Canonical Proofs in Euclid's Plane Geometry

Abel Lassalle-Casanave and Marco Panza

Abstract Since the application of Postulate I.2 in Euclid's *Elements* is not uniform, one could wonder in what way should it be applied in Euclid's plane geometry. Besides legitimizing questions like this from the perspective of a philosophy of mathematical practice, we sketch a general perspective of conceptual analysis of mathematical texts, which involves an extended notion of mathematical theory as system of authorizations, and an audience-dependent notion of proof.

7.1 Introduction

By 'Euclid's plane geometry' we will refer here to the theory presented in the first six books of *The Elements*. In those books, the application of Postulate I.2 is not uniform. This simple observation, which will be justified in Sect. 7.2, poses the following methodological question: Is that lack of uniformity to be accepted as a historical fact established on the base of textual evidence? Or does it make sense to ask oneself which of the different ways of application is the one that conform best to Euclid's plane geometry? In order to answer this question in favour of the

A slightly modified version of this paper, written in Spanish was published in *Revista Latinoamericana de Filosofía*, XLI (2), 2015. We are grateful to Oscar Esquisabel for his remarks on a preliminary version, and to Alicia Di Paolo for her valuable and indispensable linguistic help. This joint work was supported by two subsidies: CAPES/COFECUB (Number 813-14) and PDE/CNPq (Number 2200980/2015-7). The second author dedicates this paper to Roshdi, his master and friend.

A. Lassalle-Casanave (✉)
UFBA, CNPq, Salvador, Brazil
e-mail: abel.lassalle@gmail.com

M. Panza
IHPST-CNRS, Université Paris 1, Paris, France
e-mail: panzam10@gmail.com

M. Panza
Chapman University, Orange (CA), USA
e-mail: panza@chapman.edu

H. Tahiri (ed.), *The Philosophers and Mathematics*, Logic, Epistemology, and the Unity of Science 43, https://doi.org/10.1007/978-3-319-93733-5_7

second option, in Sect. 7.3 we suggest identifying a mathematical theory with a system of rules of authorization, and we distinguish between canonical proofs and enthymemathical proofs of a given theory. Finally, in Sect. 7.4, we conclude with a more general methodological discussion about the kind of reconstruction of theories that our approach implies.

7.2 A Case Study: The Application of Postulate 1.2 in *The Elements*

Euclid formulates the first three postulates of the *Elements* in this way[1]:

1. To draw a straight-line segment from any point to any point.
2. To produce a straight-line segment continuously in a straight line.
3. To describe a circle with any centre and distance.

The three postulates are all applied in the solution of the first two propositions in the *Elements*, which are problems[2]:

Prop. I.1. On a given segment to construct an equilateral triangle.

---

[1]As usual, we quote definitions, postulates, common notions and propositions of the *Elements* in Arabic characters, preceded by the book number in Roman characters, according to Heiberg's edition (Euclid, *Elementa*). We have taken into account the translation by Thomas L. Heath (Euclides *EEH*), but, we have also granted ourselves the freedom to make some changes in order to use, from our point of view, a more appropriate terminology. The most relevant cases of these changes are related to the adjective 'εὐθεῖα'. After using it as a genuine adjective in Definition 1.4 and in Postulate I.1 to coin the expression 'εὐθεῖα γραμμή', Euclid starts using it as a nominalised adjective as well. In particular, in Postulate I.2, he uses the adjective 'πεπερασμένη' to coin the expression 'πεπερασμένη εὐθεῖα'. Literally, 'εὐθεῖα γραμμή', as well as 'εὐθεῖα', when used as a nominalised adjective, should undoubtedly be translated as 'straight line', and 'πεπερασμένη εὐθεῖα' as 'finite straight line'. Still, it seems clear that in the three cases Euclid's intended reference is to straight-line segments. As this is crucial, to avoid any misunderstanding, we shall directly translate the three terms as 'straight-line segment', or only 'segment' when no confusion is possible.

[2]As usual, we distinguish the propositions in the *Elements* in problems and theorems. A problem demands the construction of a geometrical object (a point, a straight, an angle, a figure, etc.) under certain conditions, and requires a solution. A theorem states that geometrical objects of a given kind have certain properties, or that they are in a certain relation, and requires a proof. However, it is important to point out that the solution of a problem always includes, at its final stage, a proof that the object constructed in the established way satisfies the conditions of the problem. This suggests understanding the solution of a problem as the proof of a (meta-)theorem which states that the problem has a solution (which can be obtained as the solution shows). In spite of the fact that, in the versions of the *Elements* that we have received, these two kinds of propositions are not explicitly distinguished, the distinction is made perfectly clear by the fact that Euclid adopts two different canonical forms in the two cases, both for the formulation of the proposition itself and for the subsequent arguments (solutions and proofs). The classic *locus* in which the distinction between problems and theorems is explicitly drawn and widely discussed is Proclus' commentary on the first book of the *Elements* (Proclus *CEELF*, esp. pp. 77–78).

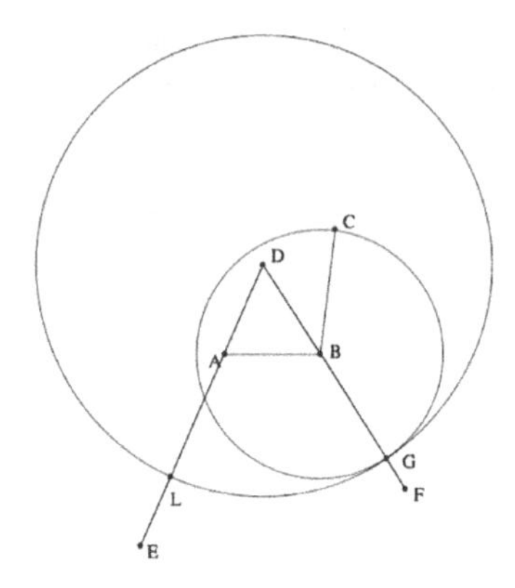

Fig. 7.1 Diagram associated with the solution of Proposition I.2

Prop. I.2. To place at a given point (as an extremity) a segment equal to a given segment.

But, while in the solution of Proposition I.1, Postulates I.1 and I.3 are applied in an easily intelligible way; in the solution of Proposition I.2 we find a less intelligible sequence of applications of Postulates I.2 and I.3 which deserves attention.

The solution to this problem is simple (Fig. 7.1): given a segment *BC* and a point *A*, Euclid constructs the equilateral triangle *ABD* on *AB*, according to Proposition I.1; and then, by Postulate I.2, segments *DB* and *DA* are produced up to two arbitrary points *F* and *E* far enough from *B* and *A*. Drawing these points far enough from *B* and *A* means that, when describing a circle with centre *B* and distance *BC* according to Postulate I.3, that circle cuts *DF* at a point, say *G*, and then, when describing a circle with centre *D* and distance *DG*, again according to Postulate 1.3, that another circle cuts *DE* at another point, say *L*. It is easy to prove that *AL* is equal to *BC*, so that its construction solves the problem.

What deserves attention is the indeterminacy of the procedure which results from applying Postulate I.2 first, and then Postulate I.3: how is it possible to know where segments *DB* and *DA* should be produced up to? Why does Euclid not apply Postulate I.3 first for drawing a circle of radius *BC* with centre *B*, so as to produce the segment *DB* up to it, and so construct *G*? And then, why is Postulate I.3 not applied again for drawing a circle with centre *D* and radius DG, so as to produce the segment *DA* up to it, and so construct *L* (Fig. 7.2)? How can Euclid's procedure, which is, so as to say, not genuinely constructive, be justified?

Certainly, the formulation of Postulate I.2 in itself licenses the two applications.[3] The problem lies in the fact that unlike as it happens with Postulates I.1 and I.3, the

[3]Our exposition of Euclid's construction with the corresponding figure is based on Heiberg's edition (Euclid, *Elementa*). The topological configuration of the diagram would be modified if the position of point *A*, in relation to segment *BC*, were different from the one that is presented here. As Proclus also observed in his commentary (Proclus *CEELF*, pp. 225–228), there would have to distinguish different cases (as it also happens with many other solutions or proofs of Euclid's propositions). In spite of the fact that the difficulty presented here would not appear in all the cases, it is not necessary to consider them (for that purpose, see Euclid *EEH*, pp. 245–246).

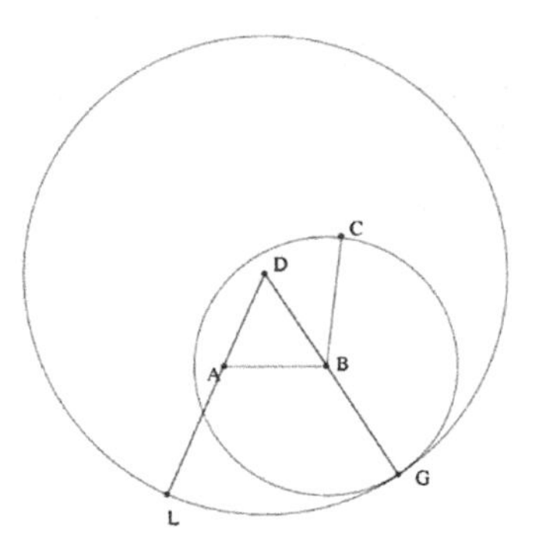

Fig. 7.2 Diagram associated with the alternative solution of Proposition I.2

output of Postulate I.2 is not univocally determined by its input. To be so, the segment would have to be produced up to meet a line, according to the second possibility. But this is not what the Postulate allows explicitly, since its formulation does not specify up to which point the segment would have to be, or could be produced. Then, even if it were applied in view of the second possibility, what would make the output univocal would not be the Postulate by itself, but something that is constructed independently of its application, that is to say, in our alternative reconstruction, the already constructed circles with centres D and B, respectively. We have referred to the first procedure as not genuinely constructive to highlight the fact that, by following it, there is no way to determine up to where the given segments should be produced in order to obtain points F and E; all what can be made is choosing arbitrarily where (or when) to stop producing them, taking into account the fact that it should not be too close (or soon), so as not to allow the construction as it is indicated.

This attitude does not seem to be consistent with the constructive rigor which characterizes most of Euclid's arguments. However, if we trust in the different versions of Euclid's text that we have received, and, particularly, in Heiberg's reconstitution, it is a fact that in the solution of Proposition I.2, Euclid proceeds according to the first possibility and not to the second one. But it is a fact, too, that Euclid does not always proceed explicitly in this way in similar situations.

For example, let us consider Proposition I.16 (this time, a theorem).

Prop. I.16. In any triangle, if one of the sides be produced, the exterior angle is greater than either of the interior and opposite angles.

In the proof, once the middle point E of the side AC of a given triangle ABC is constructed (Fig. 7.3), and the segment BE is drawn, Euclid produces it up to F making EF equal to BE.

Euclid says nothing about how point F is constructed, but he just requests BE to be produced up to F in such a way that EF is equal to BE. What procedure should be followed to carry out that construction? Should BE be produced far enough up to a point Z so as to then describe a circle with centre E and distance BE cutting BE in F? Or, should the circle with centre E and distance BE be described first, so as to then produce BE to meet it in F?

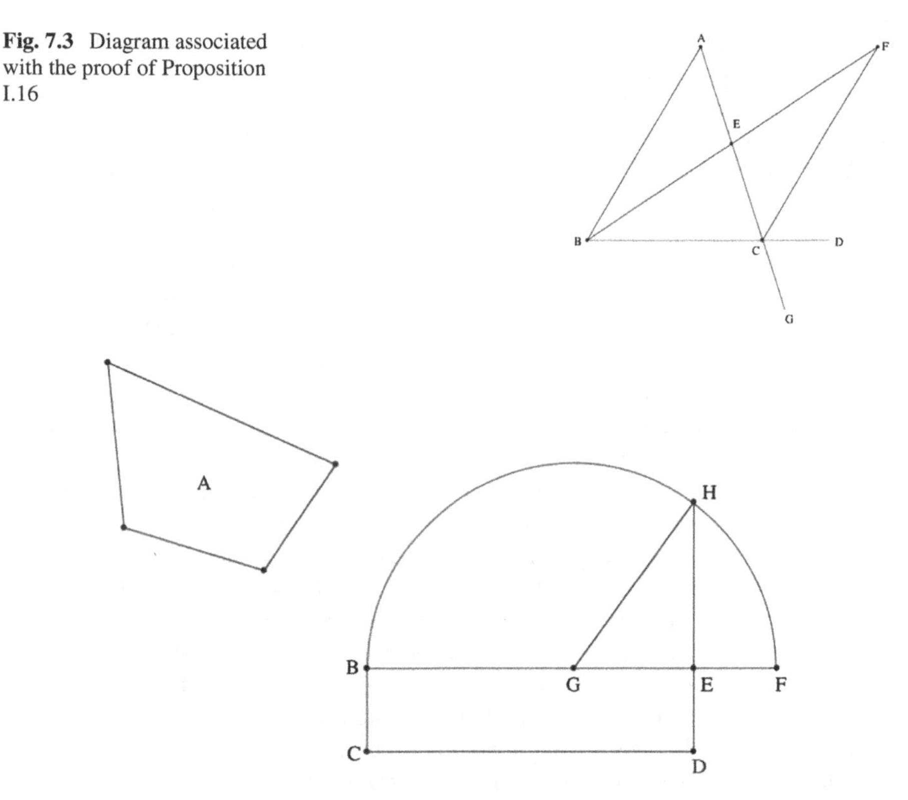

Fig. 7.3 Diagram associated with the proof of Proposition I.16

Fig. 7.4 Diagram associated with the solution of Proposition II.14

There is at least one case in which Euclid follows explicitly the second possibility. It is in the solution of Proposition II.14 (another problem, the only one in Book II):

Prop. II.14. To construct a square equal to a given rectilinear figure.

In the solution, after constructing the rectangle *BCDE* equal to the given figure *A* (Fig. 7.4), according to Proposition I.45, and having produced *BE* up to *F* making *EF* equal to *ED*, as in the proof of Proposition I.16, Euclid constructs, according to Proposition I.10, the middle point *G* of *BF*, describes a semicircle with centre *G* and distance *BF* and produces the segment *DE* up to meet this semicircle in *H*, so as to then prove that the square constructed on *EH* is equal to the rectangle *BCDE* and, in consequence, equal to figure *A*, which is what it was required to do.

In a previous paper,[4] we examined these discrepancies in Euclid's exposition, and we suggested a possible explanation of them. In a few words, we observed that the solution of Proposition II.14 does not depend on the fact that the searched for segment (the side of a square equal to *A*) let be the extension of a side of a rectangle

[4]Lassalle Casanave and Panza (2012).

equal to A, since the only relevant thing is that this segment is a cathetus of a right-angled triangle of which the other cathetus is equal to GE and the hypotenuse equal to GF.[5] From this, we inferred that Euclid allowed himself to apply Postulate I.2 as in Proposition I.2 when the solution depends on the fact that the relevant segment is got by producing a given segment in its position, but he constrained himself to apply it as in Proposition II.14 when it is not so.

In that same paper, we also observed that a way to understand Postulate I.2, which would eliminate the problem of lack of univocity of the output, would be to simply assume that this postulate allows to produce a given segment into a(n infinite) straight or half straight-line: in the first case, the output (the straight line) would be univocally determined by the given segment; in the second case, the output (the half straight-line) would be univocally determined by the given segment and the choice of the extreme as from which it extends. It could also be thought that this is the way in which Euclid conceives the postulate, though, for simplicity, he replaces the straight- or half straight-line by a long enough segment. If this conclusion were correct, it would imply that the role of straight and half straight-lines in the *Elements* would be much more important than what it is commonly thought to be. In effect, according to this interpretation, which could be called 'infinitist', Postulate I.2 would involve in itself straight- or half straight-lines, so that these would appear in all the constructions in which this postulate is applied, instead of appearing only when trying to prevent the proliferation of cases, as it happens in the solutions of Propositions I.12 and I.22.

According to the infinitist interpretation, Postulate I.2 would have by itself, as well as Postulates I.1 and I.3, an univocal output determined by its input. On the contrary, if we assume that this postulate does not let us construct a straight or half straight-line, but a segment, the only way to guarantee its output to be univocal would be to set out as condition of its application that there is another segment or circle up to which the segment can be produced. However, in the latest case, as it has already been observed, the postulate would have an univocal output only dependent on the context of its application and not by itself. Finally, it could also be simply admitted that the output of the postulate is not univocal, which would agree with conceiving it as such to allow producing a given segment far enough, as it seems to be the case in the solution of Proposition I.2.

Therefore, the problem lies on the interpretation of Postulate I.2. How should this postulate be understood? Is there a unique way of understanding it which conforms to all its applications, or does Euclid allow himself to understand it differently depending on each of them? In our previous work, we considered these questions from the point of view of the interpretation of Euclid's text. Presently, we would like to change our attitude and aim at a more general problem, which our case study of Postulate I.2 allows us to set as an example.

It seems natural to consider the *Elements* not only as a collection of solved problems and proved theorems, but also as a mathematical treatise whose main objective is to set forth a theory and to teach a mathematical practice according to (or, in better words, within) this theory. Thus, for instance, the first six books of the *Elements*

[5]Ibid., p. 113.

would set forth a theory of plane geometry and would teach to practice it according to (or, in better words, within) that theory.[6] From this point of view, the question would then be the following one: how would we have to proceed, when applying Postulate I.2, if we would like to do plane geometry according to (or within) Euclid's theory? In other words: how would we have to understand this postulate to explain it to someone who would like to learn this theory? The same questions, *mutatis mutandis*, could be asked in relation to many other mathematical texts of the past, to the theories which those texts expound, and to the different aspects of those theories which, as in the case of Euclid's Postulate I.2, are not made perfectly clear in them.

Now, in relation to these issues in general, as well as in relation to the previous questions about Postulate I.2 in particular, it can be thought that the historian would have done all he could and had to do if she/he had checked the whole content of the relevant text. Thus, in the case of Euclid's Postulate I.2, the historian could only observe that the postulate is applied in one way in the solution of Proposition I.2 and in a different way in the solution of Proposition II.14, and that, in many other cases, as in the proof of Proposition I.16, the way in which it is applied is indeterminate. Consequently, there would not be anything else to ask.

From another point of view, it could also be alleged, in a way which could be considered as complementary to the previous one, that many mathematical texts of the past cannot be considered as expositions of genuine theories, precisely because of the fact that they are ambiguous in many aspects, which can only be avoided by means of an appropriate formalization which those texts are far from presenting. If it were so, questions as the previous ones, which aim at identifying a uniform procedure which those texts do not set, would have no more sense.

In this apparent dead-end street, which would oblige to declare our questions as illegitimate, we would like to claim for the philosopher of mathematical practice, which flows between the Scylla of logical analysis and the Charybdis of the history of mathematics, the possibility of legitimizing and attempting to answer them. This is the aim we try to achieve in the next section.

7.3 Enthymemathical and Canonical Arguments

Our main idea is that the documents which are the habitual sources of the historian of mathematics, not only allow an analysis which extrapolates a sequence of acts, ideas or results, what historians teach us to do in many different ways in their daily work, but they also allow an analysis which extrapolates a system of mathematical theories, to which it is perfectly plausible to give their own life, relatively independent

[6] If the magnitudes which Book V are about are simply the same as the ones which Books I–IV deal with, that is to say, straight-line segments, polygons and rectilinear angles, Book V would be part of the exposition of plane geometry. It is, however, a quite complex question whether this is so, or Book V is rather intended to deal with whatever possible magnitudes (also if no mention of them is made in the foregoing Books), or, even, whether it is intended to deal with magnitudes in general, implicitly defined within the theory of proportions itself. We cannot consider this issue here.

of the texts which originally present them, as it is frequently done with the current mathematical theories (which are taught, in most of the cases, in a way which is absolutely independent of the consideration of its historical origin).

This idea is based on a conception of mathematical theories according to which a theory is identified with a space of possibilities of argumentation, with a reasonably clear system of authorizations or, by analogy with legal systems, with a set of power-conferring rules,[7] within which a practice is developed. This practice could be simply described as the very activity which mathematics consists of. But this extrapolation is neither a historical, nor a logical reconstruction; it is rather a reconstruction of systems of possibilities for the elaboration of acceptable arguments.

Let us take our example. The solution of Proposition I.2 by Euclid can be thought of as an argument which agrees with the rules of authorization of Euclid's plane geometry, thought as a theory in the foregoing sense. To set another example, the same can be said about a proof within any formal theory. Instead of considering a formal theory as a system determined by its language and its axioms and/or inference rules, we can consider it as a space of possible inferences from some formulae of that language to another formula of that same language. Thus, a proof within a formal theory could be thought of not as much as a sequence of formulae, but as the actualization of some of those possibilities.[8]

This observation should be enough to understand that, from the point of view we are presenting, there is no difference between a clearly informal theory (at least for the current logical criteria), such as Euclid's plane geometry, and a modern formal theory (according to the same criteria). Simply, these possibilities, or rather, the corresponding rules of authorization, present different forms in the two cases. The second case is about the possibilities of writing formulae, and the main competence required in order to allow adopting the corresponding rules is the capacity to differentiate and recognize tokens of inscription-types. The first case is about the possibilities of generating objects represented by diagrams and of attributing properties and relations to these objects, and the competences required in order to allow adopting the corresponding rules include, among other things, the capacity of appropriately using diagrams in the justification for the attribution of those properties and relations.

In both cases, there are also different ways of raising the relevant possibilities. In a formal theory these possibilities are raised in such a way that, in principle, they are immediately clear for an epistemic subject with the appropriate intellectual and cognitive competences: in principle it should be always clear for that subject if the

[7]For a brief presentation of this idea, see Lassalle Casanave (2006).

[8]This does not mean denying that a formal proof is a sequence of formulae, but conceiving that sequence as the actualization of those possibilities. From this point of view, it is the coding of the space of possibilities which a mathematical theory consists of what determines what an acceptable argument in this theory is (or establishes, at least, the conditions that it has to comply with or the criteria under which it can be recognized as such). And it is also this coding which determines which of these acceptable arguments can be considered as proofs in the theory (or establishes, at least, the relevant conditions or criteria).

writing of a certain formula in a certain situation is allowed or not.[9] On the contrary, in Euclid's plane geometry, as well as in many other important mathematical theories, it is not the same: in some cases it may not be clear, even for such a subject, if something is allowed or not.[10]

The case of the application of Postulate I.2 in the solution of Proposition I.2 is one of those cases. It could be said that if Euclid's text describes the solution of the problem in such a way that it includes the production of segments DB and DA up to two arbitrary points F and E, far enough from B and from A, then the possibility of producing a given segment far enough has to be included among those granted by the theory that the text is presenting. Although this is certainly a plausible attitude, there is another one equally plausible.

In order to approach it, let us think for a moment of the way in which a proof is exposed within a formal theory on very many occasions (it could be said it is so in most cases, except in didactical situations): all the formulae of the sequence of which the formal proof consists are not written down; on the contrary, enough instructions are given to write these formulae down. More precisely, to say that these instructions are enough means that an epistemic subject with the appropriate intellectual and cognitive competences (having enough time, pencils, paper and patience) would be able to write down the formulae in question.

Then, would it not be possible to think that Euclid's text works in the same way, too? And that this text, instead of presenting the solution of the problem in the allowed standard way within the theory in question, merely provides enough instructions so as to allow an epistemic subject with the appropriate competences, including the knowledge of the permitted possibilities, to obtain a solution in the allowed standard way? The set of those instructions (as much to one case as to the other, and to all the cases in which a similar situation is repeated) could be then thought of as an enthymemathical argument (a proof in the case of I.16, a solution as in that of I.2 or in II.14), but not in the sense of an argument which lacks some premises or other crucial components of it (in which case there would not be a valid argument, and, in particular, certainly not a proof) but in the aforementioned sense of an argument which is not required to make all its components explicit when expounded for an epistemic subject with the appropriate competences.[11] The detailed argument that

---

[9]The specification 'in principle' is used to set aside the cases in which such possibility is excluded for physical reasons, for example, if the formula is too long and complex for its structure to be identified by a subject with human cognitive limits or if it is written with ink which is too pale, etc.

[10]Evidently, there are also other important differences between the two cases. A very important one is related to the way in which the actualization of a possibility is associated to obtaining a piece of information by an epistemic subject who is operating within a theory.

[11]The notion of enthymeme is a crucial notion in Aristotle's rhetoric (*Reth* 1356b). According to Rapp (2010), for Aristotle an enthymeme is "what has the function of a proof or demonstration in the domain of public speech". If we agree with this, our use of this notion to refer to a type of mathematical argument conveys the idea that such an argument, for example a mathematical proof, is a kind of public speech, which adapts perfectly well to what we will hold further on in section IV. However, the idea of being able to speak of enthymeme in mathematics, or, in general, of enthymemathical proofs or any other sort of conclusive arguments (as Euclid's solutions of problems should be taken to be), could disconcert those who see an enthymeme as a fallacy, and

these instructions allow to obtain according to the possibilities granted by the theory would be, on the contrary, a canonical argument within this theory.[12] From this point of view, the fact that Euclid's text describes the solution of a problem or the proof of a theorem as it does, does not imply that a solution or a proof that complies with that description is the canonical one. It could be the enthymemathical, rather.[13]

For a better understanding of our perspective, let us examine the reconstruction of Euclid's plane geometry in Avigad et al. (2009, pp. 736–737). They reconstruct the solution of Proposition I.2 as follows:

Assume L is a line, b and c are distinct points on L, and a is a point distinct from b and c (Fig. 7.5).

> By Proposition I.1 applied to a and b, let d be a point such that d is distinct from a and b, and $\overline{ab} = \overline{bd}$ and $\overline{bd} = \overline{da}$.
>
> Let M be the line through d and a.
>
> Let N be the line through d and b.
>
> Let α the circle with center b passing through c.
>
> Let g the point of intersection of N and α extending the segment from d to b.
>
> Have $\overline{dg} = \overline{db} + \overline{bg}$.
>
> Hence, $\overline{dg} = \overline{da} + \overline{bg}$ [since $\overline{da} = \overline{db}$].
>
> Hence, $\overline{da} < \overline{dg}$.
>
> Let β be the circle with center d passing through g.

---

a proof or conclusive argument, in particular a mathematical one, as a valid argument. According to the perspective on mathematical theories developed here and the corresponding conception of arguments and proofs within such theories, both assumptions are questionable, or, at least, need qualifications. We would not like to take time and space here to argue that an enthymemathical proof or conclusive argument have to be thought of as a proof or a conclusive argument in a genuine sense, rather than as an argument which, despite being acceptable, has not succeeded in becoming valid. For a schematic presentation of the notion of enthymemathical proof, see Lassalle Casanave (2008).

[12]This means that in order to code a theory, in the proposed sense of system of authorizations, certain conditions that a canonical proof within this theory has to respect have to be coded.

[13]Although the idea of viewing a mathematical proof as an enthymemathical one, at least *prima facie*, is not common, it is far from being totally new. An example that proves so—also in relation to the proofs in the *Elements*—is a quote we thank Paolo Mancosu for having pointed it out to us, taken from a lecture of John Barrow (Lesson VI, given in 1664). There, Barrow arguments in favour of the thesis that mathematical proofs are causal, thesis against which it was alleged to as counter-example the solution of I.1 of the *Elements*. After his argument, in which we are not interested here, Barrow asserts (Barrow 1683, p. 105; Barrow 1734, p. 95; the first reference belongs to the first edition of the original text in Latin, the second one belongs to its first translation into English, which is quoted here):

> From which Observations most or all the Instances brought against Mathematical Demonstrations may be overthrown. And consequently, if the particular Syllogisms (brought either for the Construction, or Demonstration of the above said first Proposition of the Elements) be *Demonstrations* simply *scientific*, the Proposition ought to be reckoned as *scientifically demonstrated*. We will therefore examine them; but for brevity's sake, we will substitute *Enthymems* for Syllogisms, and insinuate the Necessity of the Consequence.

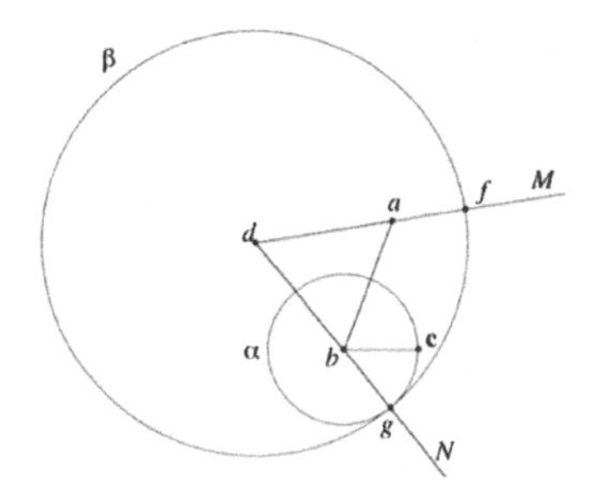

Fig. 7.5 Diagram associated to the reconstruction of the solution of I.2 by Avigad, Dean and Mumma

Hence, a is inside of β y [since d is the center and $\overline{da} < \overline{dg}$].

Let f the intersection of β and M extending the segment from d to a.

Have $\overline{df} = \overline{da} + \overline{af}$.

Have $\overline{df} = \overline{dg}$ [since they are both radii of β].

Hence, $\overline{da} + \overline{af} = \overline{da} + \overline{bg}$.

Hence, $\overline{af} = \overline{bg}$.

Have $\overline{bg} = \overline{bc}$ [since they are both radii of α].

Hence, $\overline{af} = \overline{bc}$.

This is a logically flawless reconstruction. But it is, just that: a reconstruction. And, in fact, it moves away from the original exposition by Euclid in many aspects. One of them is connected to the application of Postulate I.2, in the sense that this reconstruction makes two straight lines take part in it, which are respectively determined by points d, b and d, a,[14] and are given previously to the circles α and β, which cut them in g and f respectively.

The reconstruction of Proposition 1.2 is part of a more global reconstruction, which concerns the whole Euclid's plane geometry, and wants to show, in particular,

[14]The straight line L is not represented in the diagram which Avigad, Dean and Mumma add to their reconstruction (Fig. 7.5). That depends on the fact that the only thing that matters about this line is that two different points b and c are taken on it. It is not essential either that point a is or not on L, and actually, this is not specified. As a matter of fact, this would correspond to the distinction between cases: see n.4. The diagram reproduced in Fig. 7.5 represents a configuration in which point a is not on L. A configuration in which a is on L is represented by the diagram that follows, which shows that the construction of the segment *af* is perfectly feasible under this configuration.

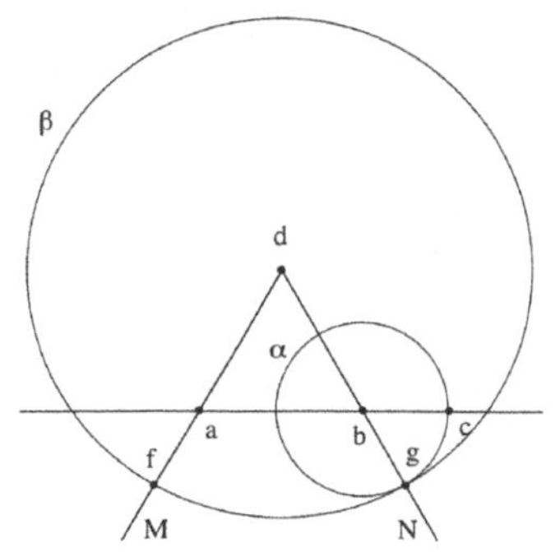

what the role of the diagrams is in that geometry. In fact, the basic idea is to reconstruct the proofs and solutions by Euclid in such a way that the reference to the diagrams is dispensable as justification to any step of them (therefore, their usefulness would simply consist in facilitating the understanding with an illustration of an argument which, as such, is independent of that illustration).[15] The changes that have to be introduced to reach that result, in particular the assumptions that have to be made (some of which appear implicitly in the preceding reconstruction of the solution of Proposition I.2), show the work which, in the original version of these arguments, was carried out by the diagrams.

It is not clear if the introduction of infinite straight lines determined by two given points is part of these changes in the authors's intention; however, that is not relevant to what we are concerned with here. The important thing is that what Avigad, Dean, and Mumma propose can be read as a reconstruction of Euclid's plane geometry, understood as a theory in our former sense, in relation to which the solutions and proofs by Euclid, in their ordinary forms, are conceived of as enthymemathical arguments in the aforementioned sense. To move from those enthymemathical arguments to the canonical ones, we must, among other things, eliminate the diagrams in favour of the application of appropriate assumptions (which Euclid would not explicit), and introduce straight lines (or, more appropriately, half straight-lines) where Euclid, for instance, applies Postulate I.2.

It is important to point out that a reconstruction as that by Avigad, Dean and Mumma is essentially different from another kind of reconstruction which intends to establish Euclid's results (or, more appropriately, a reformulation of those results) within a theory manifestly different from the original one. A very clear example to this is the reconstruction of Euclid's plane geometry within the geometrical theory

---

It is natural to ask oneself why Avigad, Dean and Mumma mention the straight line L in their reconstruction. The answer can be found in the following quote, which well shows the role that they assign to the straight lines in their reconstruction of Euclid's theory, together with their interpretation of Postulate I.2 within this reconstruction (Avigad et al. 2009, pp. 731–732; evidently, we must read "straight line" where Avigad, Dean, and Mumma write "line"):

> [...] in the *Elements*, Euclid takes lines to be line segments, although Postulate 2 ('to produce a finite straight line continuously in a straight line') allows any segment to be extended indefinitely. Distinguishing between finite segments and their extensions to lines makes it clear that at any given point in a proof, the diagrammatic information is limited to a bounded portion of the plane. But, otherwise, little is lost by taking entire lines to be basic objects of the formal system. So where Euclid writes, for example, 'let a and b be points, and extend segment ab to c,' we would write 'let a and b be distinct points, let L be the line through a and b, and let c be a point on L extending the segment from a to b.' Insofar as there is a fairly straightforward translation between Euclid's terminology and ours, we take such differences to be relatively minor.

[15]When presenting their examples of proofs within the formal system E which, according to what they explain in the summary of their article (*ibid.*, p. 700), provides "a faithful model of the proofs in Euclid's *Elements*", Avigad, Dean and Mumma write: "We include diagrams to render the proofs intelligible, but we emphasize that they play no role in the formal system" (*ibid.*, p. 734).

formulated by Hilbert in *Grundlagen der Geometrie.* Although all the theorems and problems in the *Elements* can be proved in the *Grundlagen*, Hilbert's theory neither respects nor intends to respect the nature (in particular, the heterogeneous character) or the structure of the Euclid's arguments.[16]

This distinction between a reconstruction of Euclid's plane geometry within another theory, such as Hilbert's, and reconstructions of Euclid's plane geometry such as the ones formulated by Avigad, Dean, and Mumma, would have to be enough to make it clear that the work that has to be done to obtain a reconstruction of the second type is essentially different from the one that has to be done to obtain reconstructions of the first one. A reconstruction of the first type requires some essentially mathematical work whose main objective is to promote a way of progress in mathematics. A reconstruction of the second type, consists, instead, in a conceptual analysis whose main objective is to promote a way of progress in understanding the way in which mathematics works. In other words, whereas in the first case the progress which is meant to be promoted concerns the understanding of the object itself of mathematics (whatever it is), the progress which is meant to be promoted in the second case concerns the understanding of mathematics as such, that is to say, the way in which it treats its object. Our idea is that there are different forms of mathematics and different ways of doing mathematics. On the one hand, the philosophy of mathematical practice intends to identify these forms and ways. On the other hand, it aims to understand what is common to them, that is to say, what characterizes mathematics as an intellectual activity of a specific nature, beyond these different forms and ways.

To suggest and confront different reconstructions of the second type of one same theory or of different ones just contribute to fulfil this objective. For example, to confront different reconstructions of a same theory means to confront different ways of understanding the same texts, considering they present different systems of rules of authorization. In most cases, it seems difficult, or even impossible, to identify which one among those systems is the good one, or the most adequate to the original texts, and a fortiori, to the authors' intentions. However, there are forms of argumentation in favour of and against a reconstruction instead of another. These are typical of conceptual analysis and are part of the habitual tool box of every philosopher accustomed to it.

To set an example, one could criticize Avigad, Dean and Mumma's reconstruction for it contravenes two essential features of Euclid's arguments: on the one hand, the elimination of diagrams in favour of assumptions of principles that work as supplementary axioms contravenes the heterogeneous character of these arguments;

[16]To respect that structure, still not its heterogeneous character, is, on the contrary, Avigad, Dean, and Mumma's aim. About the heterogeneous character (due to the joint utilization of language and diagrams) of Euclid's proofs, it must be mentioned here the influential "Euclid's Diagram (1995)" by K. Manders (published only thirteen years later: Manders 2008), of which Avigad, Dean and Mumma's reconstruction is in debt to, although (as already observed) the authors set as a goal the logical analysis of those proofs, showing in this way the dispensability of the diagrammatic resources. Although we will not discuss it here, the reconstructions of Euclid's plane geometry in Mumma (2006) and Miller (2007) try to justify the use of diagrams in Euclid's proofs, not to dispense them. For diagrams and geometrical proofs, see also Norman (2006).

on the other hand, the admission of infinite geometrical objects contravenes their local nature. This is not the suitable place to develop those critics, to which Avigad, Dean and Mumma could certainly respond in many ways. It is enough to observe, limiting ourselves again to the case of Postulate I.2, that one of the authors of the present paper has suggested, in the frame of a more general reconstruction of Euclid's plane geometry, another way to account for the function of this postulate, which, on the one hand, attributes a crucial role to the diagrams and, on the other hand, avoids resorting to (infinite) straight or half straight-lines, with no need to accept non-genuinely constructive procedures (Panza 2012). The idea is to account for Postulate I.2 by means of the following rule:

> If a segment is given and the concrete line representing it is such that it can be continued so as to meet a concrete line representing another given segment or a given circle, then the former segment can be produced up to meet this other segment or this circle; hence, if a segment *a* and another appropriate line *b* (either a segment, in turn, or a circle) are given, then the following other objects are susceptible of being given: two other segments, one of which, let us say *c*, extends *a* up to *b*, while the other, let's say *d*, is formed by *a* and *c* taken together; a point on *b* at which both *c* and *d* meet it; two portions of *b* having this last point as a common extremity (either two segments or two arcs of circle).[17]

Euclid's exposition of the solution of Proposition I.2, as we have seen, does not adjust to this rule.[18] But the idea is, precisely, that this exposition presents an enthymemathical argument instead of a canonical one. The reason to give an enthymematical argument in a case as the solution of Proposition I.2, in which such argument does not seem to have any advantage in comparison to the canonical one, could be the following: in this way, Euclid anticipates the exposition of some solutions or proofs for which the enthymemathical argument has evident advantages of simplicity, in comparison to the canonical one, which would imply the construction of many supplementary circles. The solution of Proposition I.22 is a pertinent example.

This proposition requires the construction of a triangle with three sides equal to three given segments, say *A*, *B*, *C*, on condition that any two of those segments taken together are not greater than the remaining one. Let *DE* be a half straight-line. Euclid's solution depends on putting three segments equal to the given ones in sequence on *DE*. Applying Proposition I.3, a segment equal to *A* can be cut off from *DE*; then, being *DE* unlimited, that same operation can be done successively for the other two segments. If *DE* were not unlimited, there would not be any guarantee, in general, that this could be done. What should be done, then, in such a case?

Let us suppose the worst possible situation, in which, instead of the half straight-line *DE*, we consider a segment *DE* lesser than each of the segments *A*, *B*, *C*; let us suppose, too, that a canonical argument within Euclid's plane geometry follows the genuinely constructive rule that we have just formulated. An enthymemathical solution could apply Postulate I.2 either by producing *DE* in a long enough segment to

[17]Panza (2012), p. 89.

[18]But there is at least a precedent of this interpretation of Postulate I.2. In Alberto Magno's commentary on the first book of the *Elements* (Lo Bello 2003, pp. 39–42) the proof of Proposition I.2 proceeds explicitly according to the mentioned rule.

be cut in sequence by segments equals to A, B, C (if we translate 'εὐθεῖα ἄπειρος' by 'unlimited segment', Euclid's arguments could be assimilated to this) or by producing it into a half straight-line (if we translate 'εὐθεῖα ἄπειρος' by 'half straight-line', Euclid's arguments could be assimilated to this, instead). The canonical solution, on the other hand, should put segment A in an extreme of DE, by Proposition I.2, then describe a circle of radius A, and produce DE up to the circumference constructing a point, say, G. Then the second segment should be put on G and the same operation should be repeated. And so again with C. Evidently, the complexity of the canonical argument would be greater.

7.4 Rethoric and Mathematical Arguments: Beyond Euclid

The analysis that has to be carried out over the available documents in order to get a reconstruction as the one we suggest is different (though, complementary rather than opposed) from logical analysis. In concluding our paper, we would like to emphasise some differences between these types of analysis of mathematical theories and arguments and the one we suggest.

What the logical analysis of a mathematical argument looks for is a unique inferential structure (at least, relative to a particular logic chosen for this aim). What a reconstruction of the kind we suggest looks for is to capture a variety of different phenomena which, as such, are independent of that structure, or, at least, are not completely determined by it. The distinction between canonical and enthymemathical arguments would have to allow, for example, to capture the phenomena which depend on giving us proofs to one another (or even to ourselves). Without going any further, it would have to allow to capture the existing differences among the proof that a mathematician gives to another one, the proof published in a specialised journal, the proof a teacher gives to his/her students or the proof given to oneself, etc. These are differences that disappear in the elegant uniformity of logical analysis (and even more, if this involves a complete formalization).

But, if we took this into account, where could we find the conceptual frame which allows an analysis of the differences among these ways to give a proof? The mentioned distinction between enthymemathical and canonical arguments is an element of such conceptual frame: the proofs that we have just distinguished are, in fact, different types of enthymemathical arguments which would correspond, perhaps, to a same canonical argument within a given theory. However, to say this is not enough. What we suggest is to add to this characterisation other tools arising from a viewpoint which could be considered as rhetorical.

The distinction proposed between enthymemathical and canonical arguments already assumes a rhetorical point of view. But it is important to make it clear that we are considering rhetoric in an Aristotelian sense, associated in consequence with valid arguments and appropriate justifications, rather than with mere persuasion (in ciceronian sense) or style.[19] Although for Aristotle rhetoric arguments do not belong to the domain of necessary knowledge, proper of the demonstrative science, our usage of rhetoric tools should not be understood as the rejection of the idea that mathematical truths are necessary and/or a priori. From Aristotelian rhetoric we would like to recover the fact that arguments depend on an audience and presuppose what is known by its members. Since, as Aristotles says,

> the enthymeme [...][is] concerned with things which may, generally speaking, be other than they are [...] being [...] a kind of syllogism, and deduced from few premises, often from fewer than the regular syllogism; if any one of these [premises] is well known, there is no need to mention it, for the hearer can add it himself. For instance, to prove that Dorieus was the victor in a contest at which the prize was a crown, it is enough to say that he won a victory at the Olympic games; there is no need to add that the prize at the Olympic games is a crown, for everybody knows it.[20]

The dependence of the audience is manifested in the enthymemathical character of the argument, but not because this would be incomplete due to a lack of premises, but for the fact that the competence of the audience makes it unnecessary to explicit them. We would like to use now the notion of audience in connection with our analysis of the concept of a mathematical argument. For short, we only consider the case of proofs. What we shall say about them might be easily generalised to other sorts of such arguments.

Formally speaking, a proof is a sequence of formulae. We submit that this result from two acts of reduction: firstly, the theory in which the proof is carried out is reduced to a theory in which the canonical proofs are required to be formal, that is, to consist, precisely, in appropriate sequences of formulae; secondly, the audience of a proof within a theory of this kind is reduced to one whose only relevant competence is taken to be that of recognizing tokens of inscription-types, and of manipulating them according to some rules of inference. Obviously, a formal proof in this sense is not, by definition, enthymemathical. But an audience like this occurs rarely, so as not to say that it is nonexistent in practice.

Our conception of enthymemathical proof within a theory, understood as a system of rules of authorization, makes possible to take into account different sorts of audiences, each of which is characterised by the competences required for conducting and/or understanding a certain sort of enthymemathical proofs. For example, the proofs that a teacher gives to his/her students, those that a mathematician gives to another mathematician (of the same or different level), those that we can give to ourselves, etc., could all be considered as reformulations of canonical proofs, each of which is adapted to a different sort of audiences. Precisely, that is one of the crucial characteristics of rhetoric arguments: the dependence on that which is known (and assumed) by all the members of the audience. Thus, for example, these different enthymemathical proofs might be regarded as consisting of more or less instructions, or of instructions of different nature, depending on the audience.

---

[19] For the notion of style in mathematics, see Mancosu (2009).

[20] *Rhetoric* 1357a (translation by J. H. Freese: Aristotle *Rhet*).

The fact of taking different audiences and different enthymemathical proofs into account does not seem to us to be incompatible at all with the idea that a theory is characterised, among other things, by the rules or conditions that a canonical proof within it has to respect. On the contrary, it seems to us that, whatever the audiences in question might be, there must be some notion of canonical proof to be used as a parameter for allowing one to properly speak of proofs in all the cases mentioned above: nothing should be legitimacy considered as a proof if it could not be taken as an enthymeme of a canonical proof proper to a certain theory. For example, in modern mathematics, at least for a large part of it, nothing should be legitimately considered as a proof if it could not be taken as an enthymeme of a formal proof within an accepted (formal) theory. In the same way, nothing should be legitimatly considered as a proof within the mathematical tradition coming from Euclid (which encompasses a large part of classical mathematics) if it could not be taken as an enthymeme of a canonical proof according to the standards fixed by the *Elements* (which makes evident the importance of fixing these standards, both from a historical and a philosophical viewpoint).

This simple observation is enough to make clear that, from our point of view, audiences not only differ for the enthymemathical versions of the same canonical proofs related to them, but also in relation to the canonical proofs which they are supposed to be able to admit. From this point of view, it is important to add that recognizing the possibility and the existence of different theories, including different types of canonical proofs admitted or admissible by different audiences with their different competences and skills, does not mean admitting that these theories and these audiences are incommensurable. Although we reject the idea of a single universal audience for mathematics *tout court*, we see no reason for excluding the possibility of mutual reinterpretation of mathematical arguments from audience to audience.

With the notion of a theory as a system of authorizations and the notion of an enthymemathical argument and its dependence on an audience, we mainly intend to legitimize a series of questions such as the ones posed in Sect. 7.2 in relation to Euclid's *Elements* (and outline their answer). These are questions which are often dismissed or ignored from other perspectives, but which acquire full sense from the perspective we have outlined. Furthermore, we intend to introduce a conceptual tool which could turn out to be fruitful for the analysis of the different forms of mathematical practice.

References

Aristóteles [*Rhet*]. (1926). *Art of Rethoric* (J. H. Freese, Trans.). Cambridge, MA: Loeb Classical Library 193, Harvard Univ. Press.

Avigad, J., Dean, E., & Mumma, J. (2009). A formal system for Euclid's elements. *Review of Symbolic Logic,* 2(4), 700–768.

Barrow, I. (1683). *Lectiones Habita in Scholis Publicis Academiæ Cantabrigiensis An. Dom, M.DC.*LXIV, Typis J. Playford, pro G. Wells in Cœmenterio D. Pauli, Londini.

Barrow, I. (1734). *The usefulness of Mathematical Learning [...]*. Printed for S. Austin at the Angel and Bible in St. Paul's Church-yard, London.
Euclid. (1883–1888). *Elementa*, vols. I–IV de *Euclidi Opera Omnia*. B. G. Teubneri, Lipsiæ, Edited by I. L. Heiberg and H. Menge. vols. 8 + suppl. 1. New edition by E. Stamatis.
Euclid [*EEH*]. (1926). *The thirteen books of the elements* (Translated with introduction and commentary by Sir Thomas L. Heath) (2nd ed., Vol. 3). Cambridge: Cambridge University Press.
Hilbert, D. (1899). *Grundlagen der Geometrie*. Leipzig: Teubner.
Lassalle Casanave, A. (2008). Entre la retórica y la dialéctica. *Manuscrito, 31*(1), 11–18.
Lasalle Casanave, A. (2006). Matemática elemental, cálculo y normatividad. *O que nos faz pensar, 20*, 67–72.
Lassalle Casanave, A., & Panza, M. (2012). Sobre el significado del Postulado 2 de los *Elementos*. *Notae Philosophicae Scientiae Formalis, 1*(2) http://gcfcf.com.br/pt/revistas/filosofia-da-pratica-matematica/.
Lo Bello, A. (2003). *The commentary of Albertius Magnus on Book I of Euclid's elements of geometry*. Boston-Leyden: Brill Academic Publishers.
Mancosu, P. (Ed.). (2008). *The philosophy of mathematical practice*. Oxford: Oxford University Press.
Mancosu, P. (2009). 'Style' in mathematics. In E. N. Zalta (Ed.), *The stanford encyclopedia of philosophy* (First published Jul 2, 2009) http://plato.stanford.edu/entries/mathematical-style/.
Manders, K. (2008). The Euclidean diagram. In: P. Mancosu (Ed.), *The philosophy of mathematical practice* (pp. 80–133). Oxford: Oxford University Press.
Mumma, J. (2006). *Intuition formalized: Ancient and modern methods of proof in elementary geometry*. Ph.D. thesis, Carnegie Mellon University.
Miller, N. (2007). *Euclid and his twentieth century rivals: Diagrams in the logic of Euclidean geometry*. Stanford CSLI Publications.
Norman, J. (2006). *After Euclid. Visual reasoning and the epistemology of diagrams*. Stanford CSLI Publications.
Panza, M. (2012). The twofold role of diagrams in Euclid's plane geometry. *Synthese, 186*(2012), 55–102.
Proclus [*CEELF*]. (1873). *In primum Euclidis Elementorum librum commentarii*. Teubner, Lipsi. Ex recognitione G. Friedlein.
Rapp, C. (2010). Aristotle's Rhetoric. In E. N. Zalta (Ed.), *The stanford encyclopedia of philosophy* (First published May 2, 2002; substantive revision Feb 1, 2010).
http://plato.stanford.edu/archives/spr2010/entries/aristotle-rhetoric/.

Chapter 8
Some Reasons to Reopen the Question of the Foundations of Probability Theory Following Gian-Carlo Rota

Carlos Lobo

Abstract Roshdi Rashed's work illustrates perfectly what can be a conscious and cautious practice of reflection, with the purpose of setting history of science (and mathematics) on renewed and deeper grounds (See the introduction, Problems of method: history of science between history and epistemology, in *Classical Mathematics from Al-Khwarizmi to Descartes*, 2014, (Rashed 2014).). This entails the methodical operations that he enumerates, such as enlargement towards undermined or ignored traditions (Chinese, Arab, Indian, etc.), careful and reasoned decompartmentalization of disciplines, correlative changes of periodization (without which the critique of scientific ideology and ideology of scientists would risk of falling back into some counter-ideological history, particular or general). (Europeocentrism for instance is twofold: promotion of the ambiguous and disputable notion of "western science" and ignorance or "minorization" of the contributions of non-western traditions. Cf. (Rashed 1984) and appendices in *The Notion of Western Science: "Science as a Western Phenomenon"* and "*Periodization in Classical Mathematics*" (Rashed 2014).) Among mathematicians, Gian-Carlo Rota is certainly both exceptional and, for this reason, exemplary. *By choosing this perspective as a tribute, I hope that Roshdi Rashed will consider my comments not too unworthy*. For any philosopher of science not insensitive to history of science, and for any historian not completely allergic to philosophical reflection, studying Rota's contribution in the fields of logic and phenomenology reveals itself instructive and fruitful. Contrary to dominant trends amongst his colleagues, in his own way, Rota showed a strong and continuous interest in logic, history of science and philosophy.

C. Lobo (✉)
College International de Philosphie, Paris, France
e-mail: carlos.lobo.ag@orange.fr

H. Tahiri (ed.), *The Philosophers and Mathematics*, Logic, Epistemology, and the Unity of Science 43, https://doi.org/10.1007/978-3-319-93733-5_8

There is no better way to get acquainted with the *Rota touch*[1]—as a mathematician *and* a philosopher—, than giving a hard look at the way he builds up his historical perspectives. While doing so, we notice at first glance, that many of Rota's papers, as it has been observed, appear with a title echoing Hilbert's challenging Paris conference of 1900 on the *21 problems of mathematics* and, beyond, the Cartesian methodological act of enumeration (*dénombrement*). Most of these papers, retrospective or prospective, as they are, take the form of lists of (old and new) (mathematical) problems, which *have been solved but remain unnoticed,* anecdotic so to speak, *problems which could and should have been, should be and will be, or, perhaps, should but won't be solved*. Among these titles and lists: "*10 things I wish I had learned....*"; "*10 mathematical problems I will never solve*";[2] "*Ten remarks on Husserl and phenomenology*", etc. The *Twelve problems in probability no one likes to bring up* we are pointing at, presented as *The Fubini Lectures* in 1998, belong obviously to this group.[3] Such a practice, at least in Rota's case, leans on a counterfactual history of science.

In the course of a piece of scientific research, this peculiar kind of historical reflection marks out a typical phase of overview—and eventually of classification—of past developments combined with a paving of new paths, an opening of new perspectives. More radically, this practice presupposes a methodological splitting between what is known (solved) and unknown (unsolved), between what is certain and what remains conjectural, problematic, likely, etc. Although many scientists and historians of science perform these methodological reflections, they rarely take the time to describe and elucidate them.

As an intermediation between the reflecting historian and the reflecting mathematician, a third member should join the company: the phenomenologist. By vocation, a phenomenology should describe these operations, and locate them in their proper frame, i.e. that of a larger reflection on the correlational a priori, and specifically that of an (intersubjectively) *universalisable* historical attitude and consequently that of a critical historical and epistemological attitude which, though it has broken through recurrent theological-political polemics, is still nowadays groping around through many ideological blockages and counterfeits. Unfortunately, very few among the phenomenologists think it practicable; or if they do, they don't think it worth devoting any effort to it; and when they acknowledge and even praise such efforts, they are rather sceptical regarding the ability of phenomenology to succeed in this task. This concerns "general history" and even more history of science. Hopefully, Rota compensates for this shortage by his own contributions with a series of phenomenological and logical investigations referring to Husserl (Rota 2000, p. 89 sq.).

---

[1]Crapo and Senato (2001). D. A. Buchsbaum, *Resolution of Weyl modules: the Rota touch*, in H. Crapo, D. Senato (Eds) *Algebraic Combinatorics and Computer Science A Tribute to Gian-Carlo Rota*, Springer, 2001, p. 97-sq.

[2]Crapo and Senato (2001: 4).

[3]Rota (2001, p. 51 sq.) *Twelve problems in probability no one likes to bring up*, in *Algebraic Combinatorics and Computer Science,* H. Crapo et alii. (Eds.), Springer-Verlag, Milano, 2001.

For any philosopher of science not insensitive to history of science, and for any historian not completely allergic to philosophical reflection, studying Rota's contribution in the fields of logic and phenomenology reveals itself instructive and fruitful. Contrary to dominant trends amongst his colleagues, in his own way, Rota showed a strong and continuous interest in logic, history of science and philosophy.[4] By "his own way", I mean that, in many instances, he practices a counterfactual history of science, but also has some view as to what the collaboration between phenomenology, logic and mathematics might be, could be and should be. This special procedure consolidates the matter-of-fact, i.e. "positive" (and positing) history. Indeed, it is now being slowly and cautiously introduced into the ordinary tool-box of some historians and philosophers. Its effectiveness is enhanced in the case of history of science because of the striving of generations of researchers directed toward supposed and expected permanently valid achievements.

A first question: in this particular case, what are the requirements and what is the meaning of a first-layer counterfactual history of science (and mathematics in particular)? Second question: what might be the motivations and the resources of a second-layer counterfactual history dealing with possible or eventually desirable relations between science itself, history of science and epistemology? The most intriguing point is certainly that, instead of vanishing into the vacuum of an empty speculation, the latter sort of reflection is engaged in advanced scientific researches. As a matter of fact, Rota scientific work is supported by a continuous historical reflection which in turn uses explicitly *leading threads*, so much so that his scientific career and his achievements could be interpreted afterwards as directed through and through by logical and philosophical concerns. And like some of his prestigious predecessors (Poincaré, Weyl), he gave us enough synoptic and insightful reflections to indicate that he was perfectly aware that this theoretical practice presupposes some kind of eidetic dimension. Such a dimension does not refer back to some spontaneous philosophy—be it the so-called Platonism of mathematicians[5]–, rather to a subtler and more complex form of idealism, the full structure of which is akin to that exhibited in Husserl's transcendental idealism (inclusively in his later works). This particular affinity is neither contradicted by Rota's unambiguous stance against mathematical Platonism, (which probably does not correspond to Plato's position

---

[4]About his indefectible interest in logic, despite his teachers in mathematics, cf. his narrative about Church's lectures in *Indiscrete Thoughts*, (Rota 1997b, p. 4–7); on his appraisal of the discipline as such, against what he calls, following Kemeny, "the mathematician's bigotry", see (Rota 1997b, pp. 7, 105, 123) (Rota 1990a).

[5]Despite and because of the strong eidetic background of his approach, Rota refutes expressly the classical problematic of mathematical existence in terms of Platonism (Frege, Gödel, etc.) in favour of an "eidetic perspectivism", which has strong affinities with Husserl's transcendental idealism. (See below, Sect. 2.) But we can anticipate on this particular point: "To the mathematician, an axiom system is a new window through which the object, be it a group, a topological space or the real line, can be viewed from a new and different angle that will reveal heretofore unsuspected possibilities. In saying this, we are acknowledging the actual practice of mathematics, and we are not arguing for Platonism: we are not in the least concerned with the problem of existence of mathematical objects, any more than the grammarian is concerned with the problem of existence of the verb or the adjective". (Rota et al. 1988, p. 382). (See Rota 2000, p. 93; Rota 1991, p. 133–138.)

relative to mathematical objects), nor by the reservations expressed on "the blend of logical insights and questionable assertions on the transcendental ego and intersubjectivity", or on the "blurring of the distinction between logic and metaphysics", and the subsequent proposal to free "genetic logic" "from all remnants of idealism".[6] For simple reasons: these reservations do not concern idealism or ideal objects *in general*, but strictly, the interference between philosophical matters and formal logic, the latter being "not concerned with the truth", but only "with the game of truth" (Rota 1991, p. 134). Nonetheless, before the activity of proving theorems, although not a "seeing the truth", mathematics is a "seeing through", which aims at trivializing what is previously given as an "object" (real line, set, probability, manifold, etc.). Yet what is still pointed at, "through" mathematical objectivities and layers of formal constructions, is something like an idea in Kantian sense.[7] Moreover the mathematical activity is necessarily referred back to a subject; and correlatively, everything "on" which this activity is exercised must be called an "ideal object", and it is by "naïve prejudice" that one is prevented from considering that "ideal objects (such as prices, poems, values, emotions, Riemann surfaces, subatomic particles, and so forth)" are as much "real" as "physical objects (such as chairs, tables, stars, and so forth)" (Rota et al. 1992, p. 167).

The eidetic structure at the core of genetic logic, as we shall see, provides us with a formal frame and a logical understanding of the reflexivity spontaneously performed by scientists in their productive phase (inseparable from that of scientific discovery proper), and is consequently a (tacit) resource for any relevant history and philosophy of science. Strikingly, as we shall see, Rota deepens Husserl's insights, by translating this eidetic structure into the language of modern logic, and highlights the blind spots of the latter. He shows reciprocally that resistance against confusion between effective and ludic formalization, requires the newer semantic-syntactic distinctions to be reworked and understood in a phenomenological perspective, in the frame of this eidetic structure. In this dynamical platonic perspective, scientific endeavour is a striving toward form, hence a passing through layers of forms, always keeping one eye on informal (although not indescribable) "phenomena" (observables, processes, etc.), and the other one on undisclosed resources of formalization, the elaboration of which aims at producing an enlargement of the old ones, and transforming them into special (eventually limit) cases of the new ones.

Rota's sensitivity to the manifold dimensions of scientific problems goes on a par with his awareness of the multi-layered and entangled eidetic dimensions involved in any historical and contingent epistemological situation. Mathematical imagination does play its role, as is evidenced by the remote and far-reaching mathematical analogies worked out by Rota. Rota's style, if any, compared to the categorist's virtuosity and the logician's liberality, is undoubtedly that of an algebraist, mitigated with an

[6] Rota 1991, p. 247.

[7] Husserl (1950a, p. 170). See full quote below. Rota (1991, p. 136), insists on the fact that this applies to mathematics too: "Thus it appears that the activity of the mathematician is not that of proving theorems, but another one: that of proving that all theorems are intuitively evident, by an evidence that shall be as close as possible to Kant's ideal of an analytic a priori statement."

acute sense of the strengths and limits of algebra.[8] Rota stands where information circulates from one field to the other, producing, through an analogical circulation, a *communication of forms*, and therefore new (upper or more complex) forms, more or less dominated, internalised and formalized (from basic automorphisms to the most general and deepest isomorphisms). In other words, the place where Rota stands is an epistemological crossroads, where intersect not only various branches of mathematics or physics, but of logic and philosophy (Rota 1990a). So much so that we are entitled to state that many of his contributions in mathematics are implicitly, when not explicitly, guided by philosophical and logical questions, before he realizes that they could be supported by Husserl's investigations. His interest in logic goes back to his formation years and is explicitly testified in his essays from 1973 to 1978 (Rota 1973a, b). Since this point might sound unpleasant for many dominant trends of philosophy of science, let us recall Rota's own words. His conversion to Husserlian phenomenology occurs in 1964.

> "I began to read the work of Edmund Husserl sometime in 1957. *I began to understand him seven years later, in 1964*. I remember the first time I succeeded in deciphering his writing: it happened on a morning in March 1964. I was on leave from MIT, and my wife and I were driving from Illinois to Indiana in an old Plymouth, taking side roads through the charming rural landscape of the Midwest. My wife did not trust my driving (with good reason), and she did most of the driving, while I lay on the back seat reading Husserl's 'Ideas for a pure phenomenology'. As I despondently reread the conclusion of a lengthy argument, suddenly it began to make sense. The experience reminded me of one of those jumbled multicolored prints that hide a picture, which you see only if you stare at the print in a certain way. My first reaction was 'So this what it is really about!', followed by 'At last I found the key'. I was young and brash. It took a few more years to begin to understand. Whenever I pick up again one of Husserl's writings and begin to read, I experience the same feeling of being out of focus, and I have learned to wait until the hidden picture emerges. Let me state a prejudice at the outset. I believe Husserl to be the greatest philosopher of all times."[9] (Rota 2000, p. 89)

[8]"Psychologists have prescribed in turn sexual release, wonder drugs and primal screams as the cure for common depression, while preachers would counter with the less expensive offer to join the hosannahing chorus of the born-again. It goes to the credit of mathematicians to have been the slowest to join this movement. Mathematics, like theology and all free creations of the Mind, obeys the inexorable laws of the imaginary, and the Pollyannas of the day are of little help in establishing the truth of a conjecture. One may pay lip service to Descartes and Grothendieck when they wish that geometry be reduced to algebra, or to Russell and Gentzen when they command that mathematics become logic, but we know that some mathematicians are more endowed with the talent of drawing pictures, others with that of juggling symbols and yet others with the ability of picking the flaw in an argument. We often hear that mathematics consists mainly in 'proving theorems.' Is a writer's job mainly that of 'writing sentences'? A mathematician's work is mostly a tangle of guesswork, analogy, wishful thinking and frustration, and proof, far from being the core of discovery, is more often than not a way of making sure that our minds are not playing tricks.» (Rota et al. 1992, p. 154).

[9]The narrative continues: "Several years ago I stated this opinion in a preprint I sent to Gödel, and he wrote me back a letter, the only letter I have ever received from him: 'Dear Professor Rota, *You are wrong. Husserl is not the greatest philosopher of all times. He is the greatest philosopher since Leibniz*. Sincerely yours,'. Years later I understood what Gödel meant, after his writings in phenomenology were published, in the third volume of his collected papers. Gödel believed Husserl to be the philosopher who brought to a successful completion the program inaugurated by Immanuel Kant, which Kant and his successors had failed to carry through. 'Husserl is the real Kant', Gödel writes." (*ibid.*)

Although his narrative should be submitted to a critical examination, it represents, historically speaking, an inevitable starting point for any one interested in determining Rota's epistemological position and profile. The conversion experience (comparable to a stereoscopic experience) is a clear description of the eidetic experience. His affinity with Husserl's reform of logic is univocally illustrated and defended by his insistent proposal for the revision of the foundations of probability theory, on which I shall now focus.

In order to avoid any excess of ambiguous generalities, we shall look more closely at how and why he outlines new prospects concerning the foundations of probability. For what reasons did he embark on reopening "some questions in the foundations of probability, and thus also a much touchier and much neglected subject, the foundations of statistics" (Rota 1998a, p. 57)? From the point of view of quantum probability, taken in its narrow sense, this might be taken as a way of coming «into play when the game was over». This is certainly true. Yet Rota's interests in the foundations of probability theory and statistics go back to the 70s. For reasons which will appear, this interest was *not* limited to the setting up of a new foundation (i.e. axiomatization) of probabilities fitting quantum mechanics and the advance towards getting rid of so-called paradoxes,[10] but something else and wider in scope, namely to *bridge* the gap between probability theory and other mathematical theories, including, some parts of first order logic. This strategy entails an explicit interest in logic—exceptional among mathematicians—, as well as a freer way to deal with logic (starting with the relations between syntax and semantics, and set theory as standard semantic basis), which goes beyond the opposition between standard and non-standard logics, and takes for granted that, being part of mathematics, mathematical logic must be freed from its remaining philosophical pretensions (for instance to posit an absolute norm of rationality). This does not exclude either the claim that there should be room for a philosophical logic, or that philosophy (under the form of transcendental phenomenology) could play its role in a reform of logic.

1. I shall start with a brief survey of the manifold dimensions of the problems concerning probability, which militates in favour of a non-Kolmogorovian probability, pointing to the logical and historical dimensions of Rota's review. Among the fundamental concepts and tools used by Rota, I shall insist on two major ones, those of *lattice* and *cryptomorphism.*
2. In doing so, my aim is not to decide whether and to what extend Rota contributed to the raise of non-kolmogorvian probability, but understand how and why he took it as an example of a semantic without syntax, and as a starting point for a reform of logic. As the case of probability theory shows, the formal general frame cannot be provided by formal logic since probability theory and logical

[10]Quantum probability followed another path, by giving the proof of the experimental necessity of non-Kolmogorovian models (Accardi 1981a, b), by showing that Bayes' definition of "conditional" was the hidden axiom of probability (an analogue to Euclide's 5th Postulate) (Accardi and Fedullo 1981) responsible for the difference between the Kolmogorov and the quantum mathematical model; by giving the list of axioms unifying classical and quantum probability and producing non-trivial models (Accardi 1982a, b) and finally by dissipating the so-called paradoxes of quantum theory (Accardi 1993, 1999, 2003), and proposing a presentation fully compatible with standard logic.

quantification are two non-trivially equivalent structures: the syntax of the latter cannot be transferred to the former. By bridging probability theory and logical quantification via a *functor*, Rota indicates the limits of the logical syntactical approach and even the fact that probability theory is lacking of any syntax. In order to provide another syntax a new logic is required, which should help to describe dynamic logical relations between syntax and semantics, following a threefold structure (syntax, semantics and "sameness").

3. On the basis of this survey, we will get a better notion of the general formal frame in which Rota builds up his historical perspectives, which is akin to Husserl's transcendental idealist historical perspective. If we search for more clarification and look for a justification of this comparison, we arrive at the core of Rota's philosophical project. This project is akin to the one he attributes, with good reason, to Husserl: a *reform of logic* (Lobo 2010). This project, compared to von Neumann and Birkhoff's, was wider in scope, since, from the start, he was not focused on and limited to the issue of quantum physics. Under the name of "*genetic logic*" Rota refers, among others, to certain dimensions of the history of science which Husserl considered an "historical a priori".

8.1 The Problems on Probability Mathematicians Don't Want to Bringup

8.1.1 The Desideratum of a Complete Formalization of Probability Theory

What is questioned here is a kind of self-enclosure and the ensuing silences on the part of the dominant mathematical tradition, which can only be explained, quasi-sociologically, on the basis of some common *credo*. In the present and recent cases of probability theory, as for the oldest one (that of Euclidian geometry), once axiomatically crystallized,[11] the credo "that the subject has been definitively closed by Kolmogorov" has become a *dogma*. There seems, Rota says, to have "been a conspiracy of silence on such foundational questions among mathematicians (not so among philosophers), who would like to believe that the subject has been definitively closed by Kolmogorov, and many of whom react unfavourably to any suggestion of revision of Kolmogorov's elegant setup" (Rota 2001, p. 57).

Rota's project is clearly to reopen "a number of questions in probability and statistics that the professionals *would rather* pass over in silence" (ibid.) (emphasis mine.) The way he does it will appear as a defence of a minority view, for the very reason that this criticism would appear as unfair and unjustified to most mathemati-

[11] Same observation in Weyl (1968, p. 705): "Each field of knowledge, when it crystallizes into a formal theory, seems to carry with it its intrinsic logic which is part of the formalized symbolic system and this logic will, generally speaking, differ in different fields." (Weyl, *Gesammelte Abhandlungen*. Band, III K. Chandrasekharan (ed.), Springer Verlag, Berlin, 1968, p. 705.)

cians. Part of them (maybe a vast majority) will see this axiomatics as a sufficient syntactic presentation. Others (another minority) will object that Kolmogorov (contrary to Ramsey, De Finetti, etc.) only presented a model for probability, but no real axiomatization, and, considering the discrepancy between this model and quantum probability, they will plead for a new axiomatization, precisely quantum probability in the strict sense.[12] Although contradictory, both critics presuppose Hilbert's conception of formalization and, consequently, of the relations between models and axiomatics. By their diverging evaluation of the scope of Kolmogorov's axiomatics, they illustrate at the same time that something is missing in the formal dual relation (syntax/semantic). In order to understand the *dynamic* of such enlargements, a third term must be introduced, which Rota calls, rather equivocally, "object", "identity" or "sameness". In order to show that Kolmogorov's axiomatization of probability (Kolmogorov 1956), provides no *sufficient syntactical presentation for probability*, the twelve problems enumerated by Rota cover three different forms of lacks in probability theory: (a) "examples of stochastic processes whose phenomenological behaviour is not fully matched in their mathematical presentation" [be it semantic or syntactical]; (b) "instances of (successful) statistical reasoning" which are "brushed aside in treatises purporting to be rigorous" [bits of syntax are not explicitly presented]; (c) examples of "new probability theories", such as "quantum probability" [not in Accardi's sense], "free probability", obtained "from relaxing some algebraic aspects of probability", and which does not fit in the classical formal frame (classical mathematical models *and* axiomatization). To anticipate somewhat our next considerations, we can notice that item (a) refers to concrete "models" without any corresponding formal model, and thus indicates rather a lack on the side of the actual formal semantic of probability; item (b) points at to *syntactical inadequacies* and thus

[12]Sharing Hilbert's conception of the relation between models and axiomatization, hence the so-called formalist approach to syntax and conception of formalization, Accardi (1981a, b) considers that Kolmogorov only presented a "mathematical model" for probability, comparable to Descartes geometry, and that one had to wait for De Finetti or Ramsey to get a first real but still incomplete axiomatization of *classical probabilities*, comparable to Hilbert's axiomatization of Euclidian geometry. In this perspective, this axiomatization remained incomplete, because some principles were neither deducible from the axioms, nor exhibited as an axiom. This was the case for *countable additivity*. Another limit is the inadequacy of Kolmogorov's model and its axiomatizations to handle probabilities in quantum physics, which required a new mathematical model and a new axiomatization. According to Accardi, Von Neumann only proposed a few steps toward a construction of a mathematical model (taking into account an analogue of space probability and random variable). The full project was only accomplished by quantum probability in his sense, by taking into account stochastic processes, quantum joint probabilities, independence, conditioning. This enlargement is compared to those of non-Euclidian geometries. — In order to avoid major misunderstandings and as it will appear clearly in the following, let us recall that, first, in the case of geometry as in the case of probability, Rota has a quite different understanding of the relation between models and axiomatics, semantics and syntax, and, secondly, that by quantum probability he does *not* aim at quantum probability in Accardi's sense, but essentially von Neumann attempts to propose a syntax (not a model) for quantic probability; thirdly, that contrary to a limited view of combinatorics, consideration of continuity are not excluded from it and it was precisely the purpose of Rota and Klain's "geometric probability" to extend enumeration to "the assigning of invariant measures" (Rota and Klain 1997).

to the syntactical incompleteness of the standard probability theory, eventually its lack of any syntax; item (c) suggests some complementary formalization whose syntax appears problematic from the point of view of the classical theory of probability (pointless probability, free probability, quantum probability, etc.). These somewhat puzzling statements—as puzzling as Gödel's theorem was in the frame of Hilbert's formalism—will become clear in point 2.

As a result, these lectures are not a mere critique of the standard axiomatization of probability, nor an introduction (an "intensive" one) to a new axiomatization, but rather an invitation to pose seriously the question: *what probability is, really?* That is, to show that the "*object*" we are talking about (in its "identity" or "sameness"), or rather that something of the object grasped (namely its "essence") lies *beyond* that which is actually "objectified", i.e. conceptually and formally grasped. But in order to avoid taking this as a form of Platonic mysticism, Rota has warned us from the start that this question (whatever its qualification: mathematical, logical, philosophical, etc.) must be rigorously posed. This means, that its formal consistency as well as that of any new concept will be submitted to the severe control of algebra. Taking Husserl at his own words, Rota knows that formalizing is "algebraizing"[13] (Husserl 1969: 48, 218) but neither blindly nor arbitrarily, and this explains how, without knowing the exact answer to each question, we are not condemned to get stuck in the mud of arbitrary symbolical games.

> I will lay my cards on the table: a revision of the notion of sample space is my ultimate concern. I hasten to add that I am not about to put forth concrete proposals for carrying out such a revision. We will, however, be guided by a belief that has been a guiding principle of the mathematics of this century. Analysis will play a second fiddle to algebra. The algebraic structure sooner or later comes to dominate, whether or not it is recognized when a subject is born. Algebra dictates the analysis. (Rota 2001, p. 57)

The twelve problems are named: (1) algebra of probability; (2) densities of random variables; (3) structure theory for Boolean σ-algebras; (4) entropy; (5) maximum entropy principle; (6) conditional probability, conditional expectation and Bayes' law; (7) justification of the univariate normal distribution; (8) probability-preserving transformations; (9) from ordinary probability to quantum probability; (10) the multivariate normal distribution and the Clifford distribution; (11) cumulants; (12) free probability theory. These problems cannot simply be dispatched on the few lines listed above, *for algebra is at work everywhere*, including in the presentation of informal instances of probability waiting for a rigorous mathematical treatment. Yet they are partially ordered in a strictly formal sense. To get a clue about this order, it is necessary to notice that from the start we are moving within the realm of algebraic forms. I shall not review the whole series of problems listed by Rota. Modulo my own limits, my aim is to highlight the major limits of the classical fundamental concepts of probability, and to understand how and in which sense formalization plays its heuristic role. For this reason, I shall insist mostly on the first problem, and survey rather rapidly the remaining ones, chiefly problems 3, 6 and 9, not because they are less technical and important than the others, but appear more "elementary".

[13] See *Formal and Transcendental logic*, English tr. p. 48 and p. 218.

8.1.2 Positing the Problem of a "Pointless" Probability in Lattice-Theory Terms

In order to get the most general formal concept of probability, it is necessary from the start to translate the usual analytical fundamental concepts of probability (measure of the probability of an event, comparing the ratio of favourable cases over the whole set of all possible cases) into those of algebra. This is the first problem named *algebra of probability*. Probability is indeed defined, in standard theory, as "the study of sets of random variables on a sample space and of their joint probability distributions", "a sample space", as "a measure space (Ω, Π, P) where Ω is a set, Π a s-subalgebra of subsets, that is, a Boolean σ-algebra of subsets of Ω, and P is a countably additive non-negative measure such that $P(\Omega)= 1$"[14]; atomic events, as "sample points", i.e. as elements of the sample space Ω; probability (P), as "a real-valued function defined on Ω, which is measurable relative to the s-subalgebra Π", etc. But being either redundant or equivocal—and this is the first difficulty—, this first formal translation is far from satisfactory. Even understood correctly, as definitions implicitly introduced by the axiom system, these definitions should be corrected: notions of *event* and *sample space* are redundant. Besides a strict algebraic presentation with pure random variables instead of sample space would be of greater interest for epistemological purposes. But instead of digging more deeply, seeking for more "hidden algebraic underpinnings of probability" (p. 62), the algebraization stops halfway, and, to make matters worse, algebraist mathematicians and mathematical statisticians seem to ignore each other's work.

> The alternative definitions of probability, by means of a sample space and by means of an algebra of random variables, are equivalent. Mathematicians opt for the definition in terms of random variables, because they do not wish to miss a chance to appeal to Her Imperial Majesty, the Theory of Commutative Rings. However, *no practicing statistician has ever felt comfortable with algebras of random variables*. The answer to a problem in statistics is likely to be a number that specifies the probability of an event. *Events are essential to statistical intuition and cannot be done without*. (Rota 2001, p. 58) (Emphasis mine)

Many introductory courses or treatises on the foundations of probability start by an algebraic presentation in terms of random variables but shift, when dealing with probability proper, to sample spaces (see Rényi 1970, p. 20–31 and p. 38–63; Moran 1968, p. 2–7; Jedrzejewski 2009, p. 21–28). This indicates according to Rota insufficiencies in the on-going process of algebraization. Statisticians (or physicists) stick, with good reason, to their semi-formal concepts of probability of events (or observables), concepts stemming from a "pre-axiomatic grasp" (see for instance: Neyman 1950, p. 9–12; Savage 1972, p. 10–21). To put it in plain logical terms: although apparently syntactically and semantically consistent and complete, the actual formal (Kolmogorov) covering of probability needs to be reworked, by getting rid of superfluous formal items and by introducing *new algebraic underpinnings tacitly*

[14] A σ-algebra of a set Ω is the collection Σ of subsets of Ω satisfying the following conditions whatever the the number (from 0 to enumerable infinite) and combinations of operations implemented. These operations, in the case of a σ-algebra, are: complement, union and intersection.

and invisibly involved in the statistical treatment of probability. In order to ensure the complete formal (algebraic) covering of the mathematical object called *probability*, some additional modifications have to be introduced into the actual algebraic expression of probability. To start with, we obtain an algebraic covering of the notion of event and the reconstruction of a Boolean σ-algebra of events by introducing an "indicator random variable $I_A(\omega)$" and a linear functional *E(X)* (expectation of X). The values of the variable w are 1 or 0, "according as ω belongs to the event A" or not. The probability of the event *A*, *P*(*A*) is expressed by the linear function $E(I_A)$ [*read*: "expectation that a random variable takes the value 0 or 1"]. The problem can now be worded in the following terms: *How can we recover the algebra of a random variable starting from a mere abstract Boolean σ-algebra, avoiding any recourse to any set-theoretical representation, "that is, when one is forbidden to mention points"*? To sum up: a fully formal translation of probability must consider probability in terms of *linear functions of random variables and be "pointless"*. (We set aside another intermediary requirement, expressed elsewhere (Rota, 1973a): that of dealing with multi-sets rather that with sets.)

Rota distinguishes two historical phases in this process. Caratheodory[15] and von Neumann (1932) contributed to the first. The main idea is from von Neumann (who invented the expression "pointless probability")[16] and we owe to Caratheodory the first, but too complicated "contrived" accomplishment, but neither Caratheodory nor von Neumann (1981) ever achieved a full algebraization of probability. The second phase involves mathematicians such as Birkhoff and Rota himself who in a paper from 1967 sketched a simpler construction and demonstration. This work was prolonged, among other works, by a treatise on *The Valuation Ring of a distributive Lattice* (from 1973a), followed by many other incentives and almost provocative papers (such as the last conference in Harvard in Memory of Garrett Birkhoff on *The many lives of Lattice Theory*).[17] This simpler construction is summed by Rota by introducing three defining identities (that of complementary event, of sum and product of events) in terms of union and join, and ideals.

---

[15]The ring *R* generates Ω and the measure function μ, which maps from *R* onto the interval $[0, +\infty]$. Caratheodory's extension theorem affirms the existence of a measure μ' from the σ-algebras generated from *R* onto the same interval $[0, +\infty]$, extending μ. And if *m* is s-finite, μ' is unique and s-finite. Cf. Caratheodory (1948), and its presentation, in Halmos (1950, p. 54); and, more recently, Kallenberg (2002, p. 455).

[16]What about "pointless topology"? According to Rota (1997, p. 220) "the term 'pointless topology' goes back to von Neumann" and led to "serious misunderstandings". Johnston (1983, p. 41) refers it back to Hausdorff's choice of the notion of "open set" as primitive "in the study of continuity properties in abstract spaces", but it is Stone [1934] which started exploiting "the connection between topology and lattice theory". Affinity between Johnstone's approach and Rota's are obvious.

[17]Published in the Notices of the AMS, 1997, Volume 44, Number 11.

Given a Boolean σ-algebra $\mathcal{B}$, we construct a ring, functorially[18] associated with $\mathcal{B}$, as follows. We begin with the commutative algebra freely generated over the reals, by elements of $\mathcal{B}$, and impose on this free algebra the identities (where A and $B \in \mathcal{B}$):

1. $A^c = \hat{\text{I}} - A$, where $\hat{\text{I}}$ is the maximal element of $\mathcal{B}$;
2. $A + B = A \wedge B + A \vee B$;
3. $A \cdot B = A \wedge B$.

The quotient algebra of the free algebra obtained by imposing these identities is called the valuation algebra of $\mathcal{B}$. (Rota 2001, p. 61)

Since this presentation makes use of lattices, let us come back to one of the staring points, and posit simply the question: What is a lattice? Answer:

A lattice L is a set endowed with two operations, called meet and join and denoted $\wedge$ and $\vee$, which satisfy some but not all the properties satisfied by 'and' and 'or' of ordinary logic, namely, each of them is idempotent, associative and commutative. What is more important, for $a, b, c \in L$ the following absorption laws are satisfied: $a \vee (b \wedge a) = a$ and $a \wedge (b \vee a) = a$. (Rota 2001, p. 59)

The starting point, as recalled by Rota, is due to Dedekind who, at the end of the nineteenth century, noticed the analogies of two fundamental operations (meet and join, $\wedge$, $\vee$), with the logical product and sum (and, or), and which as them obeyed not only the idempotent, associative and commutative laws, but also the two "more important" "absorption laws" which stated that for any a, b, c (for example subsets of a set S):

1. $\mathrm{a} \wedge (\mathrm{b} \vee \mathrm{a}) = \mathrm{a}$
2. $\mathrm{a} \wedge (\mathrm{b} \vee \mathrm{a}) = \mathrm{a}$

The analogous logical law is known as absorption rule and states the equivalence between $(p \rightarrow q) \leftrightarrow (p \rightarrow (p \wedge q))$. This property, worked out by Birkhoff, from the remaining axioms, gives the fundamental theorem mentioned by Rota, i.e. "that a lattice satisfying the distributive law, a distributive lattice, *is isomorphic to the lattice of sets with union and intersection of sets playing the role of join and meet*" (Rota 2001, p. 60).

8.1.3 *Some Reasons Why Mathematicians Should Think More Highly of Lattices*

Incidentally we stumble here on one of the reasons *why so few mathematicians were interested in such an algebraization.*

---

[18] A *functor* extends morphisms to categories themselves; a morphism being usually a function between two sets with the same structure (field, ring, group, Boolean algebra, lattices, etc.). In category theory, the function relates eventually two or more different mathematical objects (structures), without presupposing necessarily any isomorphism.

Among the various reasons that have inhibited mathematicians from pushing further the construction of a "pointless probability", one is clearly negative, i.e. that it requires the use of lattice and that, for circumstantial reasons, lattice theory has a bad press among mathematicians. One of the semi-good reasons for this is the unfortunate use von Neumann made of lattices in his attempt at a "quantum logic" (von Neumann 1936, p. 823–843; 1995, p. 105–125).

Another reason is the (almost religious) fervour surrounding category theory among mathematicians—and philosophers.[19] This enthusiasm is partially forgivable since it faced the bigotry of opponents. Yet, adopting once again a minority view and without undermining the importance of Grothendieck and MacLane, Rota does not partake of this enthusiasm for category theory[20] and still posits the problem in the perspective of universal algebra (correctly understood), refusing to dismiss a lattice-theoretic approach. The dilemma is clearly posited in different papers, in particular in his *Indiscrete thoughts*, which provide us with beautiful instances of a counterfactual historical background, which represents a reservoir of new mathematics.[21]

Another reason which explains the bad press of lattice theory is that lattices were reputed to be the weak sisters of Boolean algebras, considered as trivial and despised as such.[22] Nonetheless, lattice theory happens to be indispensable to go a step forward

[19]See quote above, *Discrete Thoughts*, p. 154; and: "One *wonders why category theory has aroused such bigoted opposition*. One reason may be that understanding *category theory requires an awareness of analogies between disparate mathematical disciplines*, and mathematicians are not interested in leaving their narrow turf." (Rota 1997b, p. 220)

[20]"The axiomatization of the notion of category, discovered by Eilenberg and Mac Lane in the forties, is an example of beauty in a definition, though a controversial one. It has given rise to a new field, category theory, which is rich in beautiful and insightful definitions and poor in elegant proofs. The basic notions of this field, such as adjoint and representable functor, derived category, and topos, have carried the day with their beauty, and their beauty has been influential in steering the course of mathematics in the latter part of this century; however, *the same cannot be said of the theorems which remain clumsy*." (Rota 1997b, p. 220)

[21]Same statement a little more passionate and anecdotic in another indiscrete thought: "Algebraic geometry has been the bottom line of mathematics for almost one hundred years; but perhaps times are changing. The second story is more sombre. One day, in my first year as an assistant professor at MIT, while walking down one of the long corridors, I met Professor Z, a respected senior mathematician with a solid international reputation. He stared at me and shouted, 'Admit it! All lattice theory is trivial!' I did not have the presence to answer that *von Neumann's work in lattice theory is deeper than anything Professor Z has done in mathematics. Those who have reached a certain age remember the visceral and widespread hatred of lattice theory from around 1940 to 1979; this has not completely disappeared*." (Rota 1997b, p. 52)

[22]Same position in 1973: "The theory of distributive lattices is richer than the better known theory of Boolean algebras; *nevertheless it has had an abnormal development, for a variety of reasons of which we shall recall two*. First, *Stone's representation theorem of 1936 for distributive lattices closely imitated his representation theorem for Boolean algebras*, and as a consequence turned out to be too contrived (I have yet to find a person who can state the entire theorem from memory.) Second, *a strange prejudice circulated among mathematicians, to the effect that distributive lattices are just Boolean algebra's weak sisters*. More recently, the picture seems to have brightened. The definitive representation theorem for distributive lattices has been proved by H. A. Priestley; it extends at long last to all distributive lattices the duality 'distributive lattice- partially ordered sets', first noticed by Birkhoff for finite lattices. Strangely, Nachbin's theory of ordered topological spaces had been

and reach a full algebraic translation of the *probabilistic intuition* (that of a pointless random variable).

By relaxing some essential set theoretical presuppositions, this construction aims at proposing a clear, consistent and adequate *syntactical presentation* of probability reasoning. For this purpose, it is essential to show that Boolean algebras of random variable are equivalent expressions of the space algebra of events. The next step establishes the isomorphism between (pointless) Boolean σ-algebras (with their homomorphisms) and "a lattice of sets with union and intersections of sets playing the role of joins and meets". The discovery of this isomorphism demonstrated by Birkhoff was the strong and deep incentive behind Rota's research into mathematics and logic. This isomorphism is an isomorphism between two categories: sample space (with their probability-preserving maps) and abstract Boolean σ-algebras (with their σ-homomorphisms). The isomorphisms between lattice operations (meet and join) and set theoretical ones of union and intersection demonstrated by G. Birkhoff, on the one hand, and between lattice operations and logical sum and product (highlighted by P. Halmos, Birkhoff, and Rota himself) on the other hand, helps to unfold new formal constituents required by an adequate pointless probability algebra and to articulate them consistently, i.e. to reveal the adequate syntactical structure behind probability theory.

8.1.4 Formalization of Probability Through Lattices and Cryptomorphisms

But we went maybe too far and too fast. A slower pace explanation of the construction could reveal useful for the philosopher and the lay reader—for us. A paper published by Garrett Birkhoff in 1949[23] puts forward a crystal-clear presentation of lattices as well as some precious indications for the reasons why lattices should play a role in a re-formalization of probability theory.

As probabilities, a lattice (L) is a partial order set or system (noted thereafter: p.o.s.). Considering an arbitrary binary relation such as $\geq$, a p.o.s. is defined by the following axioms: for every x, $x \geq x$ (reflexivity); if $x \geq y$ and $y \geq x$, then $x = y$ (antisymmetry); if $x \geq y$ and $y \geq z$, then $x \geq z$ (transitivity). A lattice is a partial order set satisfying stronger conditions such as the principle of duality, the presence of ideals, a chain form (if for any two elements x, y, necessarily the relation $x \geq y$ or $y \geq x$ takes place), etc. The ideal in a lattice L may be defined as a sub-set H of L such that (1) for x, an arbitrary element of H and a, an element of L, the meet of a

available since 1950, but nobody before Priestley had had the idea of taking a totally disconnected ordered topological space as the structure space for distributive lattices." (*Valuation Ring of a distributive Lattice,* Proceedings of Houston Lattice Theory Conference, Houston, 1973a) (Rota 1973a, p. 577).

[23] *Théorie et applications des treillis,* in *Annales de l'Institut Henri Poincaré*, Gauthier-Villars, Paris, (Tome 11, n°5 (1949), p. 227–240).

and x belongs to H and (2) for x and y elements of H, the join of x and y belongs to H. To put it plainly, let us refer to the definition of ideal (and of its dual filter) in algebraic logic. The set of provable propositions of a Boolean algebra (A) possesses a Boolean ideal (M) "if it contains $p \vee q$ whenever it contains p and q, and if it contains $p \wedge q$ whenever it contains p".[24] The fact that any demonstrably true or false proposition in a set of proposition is reducible to the same "point" or "value" (0 or 1) is just another way to express the absorption laws. The introduction of ideals entails one of the most important proprieties of lattices: the existence of *greatest lower bounds* and *least upper bounds* (*suprema* and *infima*). By introducing new conditions and/or relaxing others, various kinds of lattices result: *complete lattices, modular, semi-modular lattices, distributive or non-distributive lattices,* etc. Corollaries: any sub-algebra of an abstract algebra is a sub-lattice (p. 320) and any partial ordered set is isomorphic to a sub-set of a complete lattice.

It is worth noticing that among the applications (that is "interpretations" or "models") of modular lattices, Birkhoff mentions probability theory and projective geometry. Since: "all the gapless chains in a modular lattice of finite length have the same length", we can define "the rank r [x] of an element in a lattice as the number of elements of the longest chain between 0 and x". We arrive then at the following conclusions: If $x>y$, then $r\,[x]>r\,[y]$. And as a corollary, the following identity "which plays a key role in probability theory" is demonstrable: $r\,[x \cap y]+r\,[x \cup y]=r\,[x]+r\,[y]$ (Birkhoff 1949, p. 234). And indeed, if we reformulate this identity, we get:

$$r[x \cup y] = r[x] + r[y] - r[x \cap y];$$

i.e. the pendent of the probabilistic identity

$$P(A \cup B) = P(A) + P(B) - P(A \cap B).$$

Other consequences: Birkhoff conjectures some relation between complemented modular lattices and the "logic of quantum mechanics". Last but not least, he insists, in many occasions, on the Boolean algebra as the perfect and most common example of distributive lattices with complements. Since there is a perfect match between

[24]We are following here Halmos's definition and can justify this by its own comment: "Although it looks slightly different, this definition is, in fact, equivalent to the usual one: a Boolean algebra is, after all, a ring and a Boolean ideal in the present sense is the same as an ordinary algebraic ideal. Each of the two concepts (filter and ideal) is, in a certain sense, the Boolean dual of the other. (Some authors indicate this relation by using the term *dual ideal* instead of filter). Specifically, if P is a filter in a Boolean algebra A, and if M is a set of all those elements p of A for which $p' \in$ P [read: not p *belongs to* P], then M is an ideal in A, the reverse procedure (making a filter out of an ideal) works similarly. This comment indicates the logical role of the algebraically more common concept; just as filters arise in the theory of provability, ideals arise in the theory of refutability. (A proposition p is called refutable if its negation p' is provable. (…) Logic is usually studied from the "1" approach *i.e.,* the emphasis is on truth and provability, and consequently, on filters." (P. Halmos, *The Basic Concepts of Algebraic Logic*, 1956, p. 371–372.) A Boolean logic will be a pair (A, M) where A is a Boolean algebra and M is a Boolean ideal in A, with A (set of all Boolean propositions) and M (refutable propositions) and eventually M* (filter or dual-ideal) refutable propositions.

Boolean algebras and Boolean rings (rings in which $xx'=x$) and their ideals, and since "*every Boolean algebra is isomorphic to a field of sets and reciprocally*", he concludes by a statement which announces its famous theorem (the *Birkhoff transfom*), which "provides a systematic and useful translation of the combinatorics of partially ordered sets into the algebra of distributive lattices" (Rota 1997a, p. 1441). As concerns the notion of *ideal of events of probability zero,* it is to put it simply the Boolean analogue of the proof theoretical notion of refutable, and of impossible event.

> The setup is not as artificial as it may appear. *Among probabilists, mention of sample points in an argument has always been bad form.* A fully probabilistic argument must be pointless.
>
> To this end, consider two categories: on the one hand, the category of sample spaces and probability-preserving maps between them; on the other hand, the category of abstract Boolean σ-algebras and their s-homomorphisms. There is a contravariant functor of the first category into the second that is an isomorphism of categories (an isomorphism is an unusual occurrence in category theory). This isomorphism of categories makes it possible *to systematically translate set-theoretic concepts expressed in terms of points, sets and functions*, in terms of abstract Boolean σ-algebras and their morphisms, *thereby getting rid of points*. One advantage of this translation is that it *does away with events of probability zero*, since one can deal with the Boolean σ-algebra of events in a sample space *modulo* the ideal of events of probability zero. (Rota 1998a, p. 60)

In his first attempt (Rota 1973a, p. 575),[25] Rota constructed a *functor* called "valuation ring", which associated a torsionless ring $V(L)$ to any distributive lattice L, $V(L)$ being generated by idempotents, and he sketched at the end of the same paper the main lines of an extension of this construction to "probability spaces" and to predicate logic (Rota 1973a, p. 622). In his last paper (Rota 1998a, p. 61), Rota comes back to this functorial translation of the category of sample space (with their applications preserving probability) to the category of abstract Boolean σ-algebra (with their σ-homomorphisms). Since there is a *contravariant functor* (from the latter to the former), we get an isomorphism, which ensures the "translation" from the set theoretical classical presentation to the pointless Boolean σ-algebraic presentation. Following Rota, even if it appears rather artificial, this new presentation is finally much less unnatural than the usual one in terms of sample space and more akin to the intuitive way of probability reasoning, which is "pointless". Last but not least, pointless Boolean σ-algebras standing in-between logic and set theory, and more precisely at the junction between a Boolean algebraic logic and a Boolean σ-algebra of events in a sample space, helps to reveal the *deep syntactical structure of probability*. Following Birkhoff, Rota calls this kind of isomorphism, *cryptomorphism*.

---

[25]"Traditionally, the algebraic properties of Boolean algebras are reduced to those of Boolean rings by a well known construction. A Boolean ring, however, has the double disadvantage of having torsion, and of not being applicable to the richer domain of distributive lattices. In this paper we describe another construction, or *functor*, called the valuation ring, which associates to every distributive lattice L a torsionless ring V(L) generated by *idempotents*. The lattice L can be recovered by giving a suitable order structure to the valuation ring V(L), and thus the entire theory of distributive lattices is reduced to that of a simple class of rings. For example, the representation theory of distributive lattices is subsumed to that of valuation rings, where standard methods of commutative algebra apply." (Rota 1973a, p. 575)

A cryptomorphism works like a *criterion of syntactical equivalence*, "applicable to all axiomatic presentations", and reveals the syntactical structure of a theory whatever its completeness or incompleteness: in the present case, the syntax of "the 'pointless' version of probability, arrows point in a direction that is *opposite the one in terms of which most mathematicians still think*". But at the same time, this isomorphism works in a semi formal way. This is why Rota adds: "At our present state of understanding of formal systems, *it is not clear what such a criterion could be*, even though in mathematical practice there is never any doubt as to the equivalence of different presentations of the same theory." (*ibid.*) Conclusion: Lattices do not act as a substitute of category theory, rather as a paradigm of a syntactically and semantically complete mathematical formalization, and show on isomorphic "models" (semantics) the possibility of two equivalent syntaxes.[26]

> This result, first clearly stated by Garrett Birkhoff and Marshall Harvey Stone, is the *paradigm of formalization of a mathematical theory*. Elementary though it has become after successive presentations and simplifications, the theory of distributive lattices is the ideal instance of a mathematical theory, *where a syntax is specified together with a complete description of all models,* and what is more, a *table of semantic concepts and syntactic concepts is given*, together with a translation algorithm [the *functor* or *functorial*] between the two. (Rota et al. 1988, p. 378–379; 1997, p. 153)

Stones representation theorem of 1936 for distributive lattices is for Rota an imitation of his representation theorem for Boolean algebras (Rota 1973a, p. 577).

[26]While revising this paper, thanks to Mioara Mugur-Schächter, I got acquainted with Olivia Caramello's work, which shares with Rota a comparable interest in logic. I see also some affinity between the way she uses in a freer spirit Grothendieck toposes and the way Rota uses lattices to build "bridges" between different fields of mathematics, not to speak of their similar scepticism vis-à-vis the pretension of set theory or category theory to be the ultimate language of mathematics (semantics or syntax). "Considered the importance of building bridges between distinct mathematical branches, it would be highly desirable that Logic would not just serve as a tool for analyzing analogies already discovered in Mathematics but could instead play an active role in identifying new connections between existing fields, as well as suggesting new directions of mathematical investigation. As it happens, we now have enough mathematical tools at our disposal for trying to achieve this goal. By providing a system in which all the usual mathematical concepts can be expressed rigorously, Set Theory has represented the first serious attempt of Logic to unify Mathematics at least at the level of language. Later, Category Theory offered an alternative abstract language in which most of Mathematics can be formulated and, as such, has represented a further advancement towards the goal of 'unifying Mathematics'. Anyway, *both these systems realize a unification, which is still limited in scope, in the sense that, even though each of them provides a way of expressing and organizing Mathematics in one single language, they do not offer by themselves effective methods for an actual transfer of knowledge between distinct fields*. On the other hand, the principles that we will sketch in the present paper define a different and more substantial approach to the unification of Mathematics. Our methodologies are based on a *new view of Grothendieck toposes as unifying spaces, which can serve as 'bridges' for transferring information, ideas and results between distinct mathematical theories*. (…) In this paper, I give an outline of the fundamental principles characterizing my view of toposes as 'bridges' connecting different mathematical theories and describe the general methodologies which have arisen from such a view and which have motivated my investigations so far. (…). These principles are abstract and transversal to the various mathematical fields, and the application of them can lead to a huge amount of surprising and non-trivial results in any area of Mathematics, so we hope that the reader will get motivated to try out these methods in his or her fields of interest." (Caramello 2010, p. 3–4)

8.1.5 Other Problems, Which Should Be Posited and Could Be Solved Thanks to This Syntactical Clarification

Before moving back to the more general framework of Rota's investigation, let us survey the large field of problems that this new formalization could enlighten and help to solve.

The *second problem* focuses more sharply on the concept of *random variable* and seeks for a generalization of the concept of density of a random variable, which could fit for variables which are neither discrete nor continuous; Rota guesses that this could depend on a more generalized theory of distribution. The *third problem* points at the absence of a structural theory of the Boolean σ-algebras, and involves a whole series of lacks: there is neither a properly structural characterization of the lattice of all Boolean s-subalgebras of P (because there is no "intrinsic definition of the lattice S (family of all s-subalgebras of the σ-algebra P)); nor a structural knowledge of the individual s-subalgebras (von Neumann's theorem stating that any two atomless Boolean σ-algebras are isomorphic has not received yet a simple proof, this proof depending on the proof of the lemma stating that $P(A) = P(A^c) = ½$ etc.).

Fourth problem: the algebraic concept of entropy (in information theory) plays a similar role for s-subalgebras as the concept of probability for spaces of events, but has not received up to now an "information theoretic interpretation". This lack indicates some incompleteness of the actual axiomatic presentation of the concept, which holds for "atomic s-subalgebras" only; "all definitions attempted so far of the entropy of an atomless s-subalgebra rely on crude approximation techniques whereby an atomless s-subalgebras is approximated by a sequence of atomic s-subalgebras" (Rota 2001, p. 69). On the other hand, the physical definition of entropy in thermodynamics, in terms of random variables requires a refinement of the algebraic notion of *conditional entropy*. Yet, despite attempts (viz. Shannon entropy) to give an algebraic formal model, these attempts stumble over the continuous character of Boltzmann conditional entropy; the "abyss" between both, and the dilemma of choosing between two models (in terms of "measure of randomness of probability densities" or of "measure of randomness of s-subalgebras") could be seen as a sign of inappropriate character of the latter program. Rota interprets it rather as another symptom of the inadequacy of "the present Kolmogorov formulation of probability in terms of events (…) to the phenomenological needs of probability" (Rota 2001, p. 72), and consequently of the need for a firm revision of Kolmogorov axiomatics, and more specifically his consistency theorem (stating "that under certain technical conditions any consistent system of joint densities can be realized by a set of random variables in a sample space") and his concept of randomness.

> To all practical purposes a sample space is completely determined by a pre-assigned set of consistent joint densities of random variables. The Kolmogorov consistency theorem confirms the realistic view of randomness. What are given are densities, or rather, the histograms of certain densities. One may argue that every probabilistic notion depends on such densities. (…) It may well tum out that an extension of the notion of sample space may have to take its lead from Kolmogorov's consistency theorem, by allowing a consistent set of measures, which are not necessarily probability densities. (…). However, *I have no constructive sug-*

> *gestion on how this extension could be carried out, nor what notion may one day replace our sample spaces*, if Kolmogorov's consistency theorem is ever extended to include infinite measures. (Rota 2001, p. 72)

Incidentally Rota develops an historical view, which relies on counterfactual lines of development.

> As a matter of fact, after the publication of Kolmogorov's tract of 1932, there remained some diehard mathematicians who refused to work with random variables, and who insisted that all of probability should be done directly with densities. One of these mathematicians was Aurel Wintner of Johns Hopkins University, who published a set of notes on probability formulated entirely in terms of densities. In these notes Wintner proved some important new results, well ahead of their time; however, because he wrote in terms of densities instead of random variables, his work was ignored, and his discoveries are now attributed to others. (ibid.) (Emphasis mine)

Problem five: the "complete and utter lack of justification" of the maximum entropy principle is a scandal. This problem is closely related to the former ones, since it depends on the generalization of the concept of sample space. *Problem six* concerns the divergence between the two concepts of "conditional probability" and "conditional expectation" (see Rota 1960). After Kolmogorov, the latter has been considered as "a special case" of the former, but this subordination is nothing but obvious. Filling this new lack of justification amounts to elucidating the algebraic underpinnings of Bayses' law used by statisticians, which states that a conditional probability (probability of event A conditioned by a probability of event B) (written $P(A|B)$ is expressed by the proportion $\frac{P(B|A)P(A)}{P(B)}$, which remains mysterious unless one goes back to its original source, in Laplace (Rota 2001, p. 76). I shall leave aside Rota's reconstruction of Laplace's reasoning which is paradigmatic of the "realistic" interpretation of sample spaces and fits perfectly what one of the tenants of non-kolmogorovian probabilities and quantum probabilities calls the "urn model" (Accardi 2003). Let us go straight to the general conclusion: namely discrepancy between sound statistical reasoning and the lack of formal justification of current procedures, which may be taken "as further evidence for the need of a rethinking of the notion of sample space and of Kolmogorov's consistency theorem along the lines" (Rota 2001, p. 77) just proposed.

Jumping over *problems seven and eight*, we arrive at what was announced as cases of "deviant" probabilities, and which represents one of the strongest epistemological incentives for a deep revision of probability. Rota's stance is not the usual one, since he does not limit the prospect to providing an axiomatic system for quantum mechanics, but, he takes the so-called quantum paradoxes as indexes of limits of the formal concepts involved in probability theory and, consequently, as a motivation for a revision of fundamental formal concepts, starting with that of σ-algebra. By looking at the way Rota enters *problem nine*, we get a perfect example of the interdependence between the building of counterfactual, ideal historical perspective and the opening of new formal-logical perspectives.

> All deviant versions of probability theory that have been attempted so far take their lead from a simplification of the structure of the lattice of a s-subalgebras. Let us see how one takes

the lead from such a simplification to arrive at quantum probability. *Our motivation does not follow the historical development of quantum probability, which originated in quantum mechanics. It may instead be taken as an ideal history of how quantum probability ought to have developed. The* 'Leitfaden' *is a hard look at the concept of independence of σ-algebras, with a view to discovering a hidden inadequacy in this concept, which will lead to an enlarged and hopefully more natural definition.* (Rota 2001, p. 80) (Emphasis mine.)

Here we enter the dimension of the counterfactual history. What is named here ambiguously "quantum probability"[27] corresponds to some revised version of the somewhat infamous Birkhoff and von Neumann (Von Neumann 1936) "quantum logics",[28] rather than to the challenging perspective followed by other authors, under the title of quantum probability (Segal 1947; Pitowski 1989, 2005; Accardi 1982a, b, 1999, 2010). But although the latter has developed independently, Rota is indirectly associated with it.[29] Although Accardi asserts that Rota's own contributions to probability theory in general and *quantum probability in particular* are minor, and that among the twelve problems exhibited by Rota only a few address the foundational problem, to my view, this opposition rests on a series of misunderstandings, that will be elucidated in the following: on (a) what *founding* means; on (b) two different senses (narrow and wide) of the expression "*quantum probability*"; on (c) diverging view on what formalizing means and on (d) the freedom allowed to mathematical analogy, "that most effective breaker of barriers" (Rota 1997b, p. xi).

---

[27]See above: note 13 and 15.

[28]We learn from Rota, via S. Ulam, that von Neumann himself, without giving up his intimate conviction about the inadequacy of naive set theory to contribute to the foundation of quantum mechanics, admitted his own failure to give an alternative foundation. "Most of von Neumann's work in pure mathematics (rings of operators, continuous geometries, matrices of high finite order) is concerned with the problem of finding a suitable alternative to Boolean algebra, compatible with the uncertainty principle, upon which to found quantum theory. Nevertheless, the mystery remains, and von Neumann could not conceal in his later years a feeling of failure over this aspect of his scientific work (personal communication from S. M. Ulam)." (Rota et al. 1992, p. 168; Rota 1987a, p. 26). But Rota (1997: 1444) "Haiman's proof theory for linear lattices brings to fruition the program that was set forth in the celebrated paper "*The logic of quantum mechanics*", by Birkhoff and von Neumann. This paper argues that modular lattices provide a new logic suited to quantum mechanics. The authors did not know that the modular lattices of quantum mechanics are linear lattices. In the light of Haiman's proof theory, we may now confidently assert that Birkhoff and von Neumann's logic of quantum mechanics is indeed the long-awaited new 'logic' where meet and join are endowed with a logical meaning that is a direct descendant of 'and' and 'or' of propositional logic." (Rota 1997a, p. 1444)

[29]While working on this paper, Luigi Accardi informed me that: "Rota è legato, sia pure indirettamente alla nascita della probabilità quantistica poiché, verso la fine degli anni 1970 (o primissimi 1980) egli tenne un corso di probabilità a Pisa e mi invitò a fare un seminario in quell'ambito. Nei pochi giorni che rimasi a Pisa decisi di leggere il lavoro di Bell sulla sua nota disuguaglianza, che molti mi avevano segnalato come un contributo fondamentale ai fondamenti della MQ. Era da molti anni che riflettevo sul problema sollevato da Feynman, relativamente all'esperimento delle due fenditure, e ricordo ancora oggi l'eccitazione che provai quando, nella Biblioteca della Scuola Normale Superiore di Pisa, intuii che sia questa analisi, sia la disuguaglianza di Bell, *erano basate sulla stessa ipotesi matematica, usata implicitamente da entrambi gli autori*: *la possibilità di descrivere dati statistici, provenienti da esperimenti mutuamente incompatibili, all'interno di un unico modello Kolmogoroviano.* " (Personal communication, 6th of July, 2015).

Strangely enough, regarding "quantum probability", the program set up by Rota in 1997 departs from the usual approach by mathematicians, including Birkhoff and von Neumann, who were "hypnotized" by "modular identity" (Von Neumann 1936, p. 114–115), which was supposed to play, according to them, a central role in quantum (non-distributive) lattices. That "distracted them from the fundamental property: commutativity of conditional expectations" (Rota 2001, p. 81), as it appears from von Neumann (1955, pp. 170; 198; 201; 206–211). Starting with an algebra of commuting operators in Hilbert space, von Neumann recognize that dependence on time must not be "considered in forming the Hilbert space" (Neumann 1955, p. 198). Von Neumann is led to consider the formula (**P**) which contains "all the statistical assertions of quantum mechanics which have been made thus far" as bound to the condition or arbitrary order, i.e. as commutative, hence to attribute an inner statistical meaning to physical formulas (Neumann 1955, p. 206), and finally to adhere to the non-causal interpretation of statistical physics (Neumann 1955, p. 210). Yet what is lacking, following Rota, is *not* the construction of a non commutative algebra of observables, for "we know" (i) that such a "quantum lattice is isomorphic to a lattice of stochastically commuting σ-algebra ", and (ii) how it is possible to construct directly an algebraic analogue of the observables of quantum mechanics (a Banach algebra, hyperfinite factor); (iii) we can even guess from Connes's non-commutative geometry that there exists *other phenomena* which could fit this algebra. What is lacking in the present case is "a real-world phenomenon, *other than those of quantum mechanics*, which will turn out to be better modelled by quantum probability than by ordinary probability", and would motivate us to consider quantum probability formalism as something more than an ad hoc logical construction. More precisely the problem is: "Whether or not a phenomenon exists that *can only* be explained by quantum probability is at present a baffling open question" (Rota 2001, p. 82) (Emphasis mine). We shall leave it open as well, even though many proposals have been made in more or less recent period, by researchers from various fields and formations.[30]

[30]See the "natural" although somewhat idealized *chameleon* of L. Accardi (*Urns and Chameleons*). We have something similar with the embedment of quantum experiments into a non-differentiable and fractal space-time proposed by L. Nottale; many macro-physical and classical phenomena appeared describable in terms of periodical peaks of probabilities modelled (as a "Schrödinger's flower") (Nottale 2004), Daniel M. Dubois Editor, *American Institute of Physics Conference Proceedings*, 718, 68–95 (Nottale 2004, p. 68–95)). —Departing from quantum probabilities, see the new epistemological approach promoted by Mioara Mugur-Schächter (Mugur-Schächter 1994 and 2009), involving not only a new layer for QM, but also another deeper concept of probability (Mugur-Schächter 2014) and a new general epistemology (Mugur-Schächter 2002).—Another direction, stripped of the usual mathematical apparatus of QM, is represented by the search for anthropological or psycho-neurological quantum phenomena: M. Bitbol, *La structure quantique de la connaissance individuelle et sociale*, in *Theorie quantique et sciences humaines*, 2009, CNRS Editions, Paris). Last: Pierre Uzan, *Conscience et physique quantique*, Mathesis, Vrin 2013.

8.2 The General Formal Frame of Rota's Historical Perspectives

From the preceding survey arises a question concerning the relation between the formal logical program and ideal history. What is the connection between detection of virtual, possible or potential lines of historical development and a combinatorial construction of cryptomorphisms? This question gets heavier and more complex as soon as one considers that mathematical logic is part of the problem rather than a tool, and that the most established theory can become questionable and problematic once projected onto the background of relevant counterfactual theoretical possibilities. To stick to our example, we must recall that, seen from afar, the reform of probability theory has been associated by Rota, for a while, to the construction of an algebraic logic of the kind initiated by Halmos.

8.2.1 What Was the Nature of Rota's Interest in Formal Logic and Phenomenology?

From 1967 on, Rota never gave up this project of a reform of logic by algebraic translation and import of some new formal concepts. The 1973 paper proposes clearly to apply the cryptomorphic construction and the functor called *valuation ring*, which "associates to every distributive lattice a torsionless ring generated by idempotents". Without going into the details of this construction, one should insist on its great formal generality, since an entire class of lattices becomes a simple class of rings, and consequently, every *representation* of distributive lattices becomes an instance of a valuation ring.

> The category of valuation rings is equivalent to the category of distributive lattices. It has a generator, namely, the valuation ring of the two-element lattice; we shall see that this fact can be used to obtain a representation theorem for valuation rings. *Actually, more is true, but neither category theorists nor first-order logicians have yet invented a precise way of saying it, though the appropriate term was introduced long ago by Birkhoff*: the two categories (or first-order theories) are cryptomorphic. In other words, to every fact about one there 'naturally' corresponds a fact about the other. The algebraic structure of a valuation ring is richer than that of a ring. (Rota 1973a, p. 582–583)

Pushing forward Halmos's program of algebraic logic, Rota mentions, as a possible application and as "the most promising outcome", the *translation* of "the notion of quantifier on a Boolean algebra" into that of "linear averaging operator": "in this way, problems in first order logic can be translated into problems about commuting sets of averaging operators on commutative rings" (Rota 1973a, p. 576). The final conclusion relates this translation of predicate logic to that of probability, which gives the starting point of a contribution with D. P. Ellerman (*A Measure Theoretic Approach to Logical Quantification*) (Rota 1978), including the cryptomorphic bridge between probability measure and quantification.[31] We have here, as has been noticed by Jean Dhombres, a perfect example of Rota's synoptic ability and his quasi-polymath profile.

> Rota was a master at perceiving connections among disparate subjects and in this instance, he took a combinatorial point of view. In particular, he used canonical idempotents instead of characteristic functions, borrowing the idea from Louis Solomon's work on Burnside algebras [...]. But something else was occurring: Rota *was no longer analyzing the structure of averaging operators, but constructing averaging operators to fit the application. For the study of logic, he constructed specific 'averaging' operators on valuation rings, which simulated the properties of existential quantifier*. In this way, 'Problems of first order logic, such as the decision problem, can be shown to be equivalent to algebraic problems for valuation rings with averaging operators'. *Unfortunately, it proved quite difficult to carry through this program rigorously, as one can see in the paper [...], written with David Ellerman*. (Dhombres 2003, p. 163)

Following Birkhoff example, Rota inverts what would be the current perspective for mathematicians.[32] And the eventual commentator's surprise and deception echoes the reaction of Rota's teachers in mathematic, who did not understand his interest in formal logic, and could guess his philosophical motivations even less. As to the latter, they did not stem from the deceptive perspective of some philosophers misled by the "fake philosophical terminology of mathematical logic" "into believing" that it "deals with the truth in the philosophical sense". This variety of believers form the group of philosophers under the bad influence of mathematics, whose position represents a misunderstanding concerning mathematics as well as philosophy. They fall into that typical artificial philosophizing whose symptoms, in certain schools of philosophy, are: "casual and self-satisfied symbol-dropping of mathematical logic"

---

[31]In its presentation from 1973, the analogy between probability and predicate logic is mediated through the following steps: assignment of a probability measure m to the canonical idempotents; definition of a lattice space norm on the valuation ring by using the linear functional |Sa(e)e| = S|a(e)|m(e); complementation of the resulting normed linear space which can be seen as the space of all integrable functions; representation of the averaging operator as a conditional expectation operator, which once relaxed from the restrictive condition that "every element of the range be finite-valued", can apply universally). (Rota 1973a, p. 622, 1960). We read page 607: "An averaging operator on a valuation ring $V(L)$ is a linear operator A such that: (1) $Au = u$, $Az = z$. (2) $A(fAg) = Af\ Ag$. (3) If f is in the monotonic cone, so is Af. Sometimes these operators go by the name of Reynolds operators. In probability, they are called *conditional expectations*."

[32]Despite Rota's somewhat severe judgment about category theory, it seems that some of his insights are founding a (maybe) successful realization in the work of Olivia Caramello (Caramello 2010). Her purpose, as exposed in *The Unification of Mathematics* via *Topos Theory*, is precisely: "to present a set of principles and methodologies which may serve as foundations of a unifying theory of Mathematics", "based on a new view of Grothendieck toposes as unifying spaces being able to act as 'bridges' for transferring information, ideas and results between distinct mathematical theories". Neither this use of classifying toposes, nor Rota's construction of cryptomorphism as "bridges" are trivial and limited to translating mathematical theories into one another, but include logical and syntactical properties and mathematical properties (in Rota's case, for instance, between logical quantification and probability, and in Caramello's, between completeness and joint dip). For similar reasons, by standing at the crossroads of mathematics and logics and challenging their respective and exclusive pretention to fix the formal frame of scientific rationality, Rota and Caramello contributions catalyse negative reactions from the part of "pure" mathematicians and "pure" logicians.

and contempt for history and evidence (or intuition), in the Husserlian sense of the term.[33] Abandoning this belief was the due price that logic had to pay, in order to become "a successful and respected branch of mathematics" (Rota 1997a, p. 97) (Rota 1991, p. 133–136). Giving up any philosophical pretension means that logic can be treated as any other branch of mathematics, and that we should not inject into it any philosophical meaning. Yet that does not mean we have dismissed any research concerning a philosophical logic. On the contrary, we should keep logical questions open. More: since mathematical logic either has nothing to do with the way we think, (as some logician think) or, in the better case, "formalizes only very few of the processes by which we think", *philosophic logical* investigations should remain at the same time an open field as well as an open source for potential enrichment of formal logic (preserved from any bad influence or contamination by mathematics) (Rota 1997a, p. 92–93, 1990c). Whatever the success or failure of Rota's attempt to reconstruct algebraically the quantifiers, in order to decide if it is a success or a failure, it is worthwhile seizing precisely what his logical aims are. In doing so, we gain a better understanding of the strange transference of an operator from a mathematical theory (operator constructed in the context of fluid mechanics) to logic (predicate calculus), via algebra. I mean Reynold's operator.

Although he recognizes a larger scope for Rota's exploitation of the "idea of averaging operators,"[34] Dhombres considers that the main mathematical goal Rota had in view in using Reynolds equation (a generalized form of Navier-Stoke equations)[35] was an enlargement of the field of applications (models) and a unification of measure theory.

> The mathematical task in Rota's mind was to understand the proper role of the identity [equation], but not in an axiomatic or formal way. He thought that studying a variety of situations where Reynolds operators played a role in real world (or the phenomenological world) would bring a fuller understanding of the meaning of Reynolds identity. Although

[33]Rota 1991, p. 134 & 135, respectively.

[34]"In 1973, Rota exploited the idea of averaging operators in the radically different context of valuation rings of a distributive lattice in [50] [i.e. here (Rota 1973a)]. This paper is fascinating because as Rota wrote, 'the method of presentation is deliberately informal and discursive'. Much of the reasoning is by analogy, but not in the structuralist or Bourbakian sense, and many of the results are only heuristic. The aim is grandiose, nothing less than the 'linearization of logic'." (J. Dhombres, Reynolds and Averaging Operators, in Gian-Carlo Rota, *On Analysis and Probability*, Selected Papers, Birkhaüser 2003) (Dhombres 2003).

[35]Reynolds operator is an averaging operator, i.e. a heuristic way of simplifying Navier-Stoke equations that reveal fundamental for hydrodynamic (description of gas, fluids, etc. flows) and beyond. Their form is quite complex, combining three quantities of the material elements (mass, moment, energy: whose density is measured: r(r, t); velocity field: $\mathbf{u}(\mathbf{r}, t)$ and energy density per mass-unity $e(\mathbf{r}, t)$. *For* more details see: Rota, *On the Passages from the Navier–Stokes Equations to the Reynolds Equations* (Rota 2003, p. 137–139). From the start, Rota's interest is the representations (i.e. models) of the Reynolds operator, in ergodic theory, spectral theory, and probability theory. The simplified form writes: $\mathrm{R}\,(fg) = RfRg + \mathrm{R}\,\{(f - \mathrm{R}f)\,(g - \mathrm{R}g)\}$. The passage from N-S.'s operator to Reynold's is smoothed, so to speak, by considering the functions of Navier-Stokes equations as *random functions*. (NB. The averaging form is: $A[f(Ag)] = (Af)(Ag)$, with f, g two functions belonging to a commutative algebra.) Reynolds operator is thus seen as a *conditional expectation operator* of form: $E^{S'}\,(uE^{S'}v) = E^{S'}uE^{S'}v$.

> Rota believed that a mathematician should take inspiration from physical world to do pure mathematics, he disagreed with the point of view of those mathematicians who insisted on the primacy of problems coming from technology which had to be solved, whether or not they lead to pure mathematics results. (Dhombres 2003, p. 157).

It is true that Rota did not *only* understand applications in an empirical sense, but had *also* in view formal models (and has we saw, vice versa). Significantly (remembering Rota's words as a supervisor of his PhD thesis, in 1970), Dhombres reports: "how he insisted on concrete realizations of such operators on different function algebras *and* how he focused on the way the Reynolds and related identities could simplify known results" (Dhombres 2003, p. 157); and even how gradually the Reynolds operator disappeared from his writings, after 1974, for the reason that it might have appeared too complicated or too rich. But, at the same time, even before that, from 1964 onward, Rota did not pretend anymore to use it as a means to produce a unified measure theory, and must have taken it as no more than a game which could generate "computations which could be used or illustrated by examples in functional analysis." The major reason invoked by Dhombres is that there were alternative and just as efficient "tools" for the same purpose, "such as derivations and Baxter operators."[36] From such a shift, we should surmise either an excessively adaptive attitude or an inconstancy in Rota's position, moving from one project to another. His position in and vis-à-vis the history of mathematics should appear clearly in both senses of the term, discrete (discontinuous and anecdotic).

> For Rota, progress in mathematics could sometimes be achieved by recognizing a pattern or working by analogy. Reynolds operators, which had been a sort of exercise for the mind, remained a source of inspiration for him. This suggests that other authors, while not specifically working on Reynolds operators, might also be stimulated by Rota's methods. A crude way of measuring this is a citation search, but mathematical papers tend to cite direct technical links rather than indirect inspirational ones. (Dhombres 2003, p. 157) (Emphasis mine)

Last but not least, as a side effect, Dhombres suggests that: "the need to represent Reynolds operators in different concrete situations might have fostered the development of Rota's phenomenological point of view for mathematics. He first explained his views in 1973 in a paper [51] [i.e. (Rota 1973b), republished in (Rota et al. 1992)] in Husserl. See also [52] [i.e. (Rota 1997a)." We shall say a few words on the text referred to under [51]: "Edmund Husserl and the reform of logic", *Explorations in Phenomenology*, p. 299–305. Then published in *Indiscrete Thoughts*). Contrary to what Jean Dhombres suggests in his nonetheless enlightening article,[37] Rota does not seek in phenomenology "mere" explanations about possible applications in different contexts of the Reynolds operator. Rather this minimalist interpretation explains

---

[36]See Rota (1969, 1972, 1995, 1998a), Cartier (1972).

[37]"In 1973, Rota *exploited the idea of averaging operators* in the radically different context of valuation rings of a distributive lattice in [50] [here, under (Rota 1973a)]. This paper is fascinating because as Rota wrote, 'the method of presentation is deliberately informal and discursive'. Much of the reasoning is by analogy, but not in the structuralist or Bourbakian sense, and many of *the results are only heuristic. The aim is grandiose, nothing less than the 'linearization of logic*'", Reynolds and Averaging Operators, in *Gian-Carlo Rota, On Analysis and Probability, Selected Papers,* Birkhaüser, 2003, p. 163 (Dhombres 2003, p. 163; Rota 1990b).

instead Dhombres severe judgment and underestimation of the importance of Rota's interests for logic.[38] But this is an old story, as we learn from Rota's *souvenirs* of Fine Hall (his Princeton years). In a deep and explicit affinity with Husserl's logical investigations, Rota was seeking new resources, which could help towards a methodical understanding of an implicit but necessary moment of the history of science (and of mathematics therein), and which might operate—as appears necessary—a deeper transformation of formal logic.

On the basis of this survey, we shall recall the general formal frame inside which Rota builds up his historical positive and ideal perspectives. This frame consists in—or at least can be expressed as—a dynamical interpretation of the logical relation between syntax and semantics. Without such a frame, Rota's specific critical position and statements on past and on-going history of science would be hardly understandable. Following Rota, let us call it *ideal history* (*history as it should or could have been if...*). This counter-factual history is combined with a view on history *as it actually happened*, in order to get a more acute view of the latter as well as a richer view of what should and could happen in the future. Ideal history represents the strict opposite of an ideological history. From that perspective, Rota's repeated, sharp and witty critiques of the prejudices of different sorts of academics (mathematicians, philosophers, logicians, A.I. researchers, etc.) does not express either an indictment for intellectual sports games, or a form of socio-psychological complex. If it were to be attributed some kind of psychological disposition, it should be labelled: a *concern for the communication of forms*. From this point of view, the purpose of Rota in his generalizing of Reynold's identity for instance is a perfect example of his strategy of radical, systematic and rigorous epistemological *decompartmentalization* (between probability theory, set theory, first order logic, etc.).

The resources of ideal history have affinities to those of Husserl's historical stance, and stand in contrast with ideological exploitations from artificially shrunk perspectives, whose various individual and sociological symptoms are reductionism, snobbism, and whose general epistemological disastrous result is self-contained ignorance.

8.2.2 Historicity of Science as an Eidetic Structure in Husserl's Phenomenology

Conversely, this eidetic structure is not teleologically closed. In order to avoid or at least to prevent endless arguments, let us admit that, in Husserl as in Rota, the

[38]"I was *too young and too shy to have an opinion of my own about Church and mathematical logic. I was in love with the subject, and his course was my first graduate course*. I sensed disapproval all around me; only Roger Lyndon (the inventor of spectral sequences), who had been my freshman advisor, encouraged me. Shortly afterward he himself was encouraged to move to Michigan. Fortunately, I had met one of Church's most flamboyant former students, John Kemeny, who, having just finished his term as a mathematics instructor, was being eased—by Lefschetz's gentle hand—into the philosophy department." (Rota 1997b, p. 7).

main feature of this historical structure is its openness. In the *Krisis*, Husserl (1950b) describes it as an historical horizon-structure in which science is at work, progresses, stops, stagnates, operates and endures modifications sudden or continuous, ordinary or extraordinary, etc. such that a minimal unitary sense is preserved, transmitted, traditionalized, etc. If we look more closely at this horizon-structure, we notice three main constitutive moments, which make understandable the dynamical traits of such an historical process (adaptive, projective, anticipative, uncertain, corrective, exposed to failure, etc.). Although they should be well known to any phenomenologist, let us enumerate them.

In talking about the most general scientific horizon of natural sciences, Husserl distinguishes between three moments of an eidetic structure corresponding to the a priori of the history of natural sciences and experience in the broader sense of the term. Three kinds of ideas or eidetic moments compose this structure (it goes without saying that the idea is not be taken in a conceptualist sense, rather in a very peculiar "platonoid" one). The whole eidetic frame is perspective-like or projective-like, which sheds light on parallels repeatedly exposed by Husserl, between sense individual perception and eidetic perception.

> Singular experience, experience of individual existence, results in no assertion that can be justified objectively. But then how can singular judgments of fact be valid at all? How can the *experienced* world even be in truth? Being reveals itself to be an ideal pole for 'infinities' of presumptive self-evidences with self-evidently given sense-adumbrations ('sides', appearances) through which the same being is adumbrated in self-evident manner but is [merely] presumptive in every finite series [of adumbrations]-though this is a legitimate presumption. * (*[Note] But here it is always merely nature which is in view, and *an idea of reality* is thereby presupposed for the world, *whose correlate is the idea of a truth-in-itself, the idea of a mathematically constructible truth, even though it is [given only] in any number of approximations.*) (Husserl, *The Crisis of European Sciences and Transcendental Phenomenology, An Introduction to Phenomenology*, tr. David Carr, Northwestern University Press, Evanston) (Husserl 1970, p. 304)

(1) The idea as a pole of truth as correlate of real being, as idea-*in-infinitum,* as unifying "point" at an infinite distance, so to speak, of an open infinite variety of presentations, of "experiences"; identity as a converging point of experiences in a sense sufficiently enlarged as to embrace all possible experience.[39] This idea as a pole is totally indeterminate for us, but determinable in an endless opened process, giving rise to an infinity of partial, and "at-first-go", "first-stroke ideas" (*Anhiebe Ideen*), which are as just so many anticipative presentations of the idea at a distance, and correspond to the ordinary fund of material eidetic (actual and potential) sciences.

> *Real truth is the correlate of real being*, and just as real being is an infinitely distant idea, the idea of a pole for systematic infinities of appearances, of 'experiences' in constantly

[39]Rota's preface to P. J. Davis and R. Hersh's book on *The Mathematical Experience*, Birkhaüser, 1981, p. xix. "They have opened a discussion of the mathematical experience that is inevitable for survival. Watching from the stern of their ship, we breathe a sigh of relief as the vortex of oversimplification recedes into the distance" (Rota 1981); see Husserl, in *Formal und Transcendental Logic,* § 60 and my comment on this parallel enlargement of the concepts of *experience* and *evidence* (Lobo 2006, p. 147–148; p. 160 sq.).

legitimate presumption, so *real truth is an infinitely distant idea*, [that of] what is identical in the agreement of experiential judgments, in each of which truth 'appears', achieves legitimate subjective givenness. (Husserl 1970, p. 304)

(2) The form of the pole-idea of truth, or more briefly the form-idea, with which we ascend out of mere relativism as well as out of any closed teleology. As form of the idea at the distance, it functions as norm. Such is, since the *Prolegomena* (§ 13–14), the theoretical fundament (*Fundament*) of norms. A *logical norm* in general is issued from a conjunction of two constitutive modifications: formalization (on an eidetic basis) (*Formalizierung*) and normation (*Normierung*). We know since the *Prolegomena,* that *normation* is a specific intentional modification, which converts a (previously) objectifying act (logical in the strict or in the broad sense of the term) into a (theoretically) founded practical one. Such is, for instance, the conversion of an arithmetical theorem $(a+b) \times (a-b) = a^2 - b^2$ into a rule of calculus: "in order to multiply the sum and the difference of a and b, make the difference of the squares of a and b" (Husserl 1988, p. 4–6). In turn, this rule can be thematically considered and eventually converted into a mechanical procedure analysable in terms of more elementary operations (see Rota 1985, 1986b, 1989b). Those operations can be explicitly grasped by formalizing the stock of anticipative ideas, which keep on functioning as partial presentations, as "adumbrations" of the idea at a distance: in other words, an empty form depending *on them and revisable*, but at each stage of development of experience functioning as "absolute norm for every construction of partial and anticipating ideas."[40]

The infinitely distant idea is determinable a priori in the pure form of generality which contains all possibilities, and in accord with this form one can construct, out of the finitely closed total experience (that is, out of its relativity 'closed-off appearance', out of the realm of determined sensible things, out of the sensible experiential predicates) an anticipation of an appropriate idea required by this experience and implied in it. (ibid.)

Although pre-grasped as a form, the "infinitely distant idea" lies beyond any actual formal system, and thus escapes the traditional objection that Husserl's conception of logic, like Hilbert system, would be refuted by Gödel's incompleteness theorems. First, there is no contradiction in stating that, if a system is syntactically incomplete it is semantically incomplete; although implicit, semantic aspects are necessarily involved in the syntactic proof of incompleteness, since the non-derivable true proposition does not belong in any trivial model of such a formal system.[41] Secondly, even though syntactically complete, any formal theory (be it physical or mathematical) would remain semantically incomplete, since no actual formal system can verify that

[40]Husserl writes: "*eine Idee als zugehörige, die aber nie letzte ist, sondern Anhieb, in gewisser Weise Darstellung der im Unendlichen liegenden und unerreichbaren Idee, von der nur die Form als absolut Norm aller Konstruktion der Anhiebe gegeben ist.*"

[41]Gödel's first theorem states that there will always be, at least, arithmetic problems, which are syntactically expressible and intuitively solvable, whose truth is not demonstrable within the formal system.

a proposition (a class of propositions) is (are) true in all possible anticipated models. Against a deeply rooted legend, Gödel's theorem refutes Hilbert's and similar programs, but presupposes Husserl instead.[42] If this analysis is right and consistent, Husserl's notion of definability seems more closely related to Tarski's definability than to Hilbert's syntactic notion of completeness, and is maybe wider in scope than the former, for it *describes* the dynamic of non-trivial extensions, and correlatively the non-trivial generalizations and extensions of incompleteness proofs.

> [I]n the philosophical description of the mathematical enterprise it is phenomenology, such as we find in Husserl's Logical Investigations, and not any logic such as logic is conceived today, whether standard or not, that gives a realistic description of what goes on in mathematics. Husserl's inspired insight into the future - too early an insight perhaps - forecasts a reform that only lately has come to the fore among mathematicians, namely, the idea that set theory itself does not found mathematics, that the theory of sets, which is the last and perfect rendering of the Platonic idea of objectivity, has, after the successful formalization it received beginning with Cantor and throughout the spectacular developments in our century, revealed at last its limitations. (Rota 1991, p. 137–138) (Emphasis mine)

What is developed here is the core problem of epistemology, the relation between intention and fulfilment.

(3) The partial, one sided, first-stroke ideas, corresponding to that which is really given in experience; each experience acting as an anticipation, whose validity consists in its being adequate to the idea as pure form, i.e. in the formal consistency of its content. Each is, from the point of view of the modality, a *presumptive actualization* of a possible determination; construction of "singular predicable determinations" of the pole idea (an X); "partial", "unilateral truths", allowing indeterminately many other "sides" of the ultimate truth. They are, so to speak, "modally sensitive", likely to be modified in their validity, i.e. modalized:[43] "every experience can a priori contain elements of discrepancy which will be separated out in further experience and its synthesis"; correlatively, it is always possible that each of the partial ideas, of the determination-idea "which is to be gained from" reveals itself "not only one-sided but also in part false, though required, for the sake of truth, by this experience up to now", that is without losing the thread of the historical fundamental motivation, without which historians would be deprived of the chains and background of motivations explaining why, in such a context, researchers in good faith have believed scientific truths which were revealed afterwards to be partially false, or approximately true, etc. Let us insist by quoting again Husserl:

> *Included in the form of the idea of something real are one-sided, partial ideas*, just as, in the full truth, which determines the entity (the totality of the predicates, belonging to it which determine it as itself), there is a multiplicity of individual predicable determinations, of individual truths, which leave the being still undetermined in other directions. *Insofar as every experience can* a priori *contain elements of discrepancy which will be separated out*

[42]See Okada remarkable papers (Okada 1998 and 2000) and Mark Van Atten's comments in (Atten 2015, p. 127, note 70; and among other passages: pp. 83–86, 90–92, 127, 161)

[43]Lobo (2012, p. 172–185).

> *in further experience and its synthesis, the idea-determination which is to be gained from it is capable of being not only one-sided but also in part false, though required, for the sake of truth, by this experience up to now*. (Husserl 1970, p. 304–305) (Emphasis mine).

Openness means more than falsifiability: it means modalizability. Let us see now, how, through his readings and the mediation of others (Church, Ulam, Sokolowski, etc.), Rota transposes the whole structure described by Husserl to the mathematical experience.

8.2.3 The Dynamic of the Form-Idea Moment

Le us restrict to the second moment of the structure holding for history of science, and show that, the narrow frame of history of formal sciences (i.e. logics and mathematics) involves by necessity a similar dynamical dimension. Describing *formally* the whole structure is indeed the real task Husserl will have had to face, and that should have motivated a revisionist view on a (recently arisen) mathematical logic, a task rarely taken seriously, with the exception of Rota, whose work evolves and progresses consciously within this structure.

By embedding in this frame the most advanced *formal* researches, Rota prolongs Husserl's reform of logic (mathematical logic, algebraic logic, model theory, semantics, lattice theory, etc.), which aimed at arming epistemology (theory of science) with adequate formal concepts to think that "reality"—apart from any bad Platonism—is precisely another name for this idea-at-a-distance. The normative, somewhat "dictatorial" use of "logic" by mathematizing philosophers rests on a vicious circle or an arbitrary stipulation. And as we already saw, this attitude stems from a confusion regarding the epistemological status of mathematical logic (Rota 1991, p. 133–134). In contrast to this attitude, the major task of phenomenology in relation to mathematics is mostly to disclose the living activity of the mathematician, which is not "that of proving theorems, but another one: *that of proving that all theorems are intuitively evident*, by an evidence that shall be as close as possible to Kant's ideal of an analytic a priori statement" (Rota 1991, p. 136). Mathematics is not a matter of "inferential devices stipulated by some previously agreed-upon logic", nor even of "seeing the truth of mathematical statements", but of "seeing through". But what does this "seeing through" means, and how can it be concerned with analytic a priori statements?

> It is to Husserl that we must turn for an answer. A project of evidence is not to be confused with the fulfilment of that project of evidence. The intentional project and its fulfilling moment are not to be visualized as one-shot phenomena. They are articulated projects pertaining not to individuals, but to the community of mathematicians. (...) Thus, I hope to have made the point that *in the philosophical description of the mathematical enterprise it is phenomenology*, such as we find in Husserl's *Logical Investigations*, and not any logic such as logic is conceived today, whether standard or non-standard, *that gives a realistic description of what goes on in mathematics*. (Rota 1991, p. 136) (Emphasis mine).

In a collaboration with David Sharp and Robert Sokolowski, *Syntax, Semantics, and the Problem of the Identity of Mathematical Objects* (Rota 1988), adopting a frame, similar to Husserl's, Rota extends the scope of cryptomorphisms and creates a tension between formal concepts such as semantics and syntax. Under the name of "identity" or "object", Rota does not understand any kind of equation nor any peculiar structure, but rather some thing similar to Husserl's pole-idea. Taking an opposite (and complementary) stance to the current understanding of model theory, as well as category theory, Rota asserts that a *same* mathematical "identity" ("object"),[44] is the *itself*, the *as-such* of endless objectivizations (and equations), bearing various syntactical and semantical presentations. From an historical perspective, this pre-grasp is exposed to various series of modalizations (doubts, partial, incomplete grasps, etc.). The word "object", "mathematical object" must be understood here as synonyms for "sameness".

> If one accepts our discussion leading to the conclusion that two (or more) axiom systems for the real line (say) can be recognized to present the *same* real line, then one is forced to draw the following consequences: the real line, or any mathematical object, is not fully given by any one specific axiom system, or by any specification of a finite set of axiom systems. The totality of possible axiom systems for the real line cannot be exhaustively spelled out or foreseen. Any mathematical object allows for an open-ended sequence of presentations by ever new axiom systems; that is, for the successive development of ever new axiom systems for what is perceived as one and the same object. Each such system is meant to reveal new features of the mathematical object. Learning about the real line is not a game played with axioms whereby skill is developed in drawing consequences. Rather, the very choice of what properties are to be inferred and of how the theory is to be organized is from the start guided by a pre-axiomatic grasp of the real line. Without a guiding intuition, however unverbalized, such an axiomatic theory cannot make *sense*. Even though the concept may be initially learned by diligently working through one axiomatic approach, that particular axiomatic approach will be shed after familiarity is gained. Thus, in learning through a particular axiomatic system, a concept is revealed whose full understanding lies literally *beyond* the reach of that one system. (Rota 1988, p. 384–385)

In this paper, Rota concentrates on the pole-idea of mathematical constructions, which are neither *the*, nor *a realm* of reality, according to so-called "mathematical Platonism" or realism. Nonetheless it is an essential moment (component) of the real and of any object. Avoiding the crystallization of reality by stuffing it with mathematical constructs (or "substructing it", in Husserl's words), Rota restricts the structure of scientific experience, described above by Husserl, to the sphere of mathematical experience and exposes an a priori frame for history of mathematics, which does neither exclude its openness nor annihilates the freedom of mathematical thought. But in this peculiar case, as in the general one, to phrase it in Husserlian terms: Great "as this freedom of categorial union and formation may be, it still has its law-governed limits" (Sixth Logical Investigation, § 62), (Husserl 2001, p. 309; 1984, p. 717).

[44] With good reasons, in the footsteps of Husserl's critique of objectivism, Rota will later designate them as "items" instead of "objects" or "identities", claiming the "end of objectivity", i.e. "the end of the objectivistic conception of experience" (Rota 1991, p. 138).

This scope is fully and explicitly assumed, since the starting point of the paper is to recognize "that this duality of presentation, the syntactical and the semantical, is shared by all mathematical theories and is not found only in mathematical logic" (Rota 1988, p. 377). From the way the interplay between "identity", syntax and semantics is settled, one can infer that there is neither a unique syntax, nor an ultimate syntax, and that the dual relation explored in symbolic logic, by its transference to mathematics, is transformed into a ternary structure.

> The analysis of sameness of mathematical objects, as we shall call the constructs of mathematics, points at once to a closely related problem, one that was first recognized in the philosophy of symbolic logic, but that has far wider scope in the philosophy of mathematics, as we argue below. A presentation of a mathematical system leading up to the definition of an object is of necessity syntactical, that is, it is given by axioms and rules of inference. Nonetheless, the axioms and rules of inference are intended to characterize a class of mathematical objects consisting of sets with some additional structure (such as groups, manifolds, etc.). Any set with such an additional structure that satisfies the axioms is said to be a *model* for the axioms, and the description of all such models is customarily said to be the semantic interpretation of the theory. (Rota 1988, p. 377)

The problem is thus reformulated: "how disparate syntaxes can nonetheless have the same semantics, that is, the same models"? The third term, that of "identity", is a mere idea *at a distance*. For this very reason, it is precisely that which is the most immediately, although confusingly, given. The introduction of this third term implies a relaxing and opening of the whole logical structure, and forces the "fundamental conclusion": "that a full description of the logical role of mathematical objet lies beyond the reach of the axiomatic method *as it is understood today*" (Rota 1988, p. 384) (Emphasis mine). Among the examples of "identities" or "samenesses": the "real line" is probably the most paradigmatic of a mathematical "sameness" supporting an infinite variety of formal syntaxes and semantics, which can't be anticipated, but must each time be constructed (which correspond respectively to the second and first moment of Husserl's structure). For this reason, and, for the sake of clarity, before moving to the case of probability measure, let us jump to the presentation of the eidetic and phenomenological structure of the mathematical process of its determination, as it is sketched at the end of the article. The salient points of the structure are the following:

(1) *Identity* (*sameness*) pre-grasped in a variety of syntaxes, although pre-grasped at a distance, this identity is neither a trivial identity, nor an equation.
(2) Infinite plurality of syntactical presentations, not denumerable in advance, each new syntactical presentation revealing in a very slow step-by-step process, new formal features of the *same* object, which is not a fixed, empty closed form, but rather an inexhaustible potential of ever new forms.
(3) These new forms, new determinations motivate and lead to new constructions and are partially conceived, that is partially, informally, intuitively, tacitly, anticipatively grasped; here come into play the "leads", providing soundness and sense to mathematical activity.
(4) Although the understanding of the identical object exceeds the mastery of the axiomatic "game", a syntactical presentation provides an inevitable frame for

anyone desiring really to learn something about the mathematical object; in other and plain words, axiomatics is a necessary, but not sufficient requirement.

A question arises: *What are the operations to realize such a structure and put axiomatics in the phenomenological role of (formal) presentations and the real idea at a distance (the real line)*? For the sake of comparison, we must be reminded of what Husserl says about the objects of geometry: they are at the same time *material eidê* and *ideas in a Kantian sense* (Husserl (1988). From the start, the real line is *neither* the Euclidian line, nor, of course, the line perceived by the senses. In various ways, both are just first-stroke ideas. The central operation performed is the construction of a peculiar kind of analogy, through the -morphic (potentially anamorphic) projections in both senses, without giving any preference to one form over another. These structural analogies are precisely that which has been recognized before as cryptomorphism by Rota and, in my view, something that should be compared with Husserl's use of structural analogies (mostly in his investigations into affective and volitive intentionality, and correlatively, into axiology and ethics in the larger sense of the term).[45] The promotion and culture of this mode of circulation is underpinned by a set of presuppositions, or instead, it implies a (historically and rationally motivated) relaxation from a set of presuppositions. Among underpinning decisions, we must count: (1) integration and correction of model theory[46] and, consequently, strict reform of the Tarskian notion of truth (motivated by Paul Cohen's forcing); "[47] (2) admission at its right place of category theory;[48] (3) rehabilitation of updated projects of algebraic logic and universal algebra; "[49] (4) reasonable and

[45]For the method of analogizing for the investigation of affective and volitional intentionality (and its correlates: values and goods) as well as the extraction of their corresponding formal structures see the *Lessons on ethics and theory of values* (Husserl 1988, pp. 37–38; 41–44, 45–50 etc., 2009, pp. 111–112, 115–119, 120–126).

[46]"Every field of mathematics has its zenith and its nadir. The zenith of logic is model theory (we do not dare state what we believe will be its nadir). The sure sign that we are dealing with a zenith is that as we, ignorant and dumb non-logicians, attempt to read the stuff, we feel that the material should be rewritten for the benefit of a general audience".

[47]A new paradise was opened when Paul Cohen invented forcing, soon to be followed by the reform of the Tarskian notion of truth, which is the idea of Boolean-valued models. Of some subjects, such as this one, one feels that an unfathomable depth of applications is at hand, which will lead to an overhaul of mathematics." (Rota 1997b, p. 218).

[48]Let us give the full quote of note 22: "We were turned off category theory by the excesses of the sixties when a loud crowd pretended to rewrite mathematics in the language of categories. Their claims have been toned down, and *category theory* has taken its modest place side by side with *lattice theory*, more pretentious than the latter, but with strong support from both Western and Eastern Masters.—One wonders why category theory has aroused such bigoted opposition. One reason may be that understanding category theory requires an awareness of analogies between disparate mathematical disciplines, and mathematicians are not interested in leaving their narrow turf" (Rota 1997b, p. 220).

[49]Ever since theoretical computer scientists began to upstage traditional logicians, we have watched the resurgence of nonstandard logics. These new logics are feeding problems back to universal algebra, with salutary effects. Whoever believes that the theory of commutative rings is the central chapter of algebra will have to change his tune. The combination of logic and universal algebra will take over." (Rota 2008, p. 218–219) (see also Rota 1985, 1986b)

fair evaluation of lattice theory despite the bad press of quantum logic,[50] with an expectation of drying up the source of a perfect example of *desperate philosophy*: philosophy of quantum mechanics.[51]

In order to show the fruitfulness of this softened syntax/semantic relation, Rota proceeds to a kind of eidetic variation: enumerating examples of various syntactical presentations of a same semantic; and conversely; but also cases of mathematical theories syntactically achieved but without semantics and conversely examples of semantics deprived of syntax. Interestingly, as a paradigmatic example of fully achieved formalization of a mathematical theory, with "a table of semantic concepts and syntactic concepts, together with a translation algorithm between the two", Rota quotes once again Birkhoff's (1967) and Stone's (1937) theory of distributive lattices. He observes that examples of semantics still deprived of any clear syntax are abundant in mathematics, because the set up of a syntax requires reflection, and because, unlike logicians, mathematicians are somewhat suspicious about logical reflection. Moreover, one has to get familiar with the mathematical structures at stake, through exercises, in order to get a basis for such a reflection, and a first stroke axiomatization—such as the Euclidian geometry for the manifold involved in a certain intuition of space.[52]

Unsurprisingly, Rota gives three examples of mathematical theories, which, directly or indirectly, refer to probability theory, and show the necessity of the reform of its actual axiomatization. Let us review them briefly.

1. Despite the analogy between the "closed sub-spaces of a Hilbert space" and the "events of probability theory" (or the propositions of classical logic), the "sample spaces" constituting the "elementary semantic of quantum mechanics" are deprived of any clear syntax. Von Neumann and Birkhoff's attempts at providing such syntax failed: "the numerous attempts at developing a 'logic of quantum mechanics' have failed because no one has yet been able to develop a workable

[50]"It has always been difficult to take quantum logic seriously. A malicious algebraist dubbed it contemptuously 'poor man's von Neumann algebras'. The lattice-theoretic background made people suspicious, given the bad press that lattice theory has always had." (Rota 1997b, p. 218–219).

[51]"A more accomplished example of *Desperationsphilosophie* than the philosophy of quantum mechanics is hard to conceive. It was a child born of a marriage of misunderstandings: the myth that logic has to do with Boolean algebra and the pretence that a generalization of Boolean algebra is the notion of a modular lattice. Thousands of papers confirmed to mathematicians their worst suspicions about philosophers. Such a philosophy came to an end when someone conclusively proved that those observables, which are the quantum mechanical analogues of random variables, cannot be described by lattice-theoretic structure alone, unlike random variables.—This debacle had the salutary effect of opening up the field to some honest philosophy of quantum mechanics, at the same level of honesty as the philosophy of statistics (of which we would like to see more) or the philosophy of relativity (of which we would like to see less)." (Rota 1997b, p. 219).

[52]Rota/Baclawski give the following advice to their students: "The purpose of this course is to learn to think probabilistically. Unfortunately the *only way to learn to think probabilistically is to learn the theorems of probability*. Only later, as one has mastered the theorems, does the probabilistic point of view begin to emerge while the specific theorems fade in one's memory: much as the grin of the Cheshire cat" (Rota/Baclawski 1979, p. vii).

syntactical presentation of the mathematical structure consisting of the events of quantum mechanics" (Rota 1988, p. 380).
2. Second example: the theory of multi-set, which Rota substitutes for simple sets, and which proceeds from allowing an element to recur "more than just once". By this extension of set the notion of "probability measure emerges as an abstraction derived from the multiset concept" (Rota/Baclawski 1979, p. 1.2). As probability measure, a multi-set is a function (called "multiplicity") from the set S to the non-negative integers." Ordinary Boolean operations (product, sum) can be applied, but the elucidation of the deep-lying duality between algebra of sets and algebra of multiset requires a syntactical description, which is still missing.
3. Third and last example, that of probability theory.

> Statisticians and probabilists employ informal syntactical presentations, in thinking and speaking to each other. For example, when a probabilist thinks of a sequence of random variables that forms a Markov chain, he reasons and works directly with such a sequence of random variables, seldom appealing to the complex construction of a path space in which such a Markov chain may be realized. Another example: The statistician who computes with confidence intervals and significance levels seldom appeals to the measure-theoretic justification of his reasoning. It can be surmised that a syntactical presentation of probability will view joint probability distributions as playing a role similar to truth-values in predicate calculus. Kolmogorov's consistency theorem shows how to construct actual random variables in a sample space whose joint distributions are formalized to be a given family of consistent distributions. Thus, from this point of view, Kolmogorov's theorem *should turn* out to be the completeness theorem relating the syntactic and semantic presentations of probability theory. (Rota 1988, p. 381)

This text, which anticipates the *Fubini Lectures*, shows clearly the connection between both problems: independently of the question of quantum probabilities, the lack of a real syntactical presentation and the inadequacy of Kolmogorov syntax to ordinary statistical reasoning. This lack of syntactical adequate presentation turns out to be the source of endless miscommunications.

8.3 Reform of Formal Logic and Genetic Logic

8.3.1 A Reform of Logic

Ideal history and circulation (communication) of forms are intimately articulated. Trying to understand this articulation, we are led to another dimension of Rota's interests in formal logic, explicitly related to his reading of Husserl. Through a series of circumstances, chance encounters (that of Church in Princeton, Stanley Ulam at Los Alamos, Birkhoff, Kac, and so many others), a deeper concern motivates Rota's incursions beyond the barriers which usually stem from his mathematician colleagues' inhibitions regarding formal logic and a special kind of reflexivity: a concern for a reform of logic inspired by Husserl's own project of the reform of logic. Like very few others, and obviously under the influence of Stanley Ulam,[53] Rota has read in Husserl's many-directionnal researches a project of reform of logic.[54] From the convergence of Rota's philosophical publications (Rota: and from the phenomenological lectures given at the M. I. T. (Rota 1998b), we get a confirmation of the deep involvement of Rota in the project of reform of logic he attributes to Husserl. The fact that he is right in doing so is confirmed by a reading of Husserl's published works, and not only in far-fetched manuscripts (as one might believe). Moreover this idea should not sound as something extraordinary for someone really acquainted with the historical period. It was already at the core of the Brentano School, like many other logical schools of the turn of the 20th Century.[55] According to Rota a first sample set of results of this reform in the *Third Logical Investigation*, is already exploitable (Husserl 1982, 2001). Yet, unlike many commentators willing to reduce Husserl to the so-called "first Husserl", and eventually "half of the first", cleaned from any idealist and ecological glaze, imposed by the re-writing from 1913, Rota does not feel like playing the Occamian barber. The promotion of the *Fundierung* ("foundation") to the rank of logical concept—"which ranks among Husserl's greatest logical discoveries"—has been repeatedly recalled by authors.[56] But many other concepts are on the short logical promotion list. To mention one of them, which is indicated by Husserl: the mereological relation of *container-contained* (*Enthaltende-Enthaltene*)—not to be confused with its set-theoretical counter-parts (*inclusion, belonging*). This statement is confirmed in *Ideas I*, when Husserl promotes the conceptual distinction between *container* and *contained* to the rank of fundamental formal distinction.[57] Rota suggests some other categories such as: "b lacks of a" (to be distinguished from its counter-parts such as "a is not included in b"); "a lacks b, a is absent from b (one could describe in precise terms how this differs from the classical 'a ∈ b'), *a* reveals *b*, *a* haunts *b* (as in 'the possibility of error haunts the truth'), a is implicitly present in b, 'the horizon of a', and so on, and so on" (Rota et al. 1992, p. 171).

> The rigorous foundation of the concept of time provides further examples of such relations: as shown in Husserl's lectures on *Zeitbewusstsein*, the relations of object to past and future are irreducible to classical set theory, and lead to an entirely novel theory of impeccable rigor. (Rota et al. 1992, p. 171)

---

[53] See *Indiscrete Thoughts* (Rota 1997b, p. 58–59; 1989a). For a first general overview of the problem of logic in Rota in relation to phenomenology, see also Mugnai, in Damiani (2009, p. 246 passim) and Lanciani/Majolino in Damiani (2009, p. 229–240; Rota 1987a, p. 31).

[54] Another paper on this particular subject (*La reforme husserlienne de la logique selon Rota*), previously given at a conference on Rota, at the *Istituto Veneto*, in 2014, is to appear simultaneously in the next volume of the *Revue de Synthèse* (eds. C. Alunni & E. Brian), Springer.

[55] Cf. *Franz Brentanos Reform der Logik*, Wilhelm Enoch, in: *Philosophische Monatshefte*, 29, 1893, pp. 433–458.

[56] Albino Lanciani, *Analyse phénoménologique du concept de probabilité*, Hermann, 2012.

[57] Husserl re-baptise the key fundamental and underlying relational concepts of the mereology of the Third Logical Investigation: "container" (*das Enthaltende*) and "contained." (*das Enthaltene*), *Ideas* I, § 12 [26–27] (Husserl 1950a, p. 31–32; Palombi 2003).

Parallel to this task including bits of eidetic descriptions which are awaiting their proper formalization,[58] another task falls upon us: through the practice of eidetic descriptions, that of developing and implementing Husserl's genetic logic.

8.3.2 Genetic Logic and Ideal History of Science

For Rota, the time has come to change the way in which logicians, mathematicians and philosopher consider each other's work. Elaborated in the midst of a deep scientific crisis, genetic phenomenology as an eidetic descriptive science has raised the practice of reflexion to the level of rigorous activity. Sociologically and anthropologically understandable, it is nonetheless surprising that so little has been borrowed from this theoretical stock to reflect on our present crisis; that of the foundation of mathematics. For as the most alive parts of logic belong now to mathematics, proportionally, contributions to the "philosophical understanding of the foundations" decrease. Technical logical sophistication combined to the development of algebra and algebraic logic "contributed on their part to shadow the notion of set" (*op. cit.* p 167). The situation is even worse despite many attempts to provide a foundation, or at least a communicable understanding of quantum mechanics. Von Neumann demonstrated the inadequacy of set theory, but his own striving for a logic, "compatible with the uncertainty principle", which is the "cornerstone of quantum mechanic", failed, as he confessed himself. Yet, Rota refuses to dismiss and to promote the present situation of unintelligibility to the rank of insuperable norm of intelligibility. Past solutions, which have failed (for instance, non-commutative modular lattices), conceal hidden resources and many lives, such are *The Many lives of Lattice Theory*). For complementary reasons, present successful theories will probably fade away. In other words, the horizon of science is not closed. Such a process implies a kind of virtual or potential existence of what has been for a while invalidated, or repressed by up-to-date scientific norms, which in turn may smooth and be devaluated, otherwise epistemological changes should be ultimately just a matter of taste.

> Do you prefer lattices or categories? In the thirties, lattices were the rage: von Neumann cultivated them with passion. *Then categories came along, and lattices, like poor cousins, were shoved aside*. Now categories begin to show their slips. Abelian categories are here to stay; *topoi* are probably here to stay; *triples* were once here to stay, but 'general' categories are probably not here to stay. *Meanwhile, lattices have come back with a vengeance* in combinatorics, computer science, logic and whatnot. (Rota 1997a, p. 221) (Emphasis mine)

Although not insensible to the beauty of mathematical theories, Rota refuses any kind of epistemological aestheticism. Scientific changes are matter of historical

[58]"The examples that Husserl and other phenomenologists developed of this genetic reconstruction, admirable as they are, came before the standard of rigor later set by mathematical logic, and are therefore insufficient to meet the foundational needs of present-day science. In contemporary logic, to be is to be formal. It falls to us to develop the technical apparatus of genetic phenomenology (…) on the same or greater a standard of rigor than mathematical logic." (Rota et al. 1992, p. 171)

necessity and self-responsibility, not of taste. With assumed Husserlian accents, he declares that the present scientific crisis forces us to perform a "phenomenological reduction", a "bracketing". This non-deliberately performed "bracketing" is at work each time somebody is seriously and honestly at work, and corresponds precisely to what Husserl names a "functioning *épokhé*" by which the shift from one attitude to another is possible.[59]

> It is a boon to the phenomenologist that the notion of set, together with various other kindred notations, should have become problematic at this time (as predicted long ago by Husserl); thanks to the present *crisis* of foundations, we are allowed a rare opportunity to observe fundamental scientific concepts in the detached state technically known as 'bracketing.' As often happens, the events themselves are forcing upon us a phenomenological reduction. (Rota et al. 1992, p. 168)

The performance of phenomenological reflection is required of the philosopher and of the scientist, and implies consequently a deep reform of academic habits in both faculties, and beyond in their relations to what is outside of them. Rota picks some of them up: publishing of textbooks, and consequently, teaching.

> What is the difference between a textbook on probability and one on probability modelling? More generally, what is the difference between a textbook on X and one on X-modelling? Let me tell you. A textbook on probability is expected to carry certain subjects, *not because these subjects are useful or educational but because teachers expect them to be there*. If such a textbook omits one of the liturgic chapters, then Professor Neanderthal will not adopt it, and the textbook will go out of print. As a consequence of this situation, most elementary probability textbooks are carbon copies of each other. The only variations are colours and settings of the exercises. The only way to add a topic to the standard course is to name it 'probabilistic modelling.' In this devious way, Professor Neanderthal may be persuaded to adopt the additional text as 'supplementary reading material.' The students will find it interesting, and eventually the topic will be required." What is missing is someone willing to (1) *set his research work aside* to write an elementary modelling book (thereby risking permanent status of *déclassé*) (2) take the trouble to *rethink the subject at an expository level* (a harder task than is generally assumed), and (3) find a publisher who will take a chance on the new text (increasingly difficult). (Rota 1997b, p. 226)

Though difficult, this inner intellectual reform and the culture of those skills is still required in order to make possible a decent communication between working philosophers and working scientists, and more precisely between *working phenomenologists* and *working mathematicians*.

> Can Husserl's philosophy and method help us out of these and other foundational predicaments? It is our contention that the way of phenomenology is inevitable in the further development of the sciences. (Rota et al. 1992, p. 168)

This communication as an exchange of information is a production of a higher form. Consequently, its frame has to be set up. From the historical and phenomenological *Selbstbesinnung,* we pass thus to a formalization of the schema of the rhythmical historical progression of sciences, mediated by an eidetic moment as well and

[59]Cf. Husserl 1970, § 35. See, my first, but somewhat vague, comment in Lobo 2000, p. 69–75, and more sharply, Lobo 2009, p. 59–70.

embedded, as we saw, in a more general eidetic perspectivism. This must be read as an explication of the implicit historical structure in which reflexion moves.

> I shall sketch a possible beginning of such an enterprise, by an admittedly inadequate and schematic presentation (…) From (2) and (3) we conclude that *genetic phenomenology intervenes at two stages in the development of a science* at dawn, by circumscribing an autonomous eidetic domain with its internal laws; at dusk, by the criticism of that very autonomy, leading to an enlargement of the eidetic domain. The process can be schematically represented by the following recurring pattern:
>
> → *ideal object* → *science-* → *crisis-* → *genetic analysis-* → *ideal object* →
>
> (a) ——— (b) —— (c) ——— (d) ———— (a)
>
> It is left to us to develop this program in rigor and detail with the help of the awesome amount of material Husserl left us. Should this program turn into reality, we may then live to see the birth of a new logic, the first radical reform of logic since Aristotle. It is of course all too likely, unless communications between scientists and philosophers improve, that mathematicians and scientists themselves, unaware of Husserl's pioneering work, will independently rediscover the very same way out of their present impasse. Recall that an *inspired genetic analysis* of the concept of simultaneity, carried out by a *philosophically untrained physicist*, led to the creation of the *special theory of relativity*. (Rota et al. 1992, p. 172–173)

The transcendental *épokhè* unties us from the naive ontological credo, that ideal objects (numbers, manifolds, algebraic field, poems, sub-atomic particles) are something radically different from ordinary things (such as chairs, tables), and teaches and trains us to see that they belong (in various guises of "belonging") to distinct eidetic spheres. Simultaneously, phenomenology frees us from the anti-platonic prejudices which weights on our practice of formalizations and comprehensions of forms.

Acknowledgments I am grateful to them more than I can say to all the friends and colleagues, who take the time to read and comment this paper. This paper has benefited from the careful readings and/or the criticisms of Luigi Accardi, Françoise Balibar, Franck Jedrzejewski, Pierre Kerszberg, Didier Vaudène, Mark van Atten and Maria Villela-Petit da Penha. Last but not least, I am very thankful to Marian Hobson who helped greatly to convert my prose into proper English. Thank you to Pierre Giai-Levra who gave a final, generous and acute look at the last version of this paper. Remaining mistakes and shortcomings are attributable to me. I am also very grateful to Hassan Tahiri who organized this unforgettable conference in Lisbon at the *Centro de Filosofia das Ciências*, for his patience in waiting for the final version of this paper.

References

Accardi, L. (1981a, b). Topics in quantum probability, *Physics Report* 77, 169–192.
Accardi, L. (1982a). Foundations of quantum probability. *Rendiconti del Seminario Matematico dell'Università e del Plitecnico, Torino, 1982*, 249–273.
Accardi, L. (1982b). Some Trends and Problems in Quantum Probability, *Quantom probability and applications to the quantum theory of irreversible processes*. In: L. Accardi, A. Frigerio, & V. Gorini (Eds.), *Proceedings of the second Conference: Quantum Probability and applications to*

the quantum theory of irreversible, Processes, 6–11, 9 (1982), Villa Mondragone, Rome, Springer, 1–19.
Accardi, L. (1993). Urns and chameleons: Two metaphors for two different types of measurements. *Journal of Physics: Conference Series, 459.*
Accardi, L. (1999). The quantum probabilistic approach to the foundations of quantum theory: Urns and chamaleons, in *Language, Quantum, Music*, M. L. Dalla Chiara, R. Giuntini, & F. Laudisa, (Eds.), *Synthese Library in Epistemology, Logic, Methodology and Philosophy of science*. Berlin: Springer.
Accardi, L. (2003). *Urns and chameleons, a dialogue about reality, the laws of chance and quantum theory*. English version (Out of print).
Accardi, L. (2010). Quantum probability: New perspectives for the laws of chance. *Milan Journal of Mathematics, 78*(2010), 481–502.
Birkhoff, G. (1949). Théorie et applications des treillis, in *Annales de l'Institut Henri Poincaré*, Tome 11, n° 5 (1949), p. 227–240, Paris: Gauthier-Villars.
Bitbol, Michel. (2009). *La structure quantique de la connaissance individuelle et sociale, in Theorie quantique et sciences humaines*. Paris: CNRS Editions.
Buchsbaum, D. A. (2001). Resolution of weyl modules: The rota touch. In H. Crapo & D. Senato (Eds.), *Algebraic combinatorics and computer science a tribute to gian-carlo rota* (pp. 97–100). New York, Dordrecht, London: Springer.
Caramello, O. (2010). *The unification of Mathematics via Topos Theory*, 20 June 2010, https://arxiv.org/abs/1006.3930v1
Caratheodory, C. (1948). *Vorlesungen über reelle Funktionen*, (1st ed, Berlin: Leipzig 1918), 2nd ed., New York: Chelsea.
Cartier, P. (1972). On the structure of free Baxter algebras. *Advances in Mathematics, 9*(2), 253–265.
Damiani, E., D'Antona, O., Marra, V., Palombi, F. (Eds.) (2009). *From Combinatorics to Philosophy, The Legacy of G.-C. Rota*, Dordrecht Heidelberg London New York: Springer.
Dhombres, J. (2003). Reynolds and averaging operators. In *Gian-Carlo Rota, On Analysis and Probability*, Selected Papers: Birkhaüser.
Accardi L., Fedullo, A. (1981). On the statistical meaning of complex numbers in quantum theory. *Lettere al Nuovo Cimento 34*, 161–172. University of Salerno preprint May (1981).
Halmos, P. (1950). *Measure theory*. New York: D. van Nostrand and Co.
Halmos, P. (1956). The basic concepts of algebraic logic. In: *American mathematical Monthly, 63*, 363–387.
Husserl, E. (1950a). *Ideen zur einer reine Phänomenologie und phänomenologische Philosophie*, ed. W. Biemel, Husserliana, Band III/1. Den Haag: Martinus Nijhoff.
Husserl, E. (1950b). *Die Krisis der europäischen Wissenschaften un, die transzendentale Phänomenologie, Eine Einleitung in die phanomenologische Philosophie* (Ed.) Walter Biemel. *Husserliana,* Band VI. The Hague: Martinus Nijhoff.
Husserl, E. (1969). *Formal and Transcendental Logic*, English tr. Dorion Cairn, Den Haag: Martinus Nijhoff.
Husserl, E. (1970). *The crisis of european sciences and transcendental phenomenology, an introduction to phenomenology*. tr David Carr. Evanston: Northwestern University Press.
Husserl, E. (1984). *Logische Untersuchungen, Zweiter Band, Zweiter Teil, Untersuchungen zur Phänomenologie und Theorie der Erkenntnis*, Text der 1 und der 2 Auflage, hsg. Ursula Panzer, Husserliana Band XIX/2, Martinus Nijhoff Pub., Kluwer, The Hague, Boston, Lancaster.
Husserl, E. (1988). *Vorlesungen über Ethik und Wertlehre,* 1908–1914. In U. Melle (Ed.), *Husserliana* XXVIII, Dordrecht, Boston, London: Kluwer.
Husserl E. (2001). *Logical Investigations,* edited by Dermot Moran and translated by J. N. Findlay from the second German edition, London, New York, Routledge.
Husserl, E. (2009). *Leçons sur l'éthique et la théorie de la valeur*, French. Transl. P. Ducat, P. Lang, C. Lobo, Épiméthée: P.U.F.
Jedrzejewski, F. (2009). *Modèles aléatoires et physique probabiliste*. Paris: Berlin, New York, Springer.

Johnstone, P. T. (1983). The point of pointless topology. In *Bulletin (New Series) of the American Mathematical Society*, Volume 8, Number 1, January, 1983. 41–53.
Kolmogorov, A. N. (1956). *Foundations of the theory of probability*, 2nd English ed. Tr. N. Morrison, Un. Of Oregon, New York: Chelsea Publishing Company.
Lobo, C. (2000). *Le phénoménologue et ses exemples. Étude sur le rôle de l'exemple dans la constitution de la méthode et l'ouverture du champ de la phénoménologie transcendantale*, Paris: Kimé.
Lobo, C. (2006). *Temporalité et remplissement, in Annales de Phénoménologie*. Beauvais: APPP.
Lobo, C. (2009). De la phénoménologie considérée comme un métier, *L'Œuvre du phénomène*, P. Kerszberg, A. Mazzu, A. Schnell (Eds.), (pp. 51–70) Bruxelles: Ousia.
Lobo, C. (2010). The husserlian project of reform of logic and individuation. In *The 40th Annual Meeting of the Husserl Circle,* New School for Social Research, New York, 22 June 2010, www.husserlcircle.org/HC_NYC_Proceedings.pdf, pp. 86–103.
Lobo, C. (2012). L'idée platonicienne d'eidos selon Husserl, *Les interprétations des Idées platoniciennes dans la philosophie contemporaine.* In A. Mazzu et S. Delcomminette (éds.) Paris: Vrin. 172–185.
Moran, P. A. P. (1968). *An introduction to probability theory.* Oxford: Clarendon Press.
Mugur-Schachter, M. (1994). Quantum probabilities, kolmogorov probabilities, and informational probabilities. *International Journal of Theoretical Physics, 33*(1), 1994.
Mugur-Schachter, M. (2002). Objectivity and descriptional relativities. In *Quantum Mechanics and other Fields of Science, Foundation of Science* (Vol 7; N 1–2), 2002. Dordrecht, Kluwer: 1–86.
Mugur-Schachter, M. (2009). *Infra-Quantum Mechanics and conceptual invalidation of Bell's theorem on locality. The principles of a revolution of epistemology revealed in the descriptions of microstates* (French text with English summary) (Manuscript): http://arxiv.org/abs/0903.4976.
Mugur-Schachter, M. (2014). On the Concept of Probability. In *Mathematical Structures in Computer Science* (pp. 1–91) London: Cambridge University Press.
Neyman, Jerzi. (1950). *First course in probability and statistics*. New York: Henry Holt and Company.
Nottale, N. (2004). *The theory of scale relativity: Non-differentiable geometry and fractal spacetime.* In *Computing Anticipatory Systems. CASYS'03—Sixth International Conference* (Liège, Belgium, 11–16 August 2003).
Okada, M. (1998). Husserl and hilbert on completeness and husserl's term rewrite-based theory of multiplicity. In *24th International Conference on Rewriting Techniques and Applications* (RTA' 13), ed. Femke van Raamsdonk, pp. 4–19.
Okada, M. (2000). *Husserl's 'Concluding Theme of the Old Philosophico-Mathematical Studies' and the Role of the Notion of Multiplicity* (Draft version of Paris Conference, of the 22 of March, 2000, *Rencontre sur la logique et la philosophie de la science*).
Olav, K. (2002). *Foundations of modern probability*. New York, London: Springer.
Palombi, F. (2003). *The star & the whole: Gian-carlo rota on mathematics and phenomenology.* Torino: Bollati Boringhieri.
Pitovsky, I. (1989). *Quantum probability—quantum logic*. Dordrect, London: Springer.
Pitowsky, I. (2005). *Quantum Mechanics as a Theory of Probability.* Department of Philosophy. http://edelstein.huji.ac.il/staff/pitowsky/.
Rashed, R. (1984). *Entre Arithmétique et Algèbre. Recherches sur l'Histoire des Mathématiques Arabes.* Paris: Les Belles Lettres.
Rashed, R. (2014). *Classical Mathematics from Al-Khwārizmī to Descartes*, tr. M. H. Shank, Culture and Civilization in the Middle East Series, London: Routledge.
Rényi, A. (1970). *Foundations of probability*. New York: Dover.
Rota, G-C. (1960) Une généralisation de l'espérance mathématique conditionnelle qui se présente dans la théorie statistique de la turbulence. Paris: C. *R. Académie des Sciences,* pp. 624–626.
Rota, G.-C. (1969). Baxter algebras and combinatorial identities. I, II, Bull. Amer. Math. Soc. 75, 325.

Rota, G.-C. (1973a). The valuation ring of a distributive lattice. In *Proceedings of the University of Houston, Lattice Theory Conference* (Houston, Tex., 1973) Department of Mathematics, University Houston, Houston, Tex.: 574–628.
Rota, G.-C. (1973b). Edmund husserl and the reform of logic in D. In D. Carr & E. S. Casey (Eds.), *Exploraitions in phenomenology* (pp. 299–305). The Hague: Nijhoff.
Rota, G.-C. (1978). in N. Metropolis, Gian-Carlo Rota, Volker Strehl and Neil White, *Partitions into Chains of a Class of Partially Ordered Sets*, in *Proceedings of the American Mathematical Society,* Volume 71, Number 2, September (1978), 193–196.
Rota, G.-C. (1981). Rota's Preface. In P. J., Davis & R. Hersh (Eds.), *The Mathematical Experience*, (pp. i–xix) Switzerland: Birkhaüser.
Rota, G.-C. (1985). Mathematics. *Philosophy and artificial intelligence, a dialogue with Gian-Carlo Rota and David Sharp, Los Alamos Science, spring/summer, 1985,* 93–104.
Rota, G.-C. (1986a). In memoriam of Stan Ulam-the Barrier of Meaning. Evolution, Games and Learning (Los Alamos, N. M., 1985). Phys. D 22 (1986), no. 1–3, 1–3.
Rota, G-C. (1986b). *Remarks on Artificial Intelligence.* (Italian) Boll. Un. Mat. Ital. A (6)5 (1986), no. 1, 1–12.
Rota, G.-C. (1987a). *The lost Café. Stanislaw Ulam 1909–1984*. Los Alamos Sci. No. 15, Special Issue (1987), 23–32.
Rota, G.-C. (1987b). Stanley Ulam, *Conversations with Rota.* Transcribed and edited by Françoise Ulam. Stanislaw Ulam 1909–1984. Los Alamos Sci. No. 15, Special Issue (1987), 300–312
Rota, G.-C. (1988). David Sharp, Robert Sokolowski: 1988, Syntax, semantics, and the problem of the identity of mathematical objects. *Philosophy of Science, 55*(3), 376–386.
Rota, G-C. (1989a). The Barrier of Meaning. In *Memoriam: Stanislaw Ulam*. Notices Amer. Math. Soc. *36*(2), 141–143.
Rota, G.-C. (1989b). *Remarks on Artificial Intelligence*. Translated from the Italian by A. Nikolova. Fiz.-Mat. Spis. Bügar. Akad. Nauk. 31(64) (1989), no. 1: 19–27.
Rota, G.-C. (1990a). Mathematics and philosophy: History of a Misunderstanding, in *Boll. Un. Mat. Ital*. A (7) 4 (1990), no. 3: 295–307.
Rota, G.-C. (1990b). Les ambiguïtés de la pensée mathématique. *Gaz. Math. 45*(1990), 54–64.
Rota, G-C. (1990c). *The pernicious Influence of Mathematics upon Philosophy. New Directions in the Philosophy of Mathematics* (New Orleans, LA, 1990). Synthese 88 (1991), no. 2: 165-178.
Rota, G.-C. (1991). Mathematics and the task of phenomenology. In T. Seebohm, D. Follesdal, & J. N. Mohanty (Eds.), *Phenomenology and the formal sciences*. Dordrecht: Kluwer.
Rota, G.-C. (1995). Baxter operators, an introduction. In J. P. S. Kung (Ed.), *Gian-carlo rota on combinatorics, introductory papers and commentaries*. Birkhäuser: Contemp. Mathematicians, Boston.
Rota, G.-C. (1997a). Memory of Garrett Birkhoff: The many lives of Lattice Theory, Published in the Notices of the AMS, Volume 44, Number 11, (1997): 1440–1445.
Rota, G.-C. (1997b). *Indiscrete thoughts*. Boston, Basel, Berlin: Birkhäuser.
Rota, G.-C. (1998a). *Ten mathematics problems I will never solve*, Invited address at the joint meeting of the American Mathematical Society and the Mexican Mathematical Society, Oaxaca, Mexico, Dec. 6, 1997. DMV Mittellungen Heft 2, 45.
Rota, G.-C. (1998b). *Introduction to Heidegger's Being and Time* (draft) Gian-Carlo Rota, ed. and tr. Mark van Atten, Version Fall.
Rota, G.-C. (2000). Ten remarks on husserl and phenomenology. In O. Wiegand, R. J. Dostal, L. Embree, J. Kockelmans, & J. N. Mohanty (Eds.), *Phenomenology on Kant, German Idealism, Hermeneutics and Logic*. Dordrecht: Kluwer.
Rota, G.-C. (2001). Twelve problems in probability no one likes to bring up, Fubini Lectures, in *Algebraic Combinatorics and Computer Science*, Springer, 2001, pp. 25–96.
Rota, G-C., Smith, D. (1972). Fluctuation theory and baxter algebras. *Istituto Nazionale di Alta Matematica* IX, 179 (1972).
Rota, G.-C. & Backlawski, K. (1979). *An Introduction to Probability and Random Processes*. http://www.ellerman.org/wp-content/uploads/2012/12/Rota-Baclawski-Prob-Theory-79.pdf.

Rota, G.-C., & Ellerman, D. (1978). A Measure-theoretic approach to logical quantification. *Rend. Sem. Mat. Univ. Padova, 59*(1978), 227–246.

Rota, G.-C., Rota, G.-C., Sharp, D., Sokolowski, R. (1988). Syntax, semantics, and the problem of the identity of mathematical objects. *Philosophy of Science, 55*, 376–386.

Rota, G-C., Kac, M., Schwartz, J. T. (1992). *Discrete Thoughts. Essays on Mathematics, science, and philosophy*. Scientists of Our Time. Birkhauser, Boston, Mass., 1986. xii + 264 pp. Quoted from the edition of 1992.

Rota, G.-C., & Klain, D. A. (1997). *Introduction to Geometric Probability.* A. Luigi (Ed.) Radicati di Brozolo. Cambridge: Cambridge University Press.

Savage, L. J. (1972). *The foundations of statistics*. New York: Dover.

Segal, I. E. (1947). Postulates for general quantum mechanics. *Annals of Mathematics, 48*, 930–948.

Uzan, P. (2013). *Conscience et physique quantique, Mathesis*. Paris: Vrin.

van Atten M. (2015). *Essays on Gödel's Reception of Leibniz, Husserl, and Brouwer*, Switzerland: Springer.

von Neumann, J. (1932). *Mathematische Grundlagen der Quantenmechanik*, Springer.

von Neumann, J. (1955). *Mathematical Foundations of Quantum Mechanics*, tr. Robert T. Beyer, Princeton University Press.

von Neumann, J. (1981). *Continuous geometries with a transition probability.* In Halperin I. (Ed.), Memoirs of the American Mathematical Society, 34.

von Neumann, J., & Birkhoff, G. (1936). The logic of quantum mehanics. In *Ann. Math.*, Vol. 37: 823–843.—Reprinted in *The Neumann Compendium*, F. Brody, T. Vamos, World Scientific Series in 20th Century Mathematics, Vol. 1. New Jersey, Singapore, London, 1995: 105–125.

Weyl, H. (1968). *Gesammelte Abhandlungen*. Band, III K. Chandrasekharan (Ed.), Berlin: Springer Verlag.

Chapter 9
Mathematics and the Physical World in Aristotle

Pierre Pellegrin

Abstract I would like to start with a historical question or, more precisely, a question pertaining to the history of science itself. It is a widely accepted idea that Aristotelism has been an obstacle to the emergence of modern physical science, and this was for at least two reasons. The first one is the cognitive role Aristotle is supposed to have attributed to perception. Instead of considering perception as an origin of error (even if this may be the case in exceptional situations, if the perceiving subject is sick for instance), Aristotle thinks that our senses provide us with a reliable image of the external world. The perceptive knowledge is a kind of knowledge in its own right, and the theoretical knowledge is, in fact, the continuation of the perceptive knowledge in some way. The second reason is the presumed inability of the Aristotelian philosophers to apply mathematics to the physical world. This was a formidable obstacle because modern physics came to be but as a mathematical physics. Aristotelianism had therefore to be, so to speak, superseded by the Platonic movement that originated in Florence around Ficino in order to give modern physics the conditions of its appearance. Galileo had to say that "Nature is written with mathematical letters" and Descartes that "our senses do not teach us what things are, but to what extent they are useful or harmful to us". Alexandre Koyré is right to consider Galilean physics to be basically Platonic. The theoretical justification Aristotle offers for the impossibility of a convergence between mathematics and physics seems to be based on some fundamental features of his philosophy, i.e. he rejects the Platonic conception of a unique science, encompassing all things, and replaces it with the doctrine of the incommunicability of genera, whose corollary is that there is but one science for each genus.

Aristotle's confidence about perception resulted in a qualitative physics that was impervious to any attempt of mathematisation. I will, therefore, begin with this problem of the relationship between mathematics and the physical world. Later I will briefly return to the question of perception.

P. Pellegrin (✉)
CNRS, Paris, France
e-mail: p.a.pellegrin@wanadoo.fr

H. Tahiri (ed.), *The Philosophers and Mathematics*, Logic, Epistemology, and the Unity of Science 43, https://doi.org/10.1007/978-3-319-93733-5_9

According to the canonical definition of Aristotle, a science, when not considered as a state of the knowing a subject (in this case, science is an intellectual virtue), is a system of propositions established through demonstrative syllogisms (i.e. in which the middle term is the cause of the conclusion) some properties that belong per se to a subject which is not demonstrated, but given and established by other ways (induction, intuition, perception, etc.). This subject is a genus or belongs to the genus, the science under consideration is dealing with. Thus, if one has to demonstrate that a lunar eclipse is caused by the interposition of the Earth between the Moon and the Sun, the existence of these three celestial bodies has to be taken for granted, probably through perceptions, and then only their properties are demonstrated (interposition and by the same way their relative motions).

Sciences (I am just considering here theoretical sciences) have different extensions. Thus physics is the science of the genus consisting of the beings "having within themselves a principle of change", zoology (a term which is not an Aristotelian one) has as its genus of the animals and establishes, as a famous passage in the *Politics* say, "what all animals necessarily have", i.e. digestive, perceptive and moving organs, but ornithology (another term which is not to be found in Aristotle), taking birds as its genus, demonstrates, for instance, that a special kind of bipedy is necessarily linked with the corporeal structure of birds.

The famous chapter E,1 of the *Metaphysics* considers theoretical sciences at their maximum level of generality, distinguishing three great kinds, which are characterised with respect to the provinces of being they are concerned with. The genus of physics is the realm of mobile beings non-separated from matter. Mathematics is the science of non-separated immobile beings. Theology, the science of immobile and separated beings:

> If there is something which is eternal and immobile and separated, clearly the knowledge of it belongs to a theoretical science, however, neither to physics (for physics considers certain mobile beings) nor to mathematics, but to a science prior to both. For physics considers the things that are non-separable and non-immobile, some parts of mathematics consider immobile objects but probably non–separable and embodied in matter, while the first science deals with things which are both separable and immobile.[1] (1026a10).

Two remarks can be made on this passage. First, "some parts of mathematics" may seem strange, as what Aristotle says applies to the entire mathematics. In fact, this text does not concern mathematics in itself because its aim is to describe the "first science", i.e. theology. This is why Aristotle takes a minimal position: if some branches of mathematics consider immobile objects, this is sufficient for the current demonstration. It is the same for the "probably" of the next line: Aristotle definitely does not want to tackle this question here. Secondly, there is a famous textual problem in the text, 1026a14: the manuscripts have ἀχώριστα, which mean "non-separated (or

---

[1] εἰ δέ τί ἐστιν ἀΐδιον καὶ ἀκίνητον καὶ χωριστόν, φανερὸν ὅτι θεωρητικῆς τὸ γνῶναι, οὐ μέντοι φυσικῆς γε (περὶ κινητῶν γάρ τινων ἡ φυσική) οὐδὲ μαθηματικῆς, ἀλλὰ προτέρας ἀμφοῖν. ἡ μὲν γὰρ φυσικὴ περὶ ἀχώριστα μὲν ἀλλ' οὐκ ἀκίνητα, τῆς δὲ μαθηματικῆς ἔνια περὶ ἀκίνητα μὲν οὐ χωριστὰ δὲ ἴσως ἀλλ' ὡς ἐν ὕλῃ· ἡ δὲ πρώτη καὶ περὶ χωριστὰ καὶ ἀκίνητα.

non-separable)", which Schwegler corrected as χωριστὰ. The text would then read: "physics considers things which are separable and non-immobile". This implied that the objects of physics would exist separately. An absurd correction, since "separated" would have two different meanings within two lines. What is here at stake, in the case of the three theoretical sciences, is the separation *from matter*. I mention this problem here as we shall have to come back later on the ambiguity of the term "separated" on an important point. But what is important for us here is: how should we understand that the objects of mathematics are not separated (or separable) from matter? In any case, the property of mathematical objects of being non-separated from matter is crucial for us, as we are considering the impossibility of a convergence between physics and mathematics.

While it would be interesting to consider the ontological status of mathematical objects, I shall not do that directly, because my views on this question are not very clear (on the contrary many commentators think they have clear ideas on this point, and this is why so much has been written about it) and, on the other hand, I have an impression that Aristotle does not offer a unified doctrine on this question, which would be an effect of some chronological evolution. But perhaps not: I rather suspect that he accepts the coexistence of two different approaches to mathematical objects, without making them completely compatible with each other, just as he does *Physics*, by placing side by side two conceptions of time, which he considers successively, in book IV as "the number of movement", and in book VI as a crucial example of the continuous magnitude. According to one of these approaches, mathematical objects would be basically physical, according to the other, they would be built up by the human mind.

Now, there are two points that Aristotle seems to consider more important than the ontological status of mathematical objects. In any case, he devotes much attention to these two points, whereas he is quite short and less clear on the ontological status. Actually, for Aristotle may be and for us surely, to deal with these two points is an efficient roundabout way to come to the ontological status of mathematical objects. These two points are: (i) the substantiality of mathematical objects and (ii) the relationship between mathematical objects and physical objects. It is not surprising that Aristotle focuses on these two points, since it is in books M and N of the *Metaphysics* that most of the information on the Aristotelian conception of mathematical objects can be found. But the main aim of these two books is to criticize Platonic theses, and we know, mainly thanks to Aristotle and his commentators, that possibly Plato and certainly his successors in the Academy developed a doctrine of ideal Numbers that replaced the theory of ideas or combined with them. And among the Platonists, questions (i) and (ii) are strongly related to each other.

At the end of chapter M,2, Aristotle gives a complete list of possibilities: (i) mathematical objects do not exist, a thesis that is rejected without any argument, probably because mathematical objects have to exist in some way: cf. 1076a36: "so that the subject of our research will not be whether they exist, but how they exist" (ὥσθ' ἡ ἀμφισβήτησις ἡμῖν ἔσται οὐ περὶ τοῦ εἶναι ἀλλὰ περὶ τοῦ τρόπου); (ii) they exist within the sensible things; (iii) they exist separated from the sensible things; and (iv) they exist "in some other way" (ἄλλον τρόπον, 1076a35). To understand what this list really means, we should keep in mind that in the (ii) and (iii) hypotheses, Aristotle interprets "to exist" as "to exist as a substance". The (ii) hypothesis can be said to be a "Pythagorean" one, and the (iii) hypothesis a "Platonist" one.

For, in hypothesis (ii), if "to exist" is not understood as "to exist as a substance", it would not be absurd to say that the structure of a sensible sphere is present, but not as a substance, within this sensible sphere itself, and makes it intelligible. We shall see that this is, roughly speaking, what Aristotle does. But if this spherical structure exists substantially, there will be two substances within the same object, i.e. in the same place, which is impossible. If, on the other hand, mathematical objects are separated from the sensible ones, there will be two solids. At this point, the text is quite difficult to understand. This is what I understand: for the surfaces, which the solid is made of, there will be: (a) the surfaces of the sensible solid, (b) the surfaces of the separated solid, and (c) the separated surfaces corresponding to each of the sensible surfaces. Therefore, there will be three kinds of surfaces, four kinds of lines, and five kinds of points. "We then get a strange accumulation" (ἄτοπός τε δὴ γίγνεται ἡ σώρευσις, 1076b29).

We should notice that such a position is basically a Platonic one because mathematical realities do exist outside the framework of the sensible perception, i.e. of the matter. Even if Aristotle says that, according to the Platonists, these separated realities are "prior" to sensible realities, he does not say that they consider these realities as the causes of the sensible realities. This may reveal a disagreement between Platonists, as some of them would be in the favour of ideal Numbers as causes, while others would be not.

Such a critique of the Platonist position, which makes it ridiculous, takes possibly as its target the common opinions rather than the opinions of "specialists". This is what Aristotle himself says, according to Syrianus,

> He [Aristotle] himself admits that he has said nothing against the Platonists' hypotheses and that he does not follow the doctrine of ideal Numbers, if these are different from the mathematical ones. This can be found in the second book of his *On Philosophy*: 'thus if the Ideas are a different kind of number, and not mathematical number, we can have no knowledge of them, for the majority of us, who understand any other number?' Thus, in fact, his refutation is directed against the multitude who knows no number other than the number composed of units and did not even try to grasp the thought of these divine men.[2] (*Commentary on Metaphysics*, 159,33).

In no case, then, can mathematical objects exist as substances. What is this "other way", i.e. a non-substantial way, according to which, mathematical objects do exist? Aristotle does not define directly this "other way", but describes it through the relationship between mathematical objects and sensible objects. This means that, once

[2] ἐπεὶ ὅτι καὶ αὐτὸς ὁμολογεῖ μηδὲν εἰρηκέναι πρὸς τὰς ἐκείνων ὑποθέσεις μηδ' ὅλως παρακολουθεῖν τοῖς εἰδητικοῖς ἀριθμοῖς, εἴπερ ἕτεροι τῶν μαθηματικῶν εἶεν, μαρτυρεῖ τὰ ἐν τῷ δευτέρῳ τῶν Περὶ τῆς φιλοσοφίας ἔχοντα τοῦτον τὸν τρόπον· "ὥστε εἰ ἄλλος ἀριθμὸς αἱ ἰδέαι, μὴ μαθηματικὸς δέ, οὐδεμίαν περὶ αὐτοῦ σύνεσιν ἔχοιμεν ἄν· τίς γὰρ τῶν γε πλείστων ἡμῶν συνίησιν ἄλλον ἀριθμόν;" ὥστε καὶ νῦν ὡς πρὸς τοὺς πολλοὺς τοὺς οὐκ εἰδότας ἄλλον ἢ τὸν μοναδικὸν ἀριθμὸν πεποίηται τοὺς ἐλέγχους, τῆς δὲ τῶν θείων ἀνδρῶν διανοίας οὐδὲ τὴν ἀρχὴν ἐφήψατο.

the problem of the substantiality of mathematical objects has been settled, the question "what is the ontological status of mathematical objects?" is replaced with the question "in what sense is mathematical science concerned with sensible objects?". We, then, are brought back to the problem of *Metaphysics* E,1, that is the separation from matter.

Since this question has already been addressed to a great extent, I will directly consider the conclusion of these studies, while referring to some publications, which, if combined, seem to give an acceptable picture of what Aristotle's position is. These publications are, in chronological order, of I. Mueller, J. Lear, E. Hussey, M. Crubellier.[3]

In the first half of chapter M,3 of the *Metaphysics*, Aristotle characterises the relationship between mathematics and the sensible world in the following way. (i) Mathematics considers sensible objects. This is what it means for mathematics to study "non-separated (separable)" objects in chapter E,1 of the *Metaphysics*. (ii) But mathematics considers these objects, not *qua* sensible, but to the extent that they possess mathematical properties. (iii) There is no object with mathematical properties which is not sensible. This is a corollary of the non-substantiality of mathematical objects: as they are not substances, mathematical objects do not constitute a class of existing objects different from the class of sensible objects.

Now, the very notion of genus, understood as being the domain of a science, should be reconsidered and conceived in a non-immediate way. Birds do constitute the immediate genus of ornithology, but a bird may also be described as a walking and flying machine. This is precisely this kinematic approach that is to be found in the treatise on *The Locomotion of Animals*. This is also what Aristotle says in *Physics* II,2, in claiming that both the mathematician and the physicist consider the Sun and the Moon, and he goes further in saying that the physicist also, not the mathematician alone, is concerned with their spherical form. The spherical form seems to be a mathematical attribute *par excellence*, but there are also *physical* reasons for the spherical form of the celestial bodies and of the universe, as the beginning of the treatise *On the Heaven* shows, "being eternal, the universe must be perfect; but the sphere is the perfect form." We have, then this crucial result, that both mathematics and physics do consider the sphere as one of their objects.

Then a genus is not a collection of objects sharing some features (figures for geometry, numbers for arithmetic, birds for ornithology, etc.), since the same object may be in turn an object for physics and mathematics. We could go further in quoting the beginning of chapter III,4 of the *Physics*: "natural science is concerned with magnitudes, motion and time" (202b30: 'Εστὶν ἡ περὶ φύσεως ἐπιστήμη περὶ μεγέθη καὶ κίνησιν καὶ χρόνον). What has been said about birds appears here in its general form. Even if we let motion aside, what could be more mathematical

[3]Ian Mueller, "Aristotle on Geometrical Objects", *Archiv für Geschichte der Philosophie* 1970, pp.156–171. Jonathan Lear, "Aristotle's Philosophy of Mathematics", *The Philosophical Review*, Avril 1982, pp. 161–192. Edward Hussey, "Aristotle on Mathematical Objects", in I. Mueller (ed). *Peri Tôn Mathêmatôn*, Apeiron Décembre 1991, pp. 105–133. Michel Crubellier, "La Beauté du monde. Les sciences mathématiques et la philosophie première", *Revue internationale de Philosophie* 1997–3, pp. 307–331.

than magnitude? Concerning time, a question on which I shall come back, Aristotle defines it as "the number of the motion according to anterior and posterior". These physical objects are then obviously considered by mathematics.

But, proceeding further, we should now understand exactly what is meant by the very famous passage from *Physics* II,2:

> The mathematician too treats of things like this [the sphere], but not as the limits of natural bodies; nor does he consider their attributes as attributes of such natural bodies. That is why he separates them, for they are separable from motion in thought.[4] (193b31)

The mathematical objects are non-separated (non-separable) from matter but separated (separable) in thought from physical objects. It is because separation may be understood in two senses (separation from matter and separation as perceived by the mind) that Schwegler proposed the correction in *Metaphysics* E,1 mentioned earlier.

"Abstraction" (ἀφαιρέσις), a well-known but not well-understood concept, precisely signifies this separation in thought from physical realities. We must strongly object to a widespread and historically quite important interpretation, according to which Aristotle would have considered mathematical objects as mental realities without any natural character. This puts them at different levels of ontological reality, under the level of the objects constructed by the human mind, since these objects may have, to some extent, a "natural" character. In Plato's *Phaedrus*, for instance, the good dialectician is the one who, like the good butcher, is able to cut things where Nature disposed the joints. Therefore, genus and species are not arbitrarily constructed.

According to the interpretation that I am criticising (which has been definitively refuted in Ian Mueller's article), Aristotle would think that nothing prevents our mind to "see" a sphere within a billiard ball, just as one can "see" castles in the clouds. But for Aristotle, the sphere is really in the billiard ball, whereas there is no castle in the clouds. In fact, "to abstract" or "to separate" means to put aside (Jonathan Lear says "to filter") some characteristics of an object, which are of interest for the current research, from the other characteristics of this object. Aristotle expresses this as the fact to consider an object "*qua*" (ᾗ).

All these make intelligible the curious claim Aristotle makes almost at the end of chapter M,3 of the *Metaphysics*:

> The geometer treats him [the man] neither *qua* man nor *qua* indivisible [as the arithmetician does], but as a solid. The attributes which would have belonged to him even if he were not indivisible, can clearly belong to him without indivisibility and humanity. Thus, then, geometers speak rightly and they talk about existing things, and what their propositions affirm do exist. For existing is said in two ways, as what exists in actuality, and what exists in the way the matter does.[5] (1078a25)

---

[4] περὶ τούτων μὲν οὖν πραγματεύεται καὶ ὁ μαθηματικός, ἀλλ' οὐχ ᾗ φυσικοῦ σώματος πέρας ἕκαστον οὐδὲ τὰ συμβεβηκότα θεωρεῖ ᾗ τοιούτοις οὖσι συμβέβηκεν· διὸ καὶ χωρίζει· χωριστὰ γὰρ τῇ νοήσει κινήσεώς ἐστι.

[5] ὁ δὲ γεωμέτρης οὔθ' ᾗ ἄνθρωπος οὔθ' ᾗ ἀδιαίρετος ἀλλ' ᾗ στερεόν. ἃ γὰρ κἂν εἰ μή που ἦν ἀδιαίρετος ὑπῆρχεν αὐτῷ, δῆλον ὅτι καὶ ἄνευ τούτων ἐνδέχεται αὐτῷ ὑπάρχειν [τὸ δυνατόν], ὥστε διὰ τοῦτο ὀρθῶς οἱ γεωμέτραι λέγουσι, καὶ περὶ ὄντων διαλέγονται, καὶ ὄντα ἐστίν· διττὸν γὰρ τὸ ὄν, τὸ μὲν ἐντελεχείᾳ τὸ δ' ὑλικῶς

Aristotle, then, strongly refuses the idea that mathematical objects are arbitrary constructions of the human mind since they do exist independently of the human mind. At this point, we are facing a paradox, since, as we shall see, mathematics is closely related to formality, and Aristotle says in this passage that the mode of existence of mathematical objects is to exist "in the way matter does" (ὑλικῶς), i.e. potentially. What does this mean? Edward Hussey may be right in considering that this case is comparable with that of the matter of a statue, which exists apart from the statue: it did exist before the statue existed, and it does exist identical to itself as a matter, within the statue. But we should not go too far in such a comparison. For we cannot say that physical objects are made of mathematical objects, or, in Aristotelian terms, the mathematical sphere is not the material cause of the billiard ball, just as the bronze is the material cause of the statue. But we can say that mathematical objects can be found in concrete sensible individuals, without being themselves concrete sensible individuals. We, therefore, grasp a mathematical object in depriving a concrete sensible individual of some properties: We grasp the spherical form of the billiard ball when we separate (filter) it from the wood it is made of. Once again we see how far we are from mathematical objects as pure mental constructions.

The mathematical sphere, on the other hand, is the *formal* cause of the billiard ball. For, being a science, and more precisely a theoretical science, mathematics has an explanatory function. But, according to Aristotle, this explanatory function needs to have recourse to his system of the four causes. Now, whereas physics does use the four causes (with a possible merging of formal, final, and efficient causes, but this is another problem), mathematics is concerned only with the formal cause. This is a well-known aspect of Aristotle's epistemology that I do not want to consider here. I will just outline the connexion between the formality of mathematics and the fact that mathematics considers immobile objects. For form is the immobile factor in natural processes.[6] This is even more obvious if we consider arithmetic rather than geometry. As Michel Crubellier says "les nombres entiers et les théorèmes de l'arithmétique représentent les propriétés qui se retrouvent dans toutes les collections d'objets particuliers, qu'Aristote appelle 'nombres sensibles'".[7] And in the *Physics* (and the parallel passage in the *Metaphysics*), Aristotle gives as an example of the formal cause, "<formal cause> of the octave, the relation of 2 to 1" (II,3,194b28).

Is all this specific to mathematics or is it possible to extend it to any of the sciences and should we completely revise the notion of a genus as a field of science? Because, after all, why could not we say that, as the mathematician considers a man as a solid, the physicist (biologist) considers him as a living thing? The answer is obviously "no". An artefact, for instance, cannot be considered a physical object, except in a secondary sense, as much as it is constituted out of physical elements (e.g. the table is made of trees), or if we consider it as constituted of elements (fire,

[6]Cf. *Physics* II, 7, 198a35: "The principles that cause natural motion are two, of which one is not natural, since it has no internal principle of motion. Of this kind is whatever causes motion without being moved, such as what is completely unchangeable, the primary being [i.e. the Prime mover], and the essence of a thing, i.e. the form".

[7]Op. cit. p. 316.

air, water and earth), which are natural beings. “No abstract being can be an object of the natural science” (*Parts of Animals* I,1,641b10). The table, on the other hand, certainly belongs to the genus of mathematical objects, because we can consider it as a solid in separating in its form from its matter, but in no case can it be considered a physical object in a proper sense (since it does not have an internal principle of change).

More generally, the relationship between mathematics and physics, as I have described it, is not reversible. One cannot say that physics separates the physical properties of mathematical objects as mathematics separates the mathematical properties of physical objects. The main reason for this is that though what is called the objects of science in the true sense are not substances but the properties of substances (physics establishes propositions like “all blooded animals have lungs”, or “thunder is the extinction of the fire in the clouds”), natural substances with these properties do exist. On the contrary, the mathematical objects are instantiated in *natural* substances but they are not substances themselves, as we have seen in *Metaphysics* M,2. There is nothing like mathematical substances from which one could abstract the physical properties.

We now understand why mathematics is absolutely original. This was widely ignored in the extensive literature concerning mathematics in Aristotle. Mathematics is concerned with a genus, that of mathematical realities (something like figures for geometry and numbers for arithmetic) but it is not a genus composed of *actual* objects. In a way mathematics is concerned with everything. But I am reluctant to say, as some did, that it is concerned with most or all genera: Mathematics, for instance, is not concerned with the genus “living things”, except when living things are no longer considered as alive.

All natural bodies and all artefacts are therefore potentially mathematical objects. But this is also the case for realities like motion or time, and Aristotle clearly recognises this when he defines time as “the number of the motion according to anterior and posterior”. In the same way, some commentators (including myself) falsely claimed that Aristotle, while taking into account the general theory of proportions, which can be applied to numbers, lines, solids and times (in *Posterior Analytics* I,5,74a23, he presents this theory as a recent one), overthrows his own conception of science as limited only to one genus. This is false because mathematics can be applied to everything in the way mentioned above and, at the same time, be restricted to the genus of mathematical objects.

We should simply note that some physical realities, of which some are bodies and others are not, are more mathematisable than others. This is the case for what Aristotle calls “the more physical branches of mathematics, like optics, harmonics, astronomy” (*Physics* II,2,194a7: τὰ φυσικώτερα τῶν μαθημάτων, οἷον ὀπτικὴ καὶ ἁρμονικὴ καὶ ἀστρολογία), and for notions like time. In a famous passage following the one quoted above, Aristotle says, “while geometry considers natural lines but not *qua* natural, optics investigates mathematical lines, but *qua* natural, not *qua* mathematical” (194a9: ἡ μὲν γὰρ γεωμετρία περὶ γραμμῆς φυσικῆς σκοπεῖ, ἀλλ’ οὐχ ᾗ φυσική, ἡ δ’ ὀπτικὴ μαθηματικὴν μὲν γραμμήν, ἀλλ’ οὐχ ᾗ μαθηματικὴ ἀλλ’ ᾗ φυσική).

Here we find reversibility between mathematics and physics; of which I say that it is not possible in all cases. One can do physics in considering the physical properties of a mathematical object, e.g. a mathematical line, just as one can do mathematics while considering mathematically a physical reality, e.g. a physical line. Except that Aristotle considers optics as a branch of mathematics, which seems to imply that the physical line considered by optics, namely the luminous ray, is basically for the most part an object for mathematics.

We then see that the Aristotelian conception of mathematics confers it a special status, which has, in fact, a strong Platonic taste. Mathematics is not the science of everything, a very anti-Aristotelian concept, but mathematics, and only mathematics, is the science which can be applied to everything, including the things that are the objects of other sciences. It applies to the objects, that is it deciphers their causes, and in doing so, it increases our scientific knowledge about these objects.

Mathematics can, therefore, so to speak, be superimposed on other sciences. The proposed examples, which are those Aristotle usually has recourse to, are from subordinate sciences (astronomy, optics) relative to a dominant science (geometry). We can say it to be an easy case because optics (what is meant here is geometrical optics, not physical optics) is a kind of visible geometry. But Aristotle extends this to other domains, for example in the following frequently quoted passage from the *Posterior Analytics*:

> Mathematics is about forms, for its objects are not said of an underlying subject. For even if geometrisable[8] objects are said of an underlying subject, they are certainly not geometrisable as being said of an underlying subject. There is another science related to optics, as optics is related to geometry—the science of the rainbow. It is the job of a physicist to know the fact, while the knowledge of the cause is the job of an optician, who could either be a *simpliciter* or related to mathematics. But there are also many sciences, which are not subordinate, for which, the same happens, for instance, medicine in relation to geometry. For to know that circular wounds heal more slowly belongs to the doctor, but to know why belongs to the geometer.[9] (I,13,79a7)

At the end of chapter M,3 of the *Metaphysics*, Aristotle quite remarkably refers to the possibility of mathematisation of everything in relation to the beautiful:

> Since the good and the beautiful are not one and the same thing (…), those who claim that the mathematical sciences say nothing about the beautiful or the good are mistaken. (…) For if they do not expressly mention them, but prove their properties in theirs discourses, they nevertheless deal with them. The main forms of beauty are order and symmetry and determination, which the mathematical sciences demonstrate to a great extent. And since these things (I mean for instance order and determination) are obviously causes of many

[8] Reading γεωμετρητά at 79a9 like the ms. usually corrected as γεωμετρικά, "geometrical".

[9] τὰ γὰρ μαθήματα περὶ εἴδη ἐστίν οὐ γὰρ καθ' ὑποκειμένου τινός· εἰ γὰρ καὶ καθ' ὑποκειμένου τινὸς τὰ γεωμετρητά ἐστιν, ἀλλ' οὐχ ᾗ γε καθ' ὑποκειμένου. ἔχει δὲ καὶ πρὸς τὴν ὀπτικήν, ὡς αὕτη πρὸς τὴν γεωμετρίαν, ἄλλη πρὸς ταύτην, οἷον τὸ περὶ τῆς ἴριδος· τὸ μὲν γὰρ ὅτι φυσικοῦ εἰδέναι, τὸ δὲ διότι ὀπτικοῦ, ἢ ἁπλῶς ἢ τοῦ κατὰ τὸ μάθημα. ἡ μὲν γὰρ γεωμετρία περὶ γραμμῆς φυσικῆς σκοπεῖ, ἀλλ' οὐχ ᾗ φυσική, ἡ δ' ὀπτικὴ μαθηματικὴν μὲν γραμμήν, ἀλλ' οὐχ ᾗμαθηματικὴ ἀλλ' ᾗ φυσική. πολλαὶ δὲ καὶ τῶν μὴ ὑπ' ἀλλήλας ἐπιστημῶν ἔχουσιν οὕτως, οἷον ἰατρικὴ πρὸς γεωμετρίαν ὅτι μὲν γὰρ τὰ ἕλκη τὰ περιφερῆ βραδύτερον ὑγιάζεται, τοῦ ἰατροῦ εἰδέναι, διότι δὲ τοῦ γεωμέτρου.

> things, evidently these sciences must deal with this kind of cause also, in which the beautiful is in some way a cause.[10] (1078a31)

Let us ignore the difference between the beautiful and the good. In this excerpt, Aristotle, in proposing the beautiful as the developed version of formality, goes back to the original sense of *eidos*—a term which he uses to mean both the form and the formal cause, and which etymologically refers to the apparent visible beauty. In order to show that a compound is beautiful because there is some proportionality in its elements, we need to add something to the causal knowledge of the compound. This can be done by mathematics, and by mathematics alone.

Returning to the question I posed at the beginning, we clearly see that the relationship Aristotle establishes between mathematics and the physical reality can in no way be considered an obstacle to the constitution of a mathematical physics. In a way, Aristotle is even closer to such a form of physics than Plato, because he makes mathematics not a universal science which would include physics, but saves the autonomy of physics, and makes its objects mathematisable.

We have seen that the basic anti-Platonic requisite of Aristotelianism, according to which one science considers one genus, has been an obstacle to modern science. And Aristotle endorses a strong version of this, considering that, even within mathematics, arithmetic cannot solve geometrical problems. This has been a special obstacle to modern science. But we have also seen that this rule has been applied to mathematics in such an original way that the theoretical obstacle is no longer active. Mathematics is available to complete the causal explanation of everything.

Concerning Aristotle's confidence in the perception of sense, it could not be an obstacle either, because, in addition to the qualities grasped by the five senses, Aristotle recognises the necessity to locate these perceptions in space and time, one relatively to others, and all relatively to the perceiving subject. It does not really matter that Aristotle attributes this function to a *sense*, the "common sense", rather than to a cognitive function as Plato did: Aristotle's perception is far from being a pure phenomenism but also includes the theory of ratios.

If the obstacle to modern science was not to be found within Aristotle's "system", what would have happened? This is quite a difficult question. We may at least notice that a tradition of mathematisation of physical objects did not develop within the Lyceum. Theophrastus writes, "it seems that figures, forms, ratios have been constructed by us, whereas in themselves they have no nature" (*Metaphysics* 4a21: οἷον γὰρ εμηχανημένα δοκεῖ δι' ἡμῶν εἶναι σχήματά τε καὶ μορφὰς καὶ λόγους περιτιθέντων, αὐτὰ δὲ δι' αὑτῶν οὐδεμίαν ἔχειν φύσιν). The term translated as "constructed", μεμηχανημένα, is remarkable because precisely it does indicate the recourse to the amazing or miraculous means, apparently contrary to the laws

[10] ἐπεὶ δὲ τὸ ἀγαθὸν καὶ τὸ καλὸν ἕτερον (...), οἱ φάσκοντες οὐδὲν λέγειν τὰς μαθηματικὰς ἐπιστήμας περὶ καλοῦ ἢ ἀγαθοῦ ψεύδονται. (...) μάλιστα· οὐ γὰρ εἰ μὴ ὀνομάζουσι τὰ δ' ἔργα καὶ τοὺς λόγους δεικνύουσιν, οὐ λέγουσι περὶ αὐτῶν. τοῦ δὲ καλοῦ μέγιστα εἴδη τάξις καὶ συμμετρία καὶ τὸ ὡρισμένον, ἃ μάλιστα δεικνύουσιν αἱ μαθηματικαὶ ἐπιστῆμαι. καὶ ἐπεί γε πολλῶν αἴτια φαίνεται ταῦτα (λέγω δ' οἷον ἡ τάξις καὶ τὸ ὡρισμένον), δῆλον ὅτι λέγοιεν ἂν καὶ τὴν τοιαύτην αἰτίαν τὴν ὡς τὸ καλὸν αἴτιον τρόπον τινά. μᾶλλον δὲ γνωρίμως ἐν ἄλλοις περὶ αὐτῶν ἐροῦμεν.

of nature. Theophrastus, Aristotle's successor at the head of the Lyceum, clearly departs from his master's conception of mathematics as a way of knowing physical realities. Theophrastus' successor, Strato of Lampsacus, seems to be a follower of a strong empiricism and certainly hostile to the formalisation of physical processes by the means of mathematics. Then empiricism can triumph, and the Nature is not at all written with mathematical letters, although Galileo's expression could have been, at least to some extent, endorsed by Aristotle.

References

Aristotle. (1984). The complete works of Aristotle. In J. Barnes (Ed.), *The revised Oxford translation*, Vol. 2. Princeton University Press.
Crubellier, Michel. (1997). La Beauté du monde. Les sciences mathématiques et la philosophie première. *Revue Internationale de Philosophie, 3,* 307–331.
Hussey, E. (1991). Aristotle on mathematical objects. In I. Mueller (Ed.), *Peri Tôn Mathêmatôn*, Apeiron, pp. 105–133.
Lear, J. (1982). Aristotle's philosophy of mathematics. *The Philosophical Review*, pp. 161–192.
Mueller, I. (1970). Aristotle on geometrical objects. *Archiv für Geschichte der Philosophie*, pp. 156–171.

Chapter 10
The Axiom of Choice as Interaction Brief Remarks on the Principle of Dependent Choices in a Dialogical Setting

Shahid Rahman

Abstract The work of Roshdi Rashed has set a landmark in many senses, but perhaps the most striking one is his inexhaustible thrive to open new paths for the study of conceptual links between science and philosophy deeply rooted in the interaction of historic with systematic perspectives. In the present talk I will focus on how a framework that has its source in philosophy of logic, interacts with some new results on the foundations of mathematics. More precisely, the main objective of my brief remarks is to discuss some claims of the late Hintikka (1996, 2001) who brought forward the idea that a game-theoretical interpretation of the Axiom of Choice yields its meaning "evident". More precisely I will show that if we develop Per Martin-Löf's (1984) demonstration of the axiom within a dialogical setting, the claim of Hintikka can be upheld. However, the dialogical demonstration, shows that, contrary to the expectations of Hintikka, the meaning that the game-theoretical setting provides to the Axiom is compatible with constructivist rather than with classical tenets.

10.1 Introduction

It is rightly said that the principle of set theory known as the Axiom of Choice "is probably the most interesting and in spite of its late appearance, the most discussed axiom of mathematics, second only to Euclid's Axiom of Parallels which was introduced more than two thousand years ago" (Fraenkel et al. 1973). According to Ernst Zermelo's (1904) formulation, the Axiom of Choice amounts to the claim

---

This section is based on Rahman et al. (2018) in preparation, moreover it works out and improves the versions of Clerbout and Rahman (2015) and Rahman et al. (2015), where a complete dialogical demonstration of AC was developed but with a different dialogical setting.

The paper has been developed in the context of the researches for transversal research axis *Argumentation* (UMR 8163: STL), the research project ADA at the MESHS-Nord-pas-de-Calais and the research projects: ANR-SÊMAINÔ (UMR 8163: STL).

S. Rahman (✉)
UMR 8163 - STL-Savoirs Textes Langage Université de Lille 3, Lille, France
e-mail: shahid.rahman@univ-lille3.fr

H. Tahiri (ed.), *The Philosophers and Mathematics*, Logic, Epistemology, and the Unity of Science 43, https://doi.org/10.1007/978-3-319-93733-5_10

that, given any family A of non-empty sets, it is possible to select a single element from each member of A. The selection process is carried out by a function f with domain in M, such that for any non-empty set M in A, f(M) is an element of M. The axiom has found resistance from its very beginnings and triggered heated foundational discussions concerning, among others, mathematical existence and the notion of mathematical objects in general and of functions in particular. With time however, the foundational and philosophical reluctances faded away and were replaced with a kind of praxis-driven view by the means of which the Axiom of Choice was accepted as a kind of postulate—rather than as an axiom whose truth is manifest—necessary for the practice and development of mathematics.

Around 1980, the foundational discussions around the Axiom of Choice (AC) found an unexpected revival when Per Martin-Löf showed that in constructive logic—a logic that does not presuppose the Excluded Middle—the AC is logically valid (though only in its intensional version) and that this logical truth follows naturally (almost trivially) from the constructive meaning of the quantifiers involved—this form of "evidence" is what makes it an axiom rather than a postulate. The extensional version can also be proved but then must be assumed either the third excluded or the unicity of the function. Martin-Löf's proof, for which he was awarded the prestigious Kolmogorov prize, showed that the old discussions were rooted in an even older conceptual problem: the tension there is between intension and extension. Still more recently, Jaako Hintikka (2001) has tackled the AC with a game-theoretical interpretation[1]—though he did not take Martin-Löf's proof into account, presumably because he was not in favour of the constructivist approaches.

As opposed to Hintikka's game-theoretical approach, the dialogical take on the AC does not require an unaxiomatizable language such as the one underlying Independent Friendly logic (IF-logic). Hintikka is certainly right in stressing the aptness of the game-theoretical interpretation of the AC, yet we contend that he is mistaken in what concerns the theory of meaning such an interpretation requires. As pointed out by Jovanovic (2013, 2015), Hintikka's claim, that a game-theoretical perspective on Zermelo's AC in a first-order logic was perfectly acceptable for Constructivists, has found sound confirmation in the dialogical approach to Constructive Type Theory, albeit no underlying IF-semantics is required in this framework. Ironically, Hintikka's formulation of the AC, fully spelled out, yields Martin-Löf's own CTT-formulation, making it constructivist friendly after all. By dealing with the case of the AC in dialogues for immanent reasoning, we intend to exhibit the expressive power of this constructivist game-theoretical framework based on equality in action. Our point in bringing up this case study is thus to show the main perk of our dialogical framework:

- That the dialogical formulation of the intensional version of the AC—also called the *Principle of Dependent Choices*—plainly shows how the evidence of its logical truth amounts to developing a winning strategy in which the consequent of the relevant main implication follows from an interactive take on the meaning of the antecedent.

[1]See for example (Hintikka 1996, 2001).

- In such a perspective, the logical truth of the Principle of Dependent Choices is rooted in the equality of the reasons grounding the antecedent with the reasons of the consequent.

10.2 Two Versions of the Axiom of Choice Distinguished in CTT

The Axiom of Choice (AC) was first introduced by Zermelo in 1904 in order to prove Cantor's theorem: every set can be rendered in such a way as to be well ordered. Zermelo gave two formulations of this axiom; one in 1904, and a second one in 1908. The second formulation is relevant for our discussion, since it relates to both Martin-Löf's and the game-theoretical formalization, and is spelled out in the following fashion:

> A set S that can be decomposed into a set of disjoint parts A, B, C, … each of them containing at least one element, possesses at least one subset S_1 having exactly one element with each of the parts A, B, C, … considered. (Zermelo 1908)

The AC immediately attracted a lot of attention and both of its formulations were criticized by constructivists such as René-Louis Baire, Émile Borel, Henri-Léon Lebesgue and Luitzen Egbertus Jan Brouwer. The first objections were related to the non-predicative character of the axiom, for a certain choice function was supposed to exist without constructively showing that it does. The axiom however found its way into the ZFC set theory and was finally accepted by the majority of mathematicians because of its usefulness in different branches of mathematics.

CTT differentiates the extensional and the intensional versions of the AC

Martin-Löf's proof of the AC is produced in a constructive setting and brings together two seemingly incompatible perspectives on this axiom, namely

1. Bishop's surprising observation from (1967): A choice function exists in *constructive mathematics*, because a choice is implied by the very meaning of existence; and
2. The proof by Diaconescu (1975) and by Goodman and Myhill (1978) that the Axiom of Choice implies the Excluded Middle.

These two perspectives seem incompatible. The solution resides in distinguishing two versions of the AC: an intensional and an extensional version. In his (2006) paper, Martin-Löf thus shows that there are versions of the Axiom of Choice that are perfectly acceptable for a Constructivist, namely those in which the choice function is defined *intensionally*. But in order to be able to formulate the intensional version of the AC, the axiom must be expressed within a CTT-framework. Such a setting allows the extensional and the intensional formulations of the axiom to be compared: we find that it is in fact the *extensional* version of the AC that implies the Excluded

Middle, and that Bishop's remark is compatible with the intensional version of the AC.[2]

10.2.1 Bishop's Remark, the Meaning of Quantifiers in CTT, and the Intensional Version of the AC

In harmony with Bishop's (1967) remark on the meaning of existence entailing a choice, in CTT the truth of the AC follows rather naturally from the meaning of the quantifiers. Take for instance the proposition $(\forall x : A)\, P(x)$ where $P(x)$ is of the type proposition (**prop**), provided that x is an element of the set A. If the proposition is true in a constructive setting, then there is a proof for it. Such a proof is a function that renders a proof of $P(x)$ for every element x of A. Bishop's remark should be understood as follows: the truth of a universal amounts to the existence of a proof, and this proof is a function. Thus, the truth of a universal amounts in a constructive setting to the existence of such a function. From this CTT characteristic the proof of the AC can be developed quite straightforwardly. If we recall that in the CTT-setting

- the existence of a function from A to B amounts to the existence of a proof-object for the universal *every A is B*; and that
- the proof of the proposition $B(x)$, existentially quantified over the set A, amounts to a pair such that the first element of the pair is an element of A and the second element of the pair is a proof of $B(x)$;

then a full-fledged formulation of the AC—more precisely, of the *Principle of Dependent Choices* (PDC)—follows, in which we explicit the set over which the existential quantifiers are defined.

The intensional formulation of the Axiom of Choice:

$$(\forall x : A)(\exists y : B(x))C(x, y) \supset (\exists f : (\forall x : A)B(x))(\forall x : A)C(x, f(x))$$

10.2.2 Martin-Löf's Proof of the Intensional Version of the Axiom of Choice

The proof of Martin-Löf (1984, pp. 50–51) (1980, p. 50–51) is the following[3]:

> The usual argument in intuitionistic mathematics, based on the intuitionistic interpretation of the logical constants, is roughly as follows: to prove $(\forall x)(\exists y)C(x,y) \rightarrow (\exists f)(\forall x)C(x,f(x))$,

[2]See for instance this observation of (Martin-Löf 2006, p. 349): "[…] this is not visible within an extensional framework, like Zermelo-Fraenkel set theory, where all functions are by definition extensional."

[3]For the formal demonstration spelled out as a Natural Deduction tree, see (Clerbout and Rahman 2015, p. 78) and (Rahman et al. 2015, p. 204).

assume that we have a proof of the antecedent. This means we have a method which, applied to an arbitrary *x*, yields a proof of $(\exists y)C(x,y)$. Let f be the method which, to an arbitrarily given *x*, assigns the first component of this pair. Then $C(x,f(x))$ holds for an arbitrary *x*, and hence, so does the consequent. The same idea can be put into symbols getting a formal proof in intuitionistic type theory. Let A : set, $B(x)$: set $(x{:}\ A)$, $C(x,y)$: set $(x{:}\ A,\ y{:}\ B(x))$, and assume $z{:}\ (\Pi x{:}\ A)(\Sigma y{:}\ B(x))C(x,y)$. If *x* is an arbitrary element of *A*, *i.e.* $x{:}\ A$, then by Π-elimination we obtain

$$Ap(z, x) : (\Sigma y : B(x))C(x, y)$$

We now apply left projection to obtain

$$p(Ap(z, x)) : B(x)$$

and right projection to obtain

$$q(Ap(z, x)) : C(x, p(Ap(z, x))).$$

By λ-abstraction on *x* (or Π-introduction), discharging $x{:}\ A$, we have

$$(\lambda x)p(Ap(z, x)) : (\Pi x : A)B(x)$$

and by Π-equality

$$Ap((\lambda x)p(Ap(z, x), x) = p(Ap(z, x)) : Bx.$$

By substitution [making use of $C(x,y)$: *set* $(x{:}\ A,\ y{:}\ B(x))$,] we get

$$C(x, Ap((\lambda x)p(Ap(z, x), x) = C(x, p(Ap(z, x)))$$

[that is, $C(x,\ Ap((\lambda x)\ p\ (Ap(z,x),\ x)=C(x,\ p(Ap(z,x)))$: set]and hence by equality of sets

$$q(Ap(z, x)) : C(x, Ap((\lambda x)p(Ap(z, x), x)$$

where $((\lambda x)\ p\ (Ap(z,x))$ is independent of *x*. By abstraction on *x*

$$((\lambda x)p(Ap(z, x)) : (\Pi x : A)C(x, Ap((\lambda x)p(Ap(z, x), x).$$

We now use the rule of pairing (that is Σ-introduction) to get

$$(\lambda x)p(Ap(z, x)), (\lambda x)q(Ap(z, x)) : (\Sigma f : (\Pi x : A)B(x))(\Pi x : A)C(x, Ap(f, x))$$

(note that in the last step, the new variable f is introduced and substituted for $((\lambda x)p(Ap(z,x))$ in the right member). Finally by abstraction on *z*, we obtain

$$(\lambda z)((\lambda x)p(Ap(z, x)), ((\lambda x)q(Ap(z, x)) : (\Pi x : A)(\Sigma y : B(x))C(x, y) \\ \rightarrow (\Sigma f : (\Pi x : A)B(x))(\Pi x : A)C(x, Ap(f, x)).$$

Martin-Löf (2006) further shows that—from a constructive point of view—what is wrong with the AC is the extensional formulation of it, formulation that Hintikka seems to assume and that can be expressed in the following way:

The extensional version of the Axiom of Choice:

$$(\forall x : A)(\exists y : B(x))C(x, y) \supset (\exists f : (\forall x : A)B(x))(Ext(f) \wedge (\forall x : A)C(x, f(x)))$$

where $Ext(f)$ is defined as $(\forall i,j : A)\ (i =_A j \rightarrow f(i)=f(j))$

Thus, from the constructive point of view, what is really wrong with the classical formulation of the AC is the assumption that from the truth that all of the *A* are *B* we can obtain a function that satisfies extensionality. In fact, as shown by Martin-Löf (2006), the classical version holds, even constructively, if we assume that there is only one such choice function in the set at stake:

The constructive extensional formulation of the AC:

$$(\forall x : A)(\exists!\, y : B(x))C(x, y) \supset (\exists f : (\forall x : A)B(x))(Ext(f) \wedge (\forall x : A)C(x, f(x)))$$

10.2.3 Conclusion on the Two Formulations of the AC

Let us retain that if we take $(\forall x : A)\ (\exists y : B(x))\ C(x,y) \supset (\exists f : (\forall x : A)\ B(x))\ (\forall x : A)\ C(x, f(x))$ to be the formalization of the AC, or more precisely of the PDC, then that axiom is not only unproblematic for Constructivists but it is also a theorem. In fact, it is the explicit language of CTT that allows a fine-grained distinction between the two formulations of the AC, equivalent only from the outside. This is due to the expressive power of CTT that allows expressing at the object-language level *quantifier interdependence*. What is more, Hintikka's insight is that it is the game-theoretical approach to the meaning of the quantifiers that makes the truth of the PDC apparent. Let us proceed now to our main subject: showing how the PDC in the framework of dialogues explicit the interdependency of the quantifiers through equality in action.

10.3 The PDC Within Dialogues for Immanent Reasoning

Dialogues for immanent reasoning is a dialogical framework that incorporates the features of CTT: it is therefore at least as expressive as CTT, and can thus also differentiate between an extensional version of the AC and an intensional version (the PDC) (see Appendix 2). The particle rules provide the meaning of the logical constants; the existential and the universal quantifiers have their meaning determined notably by which player has the choice—the challenger for the universal quantification, the defender for the existential quantification. What is more, the whole structure of dialogues for immanent reasoning rests on this fundamental rule, SR4 or Socratic rule (see Appendix 2), that makes explicit the dynamics of the choices of the players in interaction: the Proponent will copy the Opponent's choices and reasons in order to provide local reasons for his own statements. The Principle of Dependent Choices embeds existential quantification within universal quantification in the antecedent of the implication, and universal quantification within existential quantification in the consequent. In dialogues for immanent reasoning, just like in the standard dialogical framework, demonstrations are obtained by providing a **P**-strategy, that is **P** must be able to win whatever be **O**'s choices. Since the thesis is stated by **P** and is an implication, **O** will challenge it by stating the antecedent, **P** will be defending it by stating the consequent:

The thesis for the PDC in dialogues for immanent reasoning:

$$\mathbf{P}!\ (\forall x : A)(\exists y : B(x))C(x, y) \supset (\exists f : (\forall x : A)B(x))(\forall x : A)C(x, f(x))$$

In this fashion, **O** will have to defend the embedding of the existential within the universal, and **P** will have to defend the embedding of the universal within the existential. In other words, **O** will have to defend the embedding of her choice (defence of the existential) within a **P**-choice (defence of a universal), and **P** will have to defend the embedding of an **O**-choice (defence of a universal) within her choice (defence of an existential). Playing his moves in an optimal fashion will thus allow **P** to ask **O** to choose first and then copy her choices in order to build an equality through interaction and be able to win. The basic idea is that **P** can copy **O**'s choice for y in the antecedent for his defence of $f(x)$ in the consequent since both are equal objects of type $B(x)$, for any $x : A$. Thus a winning strategy for the implication follows simply from the meaning of the antecedent: this meaning is defined by the dependences generated by the interaction of choices involved in the embedding of an existential quantifier in a universal one. The following two dialogue tables display the relevant plays for building a winning strategy. These two plays are triggered by **O**'s decision-options concerning the existential in the consequent (stated by **P**): she can ask for the left or for the right (move 5), **P** must be able to win in both cases.

Here is a reminder of some particle rules of dialogues relevant for our present study:

Synthesis of local reasons

	Move	Challenge	Defence
Implication	$\mathbf{X}!\ A \supset B$	$\mathbf{Y} p_1 : A$	$\mathbf{X} p_2 : B$
Existential quantification	$\mathbf{X}!\ (\exists x : A)B(x)$	$\mathbf{Y}?\ L^{\exists}$ or $\mathbf{Y}?\ R^{\exists}$	$\mathbf{X} p_1 : A$ (resp.) $\mathbf{X} p_2 : B(p_1)$
Universal quantification	$\mathbf{X}!\ (\forall x : A)B(x)$	$\mathbf{Y} p_1 : A$	$\mathbf{X} p_2 : B(p_1)$

We assume by now that the appendices make the reader familiar enough with the dialogical framework and with this kind of table presentation; the comments will therefore be more to the point in explaining what is specific to the PDC. Are highlighted the moves where **P** defends his choice of a local reason by stating an equality, that is the crucial moves defining immanent reasoning where **P** says “my reason is the same as yours”.

The PDC (left decision-option)

	O			P	
C1 C2	$C(x, y)$: **set** $[x : A, y : B(x)]$ $B(x)$: **set** $[x : A]$			$! (\forall x : A) (\exists y : B(x)) C(x,y) \supset (\exists f : (\forall x : A) B(x)) (\forall x : A) C(x, f(x))$	0
1	m:= 1			n:= 2	2
3	$d_1 : (\forall x : A) (\exists y : B(x)) C(x, y)$	0		$d_2 : (\exists f : (\forall x : A) B(x)) (\forall x : A) C(x,f(x))$	4
5	$?_L$	4		$L^{\exists}(d_2) : (\forall x : A) B(x)$	6
7	$? \text{---} / L^{\exists}(d_2)$	6		$g_1 : (\forall x : A) B(x)$	8
9	$L^{\forall}(g_1) : A$	8		$R^{\forall}(g_1) : B(a)$	24
11	$a : A$		13	$? \text{---} / L^{\forall}(g_1)$	10
17	$R^{\forall}(d_1) : (\exists y : B(a)) C(a, y)$		3	$L^{\forall}(d_1) : A$	12
13	$? \text{---} / L^{\forall}(d_1)$	12		$a : A$	14
15	$? = a$	14		$L^{\forall}(g_1) = a : A$	16
19	$v : (\exists y : B(a)) C(a, y)$		17	$? \text{---} / R^{\forall}(d_1)$	18
21	$L^{\exists}(v) : B(a)$		19	$?_L$	20
23	$v_1 : B(a)$		21	$? \text{---} / L^{\exists}(v)$	22
25	$? \text{---} / R^{\forall}(g_1)$	24		$v_1 : B(a)$	26
27	$? = v_1$	26		$L^{\exists}(v) = v_1 : B(a)$	28
29	$? = a^{B(a)}$	28		$L^{\forall}(g_1) = a : A$	30
31	$? = B(a)$	30		$! B(L^{\forall}(g_1)) = B(a)$: **set**	34
33	$! B(a)$: **set**		C_2 Subst-D	$a : A$	32

Commentary:

- **Move 3**: Once the thesis has been stated and the repetition ranks established, **O** launches an attack on the material implication.
- **Move 4**: **P** states the right component of the material implication.
- **Moves 5 and 6**: **O** has here the choice between asking for the left or for the right component of the existential. The present play describes the development of the play triggered by the left choice.
- **Moves 7–13**: follow from a straightforward application of the dialogical rules.
- **Move 15**: **O** asks for the local reason that corresponds to the instruction stated by **P** at move 14.
- **Move 16**: **P** answers by recalling that **O** used the same reason, namely a, in defence of the claim that A holds (that it is not empty). And indeed this is what **O** did when she resolved $L^{\forall}(g_1)$ with a. This led **P** to implement the Socratic rule and state the equality $L^{\forall}(g_1) = a : A$.
- **Moves 27–28 and 29–30**: deal with the same kind of situation.

- **Moves 30–31**: Once move 30 has established that a—occurring in $B(a)$—constitutes a local reason for holding A, according to the Socratic rule S4.1c, **O** can launch an attack requesting **P** to show that $B(a)$ and $B(L^{\forall}(g_1))$ are equal propositions.
- **Moves 32–34**: **P** applies the rule for substitution within dependent statements (Subst-D) on the first concession, forcing **O** to state the condition that allows **P** to answer to the attack of move 31 and win the play by applying the Socratic rule to **O**'s move 25.[4]

Let us now, develop the second play, in which **O** went for the right decision-option at move 5.

The PDC (right decision-option)

O			P		
C1	$C(x, y)$: **set** $[x : A, y : B(x)]$			$!\ d : (\forall x : A)\ (\exists y : B(x))\ C(x,y) \supset (\exists f : (\forall x : A)\ B(x))\ (\forall x : A)\ C(x, f(x))$	0
C2	$B(x)$: **set** $[x : A]$				
1	m:= 1			n:= 2	2
3	$d_1 : (\forall x : A)\ (\exists y : B(x))\ C(x, y)$	0		$d_2 :\ (\exists f : (\forall x : A)\ B(x))\ (\forall x : A)\ C(x, f(x))$	4
5	$?_R$	4		$R^{\exists}(d_2) : (\forall x : A)\ C(x, L^{\exists}(d_2)(x)$	6
7	$?\ \text{---}\ /\ R^{\exists}(d_2)$	6		$g_2 : (\forall x : A)\ C(x, g_1(x))$	8
9	$L^{\forall}(g_2) : A$	8		$R^{\forall}(g_2) : C(x, g_1(a))$	30
11	$a : A$		11	$?\ \text{---}\ /\ L^{\forall}(g_2)$	10
17	$R^{\forall}(d_1) : (\exists y : B(a))\ C(a, y)$		3	$L^{\forall}(d_1) : A$	12
13	$?\ \text{---}\ /\ L^{\forall}(d_1)$	12		$a : A$	14
15	$? = a$	16		$!\ L^{\forall}(g_2) = a : A$	16
19	$v : (\exists y : B(a))\ C(a ,y)$		17	$?\ \text{---}\ /\ R^{\forall}(d_1)$	18
21	$L^{\exists}(v) : B(a)$		19	$?_L$	20
23	$t_1 : B(a)$		21	$?\ \text{---}\ /\ L^{\exists}(v)$	22
25	$R^{\exists}(v) : C(a, L^{\exists}(v))$		19	$?_R$	24
27	$R^{\exists}(v) : C(a, t_1)$		25	$?\ t_1\ /\ L^{\exists}(v)$	26
29	$t_2 : C(a, t_1)$		27	$?\ \text{---}\ /\ R^{\exists}(v_2)$	28
31	$?\ \text{---}\ /\ R^{\forall}(g_2)$	30		$t_2 : C(a, g_1(a))$	32
33	$?\ \text{---}\ /\ g_1(a)$	32		$t_2 : C(a, t_1)$	34
35	$? = t_2$	34		$R^{\exists}(v) = t_2 : C(a, t_1)$	36
37	$? = a^{C(a, t1)}$	36		$L^{\exists}(v) = t_1 : B(a)$	38
39	$? = C(a, t_1)$	38		$C(a, L^{\exists}(v)) = C(a, t_1)$: **set**	44
41	$C(a, y)$: **set** $[y : B(x)]$		C_1 Subst-D	$a : A$	40
43	$C(a, t_1)$: **set**		41 Subst-D		42

Remark: The only difference in the procedure in relation to the preceding play are the moves 40–44. Indeed, since $C(x, y)$ is a dyadic predicate, **P** must apply

[4] *Notice that when **P** attacks **O**'s initial concession **C2** he has to state a : A (move 32); according to the provisos of the Socratic rule however (see SR4.1, p. 210), **O** cannot attack this statement again: move 16 establishes **P**'s right to state this equality.*

substitution twice before obtaining of **O** the right to state the winning equality at move 44.

10.4 Conclusion on the Axiom of Choice

From the constructive point of view, functions are rules of correspondence. Rules of correspondence only make sense if we know how the correspondences are to be carried out. From the dialogical (and more generally from the game theoretical) point of view a function is the result of a player choosing an object of the domain and the defender choosing the suitable match. This can be seen as carrying out a rule of correspondence. However, the problem is that more has to be said to make an intensional function out of these interactions, though extensionality is also brought forward by interaction. Moreover, the dialogical take on constructivism is strongly linked to the view that in order to understand a play, and a winning strategy constituted by the relevant plays, it is not enough to know the rules of the game; it is not even enough to believe that there is a winning strategy behind the moves: what we need is to be able to describe the moves in such a way that it makes their contribution to the winning strategy understandable, that is, how the content involved in each moves constitutes that strategy. A proof beyond our capacities to describe it does not produce knowledge at all. This is what Hintikka's use of Wittgenstein's notion of *human-playable games* amounts to in his (2001) work; it is what the dialogical demonstration of the PDC amounts to. Let us conclude by quoting some beautiful lines of Poincaré, his response to what he considered a purely "formalistic" approach to mathematics:

> Si vous assistez à une partie d'échecs, il ne vous suffira pas, pour comprendre la partie, de savoir les règles de la marche des pièces. Cela vous permettrait seulement de reconnaître que chaque coup a été joué conformément à ces règles et cet avantage aurait vraiment bien peu de prix. C'est pourtant ce que ferait le lecteur d'un livre de Mathématiques, s'il n'était que logicien. Comprendre la partie, c'est toute autre chose; c'est savoir pourquoi le joueur avance telle pièce plutôt que telle autre qu'il aurait pu faire mouvoir sans violer les règles du jeu. C'est apercevoir la raison intime qui fait de fait de cette série de coups successifs une sorte de tout organisé. A plus forte raison, cette faculté est-elle nécessaire au joueur lui-même, c'est-à-dire à l'inventeur. (Poincaré 1905, pp. 30–31).[5]

[5]If you are present at a game of chess, it will not suffice, for the understanding of the game, to know the rules for moving the pieces. That will only enable you to recognize that each move has been made in conformity with these rules, and this knowledge will truly have very little value. Yet this is what the reader of a book on mathematics would do if he were a logician only. To understand the game is wholly another matter; it is to know why the player moves this piece rather than that other which he could have moved without breaking the rules of the game. It is to perceive the inward reason which makes of this series of successive moves a sort of organized whole. This faculty is still more necessary for the player himself, that is, for the inventor (Poincaré 2014, pp. 23–24).

Appendix 1: Basic Notions for Dialogical Logic[6]

The dialogical approach to logic is not a specific logical system; it is rather a general framework having a rule-based approach to meaning (instead of a truth-functional or a model-theoretical approach) which allows different logics to be developed, combined and compared within it. The main philosophical idea behind this framework is that meaning and rationality are constituted by argumentative interaction between epistemic subjects; it has proved particularly fruitful in history of philosophy and logic. We shall here provide a brief overview of dialogues in a more intuitive approach than what is found in the rest of the book in order to give a feeling of what the dialogical framework can do and what it is aiming at.

Dialogues and interaction
As hinted by its name, this framework studies dialogues; but it also takes the form of dialogues. In a dialogue, two parties (players) argue on a thesis (a certain statement that is the subject of the whole argument) and follow certain fixed rules in their argument. The player who states the thesis is the Proponent, called **P**, and his rival, the player who challenges the thesis, is the Opponent, called **O**. By convention, we refer to **P** as he and to **O** as she. In challenging the Proponent's thesis, the Opponent is requiring of the Proponent that he defends his statement.
The interaction between the two players **P** and **O** is spelled out by challenges and defences, Actions in a dialogue are called moves; they are often understood as speech-acts involving declarative utterances (*statements*) and interrogative utterances (*requests*).[7] The rules for dialogues thus never deal with expressions isolated from the act of uttering them. The rules in the dialogical framework are divided into two kinds of rules: particle rules, and structural rules.

Particle rules
Particle rules (*Partikelregeln*), or rules for logical constants, determine the legal moves in a play and regulate interaction by establishing the relevant moves constituting *challenges*: moves that are an appropriate attack to a previous move (a statement) and thus require that the challenged player play the appropriate defence to the attack. If the challenged player defends his statement, he has answered the challenge.
Particle rules determine how reasons are asked for and are given for each kind of statement, thus providing the meaning of that statement. In other words, the appro-

[6]The three appendices are based on the book in progress *Immanent Reasoning* by Rahman et al. (2017). The original historical sources of the origins of dialogical logic and their reprintings are to be found in Lorenzen (1969), Lorenzen and Lorenz (1978), Lorenzen and Schwemmer (1975), Lorenz (2010a, b). For an overview see Rahman and Keiff (2005), Krabbe (1982, 1985, 2006), Keiff (2007, 2009), Clerbout (2014a, b, c).

[7]Literature pertaining to the dialogical framework also uses the terms posits and assertions to designate what we will here call statements, that is, the act of stating a proposition within a game of giving and asking for reasons; the meaning of a statement is defined by an appropriate challenge and defence, or, in other words, how reasons for this statement can be requested, what constitutes reasons for this statement and how these reasons can be provided (see Keiff (2009)).

priate attacks and defences—that is, the appropriate ways of asking for and giving reasons—for each statement (or move) gives the meaning of these statements: a conjunction, a disjunction, or a universal quantification, for instance, receive their meaning through the appropriate interaction in a dialogical game, spelled out by the particle rules.
The particle rules provide the meaning of the different logical connectives, which they provide in a dynamic way through appropriate challenges and answers. This feature of dialogues is fundamental for immanent reasoning: the meaning of the moves in a dialogue does not lie in some external semantic, but is immanent to the dialogue itself, that is, in the specific and appropriate way the players interact; we here join the Wittgensteinian conception of meaning as use. The particle rules are spelled out in an anonymous way, that is, without mentioning if it is **P** or **O** who is attacking or defending: the rules are the same for the two players; the meaning of the connectives is therefore independent of who uses them.[8]

Structural rules
Structural rules (*Rahmenregeln*) on the other hand determine the general course of a dialogue game, such as how a game is initiated, how to play it, how it ends, and so on. The point of these rules is not so much to spell out the meaning of the logical constants by specifying how to act in an appropriate way—this is the role of the particle rules—; it is rather to specify according to what structure interactions will take place. It is one thing to determine the meaning of the logical constants as a set of appropriate challenges and defences, it is another to define whose turn it is to play and when a player is allowed to play a move. One could thus have the same local meaning and change a structural rule, saying for instance that one of the players is allowed to play two moves at a time instead of simply one: this would considerably change the game without changing the local meaning of what is said.

One of the most important structural rules for the present study on immanent reasoning is the Copy-cat rule (or Socratic rule when introducing CTT features in the dialogical context). This rule is not anonymous, it is a restriction on the moves the Proponent is allowed to play: the Proponent is allowed to assert an elementary judgement only if the Opponent has already asserted it. So the Opponent is not concerned by the same exact rules as the Proponent.

The Copy-cat rule accounts for analyticity: the Proponent, who brings forward the thesis, will have to defend it without bringing any element of his own in the play: his defence of the thesis will have to rely only on what the Opponent has conceded, and everything the Opponent concedes comes only from the meaning of the thesis. The Opponent will be challenging the thesis, and challenging and defending the subsequent moves made by the Proponent in reaction to her initial challenge of the thesis; but all these challenges and defences are made according to the particle rules. So everything the Opponent will concede during a play stems from an application of the particle rules starting with the thesis. The only elements whose meaning is left unspecified, in formal plays, are the elementary statements (specifying their meaning

[8] In this sense, the particle rules are said to be *symmetric*. This is imperative to preserve the dialogical framework from connectives as Prior's (1960) *tonk*. See Redmond and Rahman (2016).

is the point of material plays). The Copy-cat rule makes sure that the Proponent is not bringing in any elementary statement to back his thesis that the Opponent might not agree with: the Proponent can only back his thesis with elementary statements that the Opponent herself has already conceded.

Preliminary notions

The language

Let L be a first-order language built as usual upon the propositional connectives, the quantifiers, a denumerable set of individual variables, a denumerable set of individual constants and a denumerable set of predicate symbols (each with a fixed arity). We extend the language L with two labels O and P, standing for the players of the game, and the two symbols '!' and '?' standing respectively for statements and requests. When the identity of the player does not matter, we use the variables **X** or **Y** (with $\mathbf{X} \neq \mathbf{Y}$).[9]

Plays

A *play* is a legal sequence of moves, that is, a sequence of moves which observes the game rules. Particle rules are not the only rules which must be observed in this respect: the second kind of rules, the *structural rules*, are the rules providing the precise conditions under which a given sequence is a play.

Dialogical games

The *dialogical game* for a statement is the set of all plays from a given *thesis* (initial statement, see below the Starting rule, SR0).

A move in a play

A move M is an expression of the form '**X**-*e*', where *e* is either

- of the form '! *A*' (read: *the player **X** states A*), for some proposition *A* of L; we say it is an elementary statement, or
- of one of the forms specified by the particle rules (see below).

Challenges and defences

The words 'attack' and 'defence' are convenient to name certain moves according to their relation to other moves which can be defined in the following way.

- Let σ be a sequence of moves. The function ρ_σ assigns a position to each move in σ, starting with 0.
- Let σ be a sequence of moves. The function ρ_σ assigns a position to each move in σ, starting with 0.
- The function F_σ assigns a pair $[m, Z]$ to certain moves M in σ, where m denotes a position smaller than $\rho_\sigma(M)$ and Z is either A or D, standing respectively for 'attack' and 'defence'. That is, the function F_σ keeps track of the relations of attack and defence as they are given by the particle rules.

Let us point that at the local level (the level of the particle rules), this terminology should be bereft of any strategic undertone.

[9]This aspect (player independence) is fundamental for the symmetry of the rules. See section 0.

Terminological note: challenge, attack and defence
The standard terminology uses the terms *challenge*, or *attack,* and *defence* (sometimes *answer* in respect of challenges). We shall here make a (subtle) distinction between challenge and attack: a challenge is initiated by an attack and needs this attack to be defended against in order to be answered to. So a challenge requires a defence to be settled, whereas an attack is simply the move that opens the challenge. For instance, using the particle rules exposed below, an attack on an implication will be simply to state the antecedent, and challenging an implication will be to attack it and thus demanding that the player who stated the implication defends her posit by positing the consequent, knowing that the challenger stated the antecedent. As one can see, the difference between challenge and attack is slim, and they may oftentimes be taken as synonymous.

Local meaning of logical constants
Particle rules:
In the dialogical framework, the particle rules state the *local semantics*: only challenges and the corresponding defences for a given logical constant are at stake here, that is, we only take the main logical constant of the proposition into account. Particle rules provide a decontextualized description of how the game can proceed locally: they specify the way a statement can be challenged and defended according to its main logical constant. In this way the particle rules govern the local level of meaning (Fig. 10.1).
The particle rules for quantifiers has not been introduced, so we will be commenting these rules briefly.
The rules for universal quantification are similar to those for conjunction: stating a universally quantified proposition means that the challenger may choose any individual constant a_i and request of the utterer to make his statement by instantiating every free occurrence of x with a_i. That is, the challenger chooses which proposition he wants the utterer to state.
Properties of universal quantification

- the challenge is a request;

	Conjunction	Disjunction	Implication	Negation
Move	$X ! A \wedge B$	$X ! A \vee B$	$X ! A \supset B$	$X ! \neg A$
Challenge	$Y ? L^{\wedge}$ or $Y ? R^{\wedge}$	$Y ?_{\vee}$	$Y ! A$	$Y ! A$
Defence	$X ! A$ (resp.) $X ! B$	$X ! A$ or $X ! B$	$X ! B$	—

Fig. 10.1 Particle rules for dialogical games: propositional connectives

Fig. 10.2 Particle rules for dialogical games: quantifiers

	Universal quantification	Existential quantification
Move	$X\,!\,\forall x B(x)$	$X\,!\,\exists x B(x)$
Challenge	$Y\,?\,[x/a_i]$	$X\,?_{\exists}$
Defence	$X\,!\,B(x/a_i)$	$X\,!\,B(x/a_i)$ with $1 \leq i \leq n$

- the challenger has the choice;
- the defender must state the requested proposition.

The rules for existential quantification are similar to those for disjunction: it is the defender who chooses the proposition he wants to state in response to the challenge.
Properties of existential quantification:

- the challenge is a request;
- the defender has the choice;
- the defender chooses which proposition to state (Fig. 10.2).

Symmetry and harmony
In providing the properties of the particle rules, a central feature we have distinguished is who has the choice: is it the challenger or the defender? The meaning of the logical constants is largely determined by who has the choice in the interaction. But notice that in formulating the particle rules, the players' identities are not specified: we do not use **O** and **P** but we use **X** and **Y** instead, thus only specifying who is the challenger and who is the defender *for this particular statement*. That is, we simply provide the appropriate challenge and defence for certain logical constants and determine in this way who has the choice: we provide their meaning in terms of interaction within a dialogue (a game of giving and asking for reasons).

It would not be reasonable to base a game-theoretical approach to the meaning of logical constants in which the meaning differs according to which players utters it: this approach would make interaction senseless, for each player would be meaning something different when uttering the same thing. Equality in action is precisely based on the possibility for a player to say *the same thing* as the other player, and by that to be meaning also the same thing. Equality in action is in this regard the idea that a statement made by a player can be made by another player in a game of giving and asking for reasons (a dialogue) with the exact same meaning as the statement made by the first player, that is with the same particle rules for challenging and defending it. It is thus the interaction based on player-independent rules that allows two different players to be speaking of the same thing: equality between different statements emerges from the interaction itself.

Since the rules for the logical constants are independent of the player's identities—the rules are exactly the same for the two players—we say that these rules are symmetric. This feature captures one of the strengths of the dialogical approach to meaning: the dialogical approach is in this way immune to a wide range of trivializing connectives such as Prior's *tonk*.[10] Symmetry, or player-independence in the particle rules, must be contrasted with the dialogical rendering of harmony, which concerns the structural rules and the strategy level, not the particle rules. The structural rules, which will be introduced in the next section, are not player independent: the first rule (SR0) specifies who the players are (Proponent or Opponent) according to who plays the first move, that is who states the thesis; that player will be the Proponent. But the rule that matters most in regard to immanent reasoning is the Copy-cat rule (or Socratic rule in a CTT framework); this rule puts a restriction on the Proponent's moves, while those of the Opponent are left unrestricted: which the Proponent cannot play an elementary statement that has not been previously stated by the Opponent. The purpose of this restriction is to insure that the thesis will be grounded only on what the Opponent has conceded, and thereby secure a form of analyticity that we call immanent reasoning: the Proponent has to ground his thesis on what the Opponent brings forward in the course of their interaction (the dialogue), an interaction that is initiated by the Proponent stating a thesis, which the Opponent challenges with the ensuing series of challenges and defences defined by the particle rules and constituting the dialogue. Thus the Opponent will not bring forward anything that does not stem from the meaning (defined by the particle rules) of the thesis, and the Proponent will not bring any elementary proposition into the game that he cannot justify within that very same dialogue by referring to the Opponent's own statements ("I am entitled to state this because you have stated it yourself"). Symmetry and harmony are two essential aspects of the dialogical framework and are the principles for immanent reasoning. Dialogical harmony thus coordinates a player-independent level (the local meaning) and a player-dependent level (the global meaning and the strategy level). This aspect contrasts with the Constructive Type Theory notion of harmony which belongs to proof-theory and stays only at the level of strategies.[11] Immanent reasoning and equality in action emerge from taking the specific aspects of the three levels (local, global and strategic) into account and considering how they intertwine to build these complex and dynamic frameworks that are dialogues.

Global meaning:

The global meaning—as opposed to the local meaning defined by the particle rules—is defined by means of *structural rules* which specify the general way plays unravel by specifying who starts in a play, what moves are allowed and in which order, when a play ends and who wins it.

---

[10]See Rahman and Keiff (2005), Rahman et al. (2009) and Rahman and Redmond (2016).

[11]Rahman and Redmond (2016) and Rahman et al. (2017).

Preliminary terminology
Terminal plays:
A play is called *terminal* when it cannot be extended by further moves in compliance with the rules.
X-terminal plays:
A play is ***X****-terminal* when the play is terminal and the last move in the play is an **X**-move.

The structural rules
SR0 (Starting rule)
Any dialogue starts with the Opponent stating initial concessions (if any) and the Proponent stating the thesis (labelled move 0). After that, each player chooses in turn a positive integer called the *repetition rank* which determines the upper boundary for the number of attacks and of defences each player can make in reaction to each move during the play.
Example: if the repetition rank of ***O*** *is* $m := 1$*, then* ***O*** *may attack or defend against* at most *once each move of* ***P****. If* ***P****'s repetition rank is* $n := 2$*, then* ***P*** *may attack or defend against* at most *twice each move of* ***O****.*
SR1: Development rule
The Development rule depends on what kind of logic is chosen: if the game uses classical logic, then it is SR1c that should be used; but if intuitionistic logic is used, then SR1i must be used.

SR1c (Classical Development rule)

Players move alternately. Once the repetition ranks have been chosen, each move is either attacking or defending a move made by the other player, in accordance with the particle rules.

SR1i (Intuitionisitic Development rule)

Players move alternately. Once the repetition ranks have been chosen, each move is either attacking or defending a move made by the other player, in accordance with the particle rules.
Players can respond only to the *last non-answered* challenge of the other player.
Note: This last clause is known as the Last Duty First *condition, and makes dialogical games suitable for intuitionistic logic (hence this rule's name).*
SR2 (Copy-cat rule)
P may not play an elementary statement unless **O** has stated it first.
Elementary propositions cannot be challenged.
Note: The formulation of this rule has a downside: the thesis of a dialogical game cannot be an elementary statement. For a special version of the Copy-cat rule allowing plays on elementary statements, see below (Sect. 0) where we link this rule to equality.
SR3 (Winning rule)
Player **X** wins the play ς only if it is **X**-terminal.

Linking the Copy-cat rule (SR2) and equality

The Copy-cat rule[12] is one of the most salient characteristics of dialogical logic. As discussed by Marion and Rückert (2015), it can be traced back to Aristotle's reconstruction of the Platonic dialectics. A purely argumentative point of view can be defined within dialectics as refraining from calling on some authority beyond what has actually been brought forward during the current argumentative interaction, the ultimate authority being the fact that the other person has said it, any other consideration being set aside for the time of the dialectical exchange (in this argumentative perspective). Thus, when an elementary statement is challenged, the challenge can be answered only by invoking the challenger's own concessions. In such a context, the Copy-cat rule can be understood in the following way, when a player plays an elementary statement:

> my grounds for stating the proposition you are challenging are exactly the same as the ones you brought forward when you yourself stated that very same proposition.[13]

In this regard, elementary statements actually can be challenged (as opposed to the SR2 formulation above), the answer then being of the form "but you have said it yourself". A special formulation of the Copy-cat rule SR2 addresses this problem.

Special Copy-cat rule

O's elementary statements cannot be challenged. However, **O** can challenge an elementary statement played by **P**. The challenge and corresponding defence is determined by the following table. Notice that this (structural) rule is not player-independent and uses the names of the players.

<u>Copy-cat rule</u>

	Move	Challenge	Defence
Special Copy-cat rule (Structural Rule 2)	$\mathbf{P}!\,A$ For elementary A	$\mathbf{O}?_A$	$\mathbf{P}!\,sic(n)$ **P** indicates that **O** stated A at move n

The Copy-cat rule, and even more in this special formulation, introduces an asymmetry between the two players (the Proponent's moves are restricted in a way the Opponent's are not).

Examples of plays

These examples should allow the reader to fully understand the rules given above and their implications, especially the difference between SR1c (classical Development

[12]In previous literature on dialogical logic this rule has been called the *Formal rule* (see Lorenzen and Lorenz (1978)). Since here we will distinguish different formulations of this rule that yield different kind of dialogues we will use the term *Copy-cat rule* when we speak of the rule in standard contexts (such as in the present section)—contexts in which the constitution of the elementary propositions involved in a play is not rendered explicit. When we use the rule in a dialogical framework for CTT, as in the next chapter, we speak of the *Socratic rule*. However, we will continue to use the expression Copy-cat *move* in order to characterize moves of **P** that copies moves of **O**.

[13]For the Platonic origins of this rule see Rahman and Keiff (2010).

rule) and SR1i (intuitionistic Development rule). In the next chapter (V), strategies will be introduced, which allow to compare different plays (with different choice sequences of the players) and build the best possible way of playing for one of the players.

First example, the third excluded: $A \vee \neg A$

The third excluded (*tertium non datur*) is a principle stating that a proposition either is (A) or is not ($\neg A$), without any third possible option. This principle is much discussed in philosophy and logic, it is a valid principle in classical logic, but is not accepted in intuitionistic logic. If this principle is accepted, the principle of non-contradiction ($\neg(A \wedge \neg A)$) follows, but the reverse is not the case (and intuitionistic logic accepts the principle of non-contradiction but not the principle of third excluded). We will here give a play according to the classical (structural) rules, and then a play according to the intuitionistic (structural) rules.

Play 1: the third excluded—classical rules

O			**P**		
				$!A \vee \neg A$	0
1	$n := 1$			$m := 2$	2
3	$?_{\vee}$	0		$!\neg A$	4
5	$!A$	2		—	
{3}	$\{?_{\vee}\}$	{0}		$!A$	6

P wins (classical rules).

*Note: the curly brackets are inserted to stress the fact that **O** is not actually making a move, but that **P** is using his repetition rank of 2 in order to defend twice **O**'s challenge (move 3). We repeat that challenge (in brackets) in order to know where **P**'s move 6 comes from.*

Notice that **P** would not have won without a repetition rank higher than 1: he would not have been allowed to answer twice to O's challenge (move 3), and thus use her own assertion of A (move 5) triggered by P's first defence to O's challenge (move 3). This example is a good illustration for the Copy-cat rule and for the use of repetition ranks. Notice also that P's move 6 is an answer to the challenge of move 3, that is a challenge preceding the last unanswered challenge, which is move 5. This challenge of move 5 will never be answered, because an attack on the negation cannot be defended. So P wins because the classical rules for dialogues do not restrict P's answers only to the last unanswered challenge. This fact is the key to understand the outcome of the next play, which uses the intuitionistic rules.

Play 2: the third excluded—intuitionistic rules

O			**P**		
				$! A \vee \neg A$	0
1	$n := 1$			$m := 2$	2
3	$?_{\vee}$	0		$! \neg A$	4
5	$! A$	4		—	

O wins (intuitionistic rules).

Second example, the double negation elimination: $\neg\neg A \supset A$
The elimination of double negation is another example of a principle accepted in classical logic but rejected in intuitionistic logic. This principle is at the core of classical mathematics, for it is what is used in indirect proofs (concluding A from the demonstration that the negation of A leads to a contradiction, that is from the fact that $\neg\neg A$ holds). This principle is closely linked to the principle of excluded middle. Once again, we give a play with classical rules (**P** wins) and a play with intuitionistic rules (**P** loses).

Play 3: the elimination of double negation—classical rules

O			**P**		
				$! \neg\neg A \supset A$	0
1	$m := 1$			$n := 2$	2
3	$! \neg\neg A$	0		$! A$	6
	—		3	$! \neg A$	4
5	$! A$	4		—	

P wins (classical rules).

Notice that as for the third excluded, **P** wins here because he does not have to answer only to the last unanswered challenge (which is move 5) but answers a previous challenge (his move 6 is an answer to the challenge of move 3). This move is forbidden by the intuitionistic (structural) rules ("Last Duty First") illustrated in the next play: P should play his move 6, but is not allowed to; it is his turn and he cannot play, so he loses.

Play 4: the elimination of double negation—intuitionistic rules

O			P		
				$! \neg\neg A \supset A$	0
1	$m := 1$			$n := 2$	2
3	$! \neg\neg A$	0			
	—		3	$! \neg A$	4
5	$! A$	4		—	

It should be clear from these two examples that the intuitionistic rules for dialogues only concern the *structural* rules, namely *when* (in what conditions) a move (challenge or defence) is allowed, but not the *particle* rules which determine *how* to challenge a move or *how* to answer a challenge. The intuitionistic rules are only a *restriction* imposed on the classical rules, so if **P** wins a play according to the intuitionistic rules, *a fortiori* he should win according to the classical rules.

Third example, the double negation of the third excluded: $\neg\neg(A \vee \neg A)$

This example is a combination of the previous two. But whereas the principle of third excluded and the principle of double negation elimination are not intuitionistic principles (**P** loses), the double negation of the third excluded $\neg\neg(A \vee \neg A)$ does actually hold with intuitionistic rules. This clearly shows that, for intuitionistic logic, an expression is not equivalent to its double negation (the elimination of the double negation of the third excluded would not yield the third excluded, which would contradict the first example).

Play 5: double negation of third excluded—intuitionistic rules

O			P		
				$! \neg\neg(A \vee \neg A)$	0
1	$m := 1$			$n := 2$	2
3	$! \neg(A \vee \neg A)$	0		—	
	—		3	$! A \vee \neg A$	4
5	$?_{\vee}$	4		$! \neg A$	6
7	$! A$	6		—	
	—		3	$! A \vee \neg A$	8
9	$?_{\vee}$	8		$! A$	10

In this play, **P** also uses his repetition rank of 2 (move 4 and move 8), but this time to challenge move 3 (instead of defending a move). As opposed to the previous examples, he does not need to defend a move preceding the last unanswered challenge, so this play in winnable by **P** in the intuitionistic and in the classical contexts.

Strategy Level

The strategy standpoint is but a generalisation of the procedure which is implemented at the play level; it is a systematic exposition of all the relevant variants of a game—the relevancy of the variants being determined from the viewpoint of one of the two players.

Preliminary notions
Definitions
Extensive form of a dialogical game
The *extensive form* E(ϕ) of the dialogical game D(ϕ) is simply its tree presentation, also called the game-tree. Nodes are labelled with moves so that the root is labelled with the thesis and paths in E(ϕ) are linear representations of plays and maximal paths represent terminal plays in D(ϕ).
That is, the extensive form of a dialogical game is an infinitely generated tree in which each branch is of a finite length.
Strategy
A *strategy* for player **X** in D(ϕ) is a function which assigns an **X**-move M to every non terminal play ς having a **Y**-move as last member, such that extending ς with M results in a play.
X-winning strategy (or X-strategy)
An *X-strategy* is *winning* if playing according to it leads to **X**-terminal plays no matter how **Y** moves.
That is, a winning strategy for player **X** defines the situation in which, for any move choice made by player **Y**, **X** has at least one possible move at his disposal allowing him to win.
Extensive form of an X-strategy
Let sx be a strategy of player **X** in *D(ϕ)* of extensive form E*(ϕ)*. The extensive form of sx is the fragment Sx of E*(ϕ)* such that:

1. The root of E*(ϕ)* is the root of Sx,
2. For any node t which is associated with an **X**-move in E*(ϕ)*, any immediate successor of t in E*(ϕ)* is an immediate successor of t in Sx,
3. For any node t which is associated with a **Y**-move in E*(ϕ)*, if t has at least one immediate successor in E*(ϕ)* then t has exactly one immediate successor in Sx, namely the one labelled with the **X**-move prescribed by sx.

Validity
A proposition is *valid* in a certain dialogical system if and only if **P** has a winning strategy for it.
Some results from existing literature on the strategy level
The following three results—extracted from existing literature on the subject—establish the correspondence between the dialogical framework and other frameworks involving classical and intuitionistic logics. We will be using them in order to facilitate the building of dialogical demonstrations: the procedure presented in the next section presupposes them. For more details and recent new proofs of these results, see Clerbout (2014a, b).

P-winning strategies and leaves
Let w be a **P**-winning strategy in D(ϕ). Then every leaf in the extensive form W_{ϕ} of w is labelled with an elementary **P**-sentence.
Determinacy
There is an **X**-winning strategy in D(ϕ) if and only if there is no **Y**-winning strategy in D(ϕ).
Soundness and Completeness of Tableaux
Consider first-order tableaux and first-order dialogical games. There is a tableau proof for φ if and only if there is a ***P***-winning strategy in D(ϕ).
The fact that the existence of a ***P***-winning strategy coincides with validity (*there is a **P**-winning strategy in D(ϕ) if and only if φ is valid)* follows from the soundness and completeness of the tableau method with respect to model-theoretical semantics.
This third metalogical result for the standard dialogical framework will be taken here for granted (the proof is given for instance in Clerbout (2014b).

Appendix 2: Local Reasons and Dialogues for Immanent Reasoning

Introductory remarks on the choice of CTT
Recent developments in dialogical logic show that the Constructive Type Theory approach to meaning is very natural to the game-theoretical approaches in which (standard) metalogical features are explicitly displayed at the object language-level.[14] This vindicates, albeit in quite a different fashion, Hintikka's plea for the fruitfulness of game-theoretical semantics in the context of epistemic approaches to logic, semantics, and the foundations of mathematics.[15] From the dialogical point of view, the actions—such as choices—that the particle rules associate with the use of logical constants are crucial elements of their full-fledged (local) meaning: if meaning is conceived as constituted during interaction, then all of the actions involved in the constitution of the meaning of an expression should be made explicit; that is, they should all be part of the object-language. This perspective roots itself in Wittgenstein's remark according to which one cannot position oneself outside language in order to determine the meaning of something and how it is linked to syntax; in other words, language is unavoidable: this is his Unhintergehbarkeit der Sprache, one of Wittgenstein's tenets that Hintikka explicitly rejects.[16] According to this perspective of Wittgenstein, language-games are supposed to accomplish the task of studying language from a perspective that acknowledges its *internalized feature*. This is what underlies the approach to meaning and syntax of the dialogical framework in which

[14]See for instance Clerbout and Rahman (2015), Rahman et al. (2015).

[15]Cf. Hintikka (1973).

[16]Hintikka (1996) shares this rejection with all those who endorse model-theoretical approaches to meaning.

all the speech-acts that are relevant for rendering the meaning and the "formation" of an expression are made explicit. In this respect, the metalogical perspective which is so crucial for model-theoretic conceptions of meaning does not provide a way out. It is in such a context that Lorenz writes:

> Also propositions of the metalanguage require the understanding of propositions, [...] and thus cannot in a sensible way have this same understanding as their proper object. The thesis that a property of a propositional sentence must always be internal, therefore amounts to articulating the insight that in propositions about a propositional sentence this same propositional sentence does not express a meaningful proposition anymore, since in this case it is not the propositional sentence that is asserted but something about it. Thus, if the original assertion (i.e., the proposition of the ground-level) should not be abrogated, then this same proposition should not be the object of a metaproposition [...].[17] While originally the semantics developed by the picture theory of language aimed at determining unambiguously the rules of "logical syntax" (i.e. the logical form of linguistic expressions) and thus to justify them [...]—now language use itself, without the mediation of theoretic constructions, merely via "language games", should be sufficient to introduce the talk about "meanings" in such a way that they supplement the syntactic rules for the use of ordinary language expressions (superficial grammar) with semantic rules that capture the understanding of these expressions (deep grammar).[18]

Similar criticism to the metalogical approach to meaning has been raised by Göran Sundholm (1997, 1998, 2001, 2009, 2012, 2013) who points out that the standard model-theoretical semantic turns semantics into a meta-mathematical formal object in which syntax is linked to meaning by the assignation of truth values to uninterpreted strings of signs (formulae). Language does not express content anymore, but it is rather conceived as a system of signs that speak about the world—provided a suitable metalogical link between the signs and the world has been fixed. Moreover, Sundholm (2016) shows that the cases of quantifier-dependences motivating Hintikka's IF-logic can be rendered in the CTT framework. What we will here add to Sundholm's observation is that even the interactive features of these dependences can be given a CTT formulation, provided the latter is developed within a dialogical setting.
Ranta (1988) was the first to link game-theoretical approaches with CTT. Ranta took Hintikka's (1973) Game-Theoretical Semantics (GTS) as a case study, though his point does not depend on that particular framework: in game-based approaches, a proposition is a set of winning strategies for the player stating the proposition.[19] In game-based approaches, the notion of truth is at the level of such winning strategies. Ranta's idea should therefore in principle allow us to apply, safely and directly, instances of game-based methods taken from CTT to the pragmatist approach of the dialogical framework. From the perspective of a general game-theoretical approach to meaning however, reducing a proposition to a set of winning strategies is quite unsatisfactory. This is particularly clear in the dialogical approach in which different levels of meaning are carefully distinguished: there is indeed the level of strategies, but there is also the level of plays in the analysis of meaning which can be further

---

[17]Lorenz (1970, p. 75), translated from the German by Shahid Rahman.

[18]Lorenz (1970, p. 109), translated from the German by Shahid Rahman.

[19]That player can be called Player 1, Myself or Proponent.

analysed into local, global and material levels. The constitutive role of the play level for developing a meaning explanation has been stressed by Kuno Lorenz in his (2001) paper:

> Fully spelled out it means that for an entity to be a proposition there must exist a dialogue game associated with this entity, i.e., the proposition A, such that an individual play of the game where A occupies the initial position, i.e., a dialogue D(A) about A, reaches a final position with either win or loss after a finite number of moves according to definite rules: the dialogue game is defined as a finitary open two-person zero-sum game. Thus, propositions will in general be dialogue-definite, and only in special cases be either proof-definite or refutation-definite or even both which implies their being value-definite. Within this game-theoretic framework […] truth of A is defined as existence of a winning strategy for A in a dialogue game about A; falsehood of A respectively as existence of a winning strategy against A.[20]

Given the distinction between the play level and the strategy level, and deploying within the dialogical framework the CTT-explicitation program, it seems natural to distinguish between *local* reasons and *strategic* reasons: only the latter correspond to the notion of *proof-object* in CTT and to the notion of *strategic-object* of Ranta. In order to develop such a project we enrich the language of the dialogical framework with statements of the form "$p : A$". In such expressions, what stands on the left-hand side of the colon (here p) is what we call a *local reason*; what stands on the right-hand side of the colon (here A) is a proposition (or set).[21] The *local* meaning of such statements results from the rules describing how to compose (*synthesis*) within a play the suitable local reasons for the proposition A and how to separate (*analysis*) a complex local reason into the elements required by the composition rules for A. The synthesis and analysis processes of A are built on the formation rules for A.

Local reasons and material truth

The most basic contribution of a local reason is its contribution to a material dialogue involving an elementary proposition. Informally, we can say that if the Proponent **P** states the elementary proposition A, it is because P claims that he can bring forward a reason in defence of his statement. It is the Socratic rule that determines the precise form of that local reason, specific to A.[22] Our study focuses on formal—not material—dialogues, but we will still provide some basic elements on material truth in regard to local reasons so as to render in a clearer fashion the limits of our study and its philosophical background, the meaning of formal plays by contrast with what they are not, and the further work that can be carried out from this presentation of dialogues for immanent reasoning.

Approaching material truth

Assume the Proponent states that 1 is an odd number:

$$\mathbf{P}!\, 1\, is\ an\ odd\ number$$

---

[20]Lorenz (2001, p. 258).

[21]See Rahman et al. (2017), Clerbout and Rahman (2015), Rahman and Clerbout (2013, 2015).

[22]Recall that the Socratic rule does not prohibit the Opponent **O** from challenging an elementary proposition of **P**; the rule only restricts **P**'s authorized moves.

the Opponent can then express the following demand, asking **P** for reasons for his statement:

$$\mathbf{O}!\ find\ a\ natural\ number\ n\ such\ that\ 1 = 2.n + 1$$

Because of the restriction the Socratic rule imposes on **P**, he can defend his statement by choosing "0", provided that **O** has already endorsed the statement "0 is a natural number" ($0 : \mathbb{N}$). This produces *material truth*.
Material truth can then be described in the following way: the statement that a given proposition is materially true requires displaying a local reason *specific to that very proposition*.

Material truth and local reasons
A local reason adduced in defence of a proposition thus prefigures a material dialogue displaying the specific content of that proposition. This constitutes the bottom of the normative approach to meaning of the dialogical framework: *use* (dialogical interaction) is to be understood as *use prescribed by a rule of dialogical interaction*. This applies not only to the meaning of logical constants, but also to the meaning of elementary propositions. This is what Jaroslav Peregrin (2014, pp. 2–3) calls the *role* of a linguistic statement: according to this terminology, and if we place his suggestion in our dialogical setting, we can say that the meaning of an elementary proposition amounts to its *role* in that form of interaction that the Socratic rule for a material dialogue prescribes for that specific proposition. It follows from such a perspective that material dialogues are important not only for the general question of the normativity of logic, but also for the elaboration of a language with content.

Material dialogues and formal dialogues
Summing up, what distinguishes formal dialogues from material dialogues resides in the following:

- The formulation of the Socratic rule of a formal dialogue prescribes a form of interaction based only on the meaning of the logical constant(s) involved, irrespective of the meaning of the elementary propositions in the scope of that constant.
- The choice of the local reason for the elementary propositions involved is left to the authority of the Opponent.

In other words, in a formal dialogue the Socratic rule is *not* specific to any elementary proposition in particular, but it is *general*; definitions that distinguish one proposition from another are introduced during the game according to the local meaning of the logical constant involved: formal dialogues are the purest kind of immanent reasoning. The synthesis and analysis of local reasons for a proposition *A* are determined by the actions prescribed by the Socratic rule specific to the kind of play in which *A* has been stated:

- If the play is *material*, the Socratic rule will describe a kind of action specific to the formation of *A*.

- If the play is *formal*, as assumed in the main body of our study, the Socratic rule will allow **O** to bring forward the relevant local reasons during the development of the play.

The point is that in formal dialogues, when the Opponent challenges the thesis, the thesis is assumed to be well-formed up to the logical constants, so the formation of the elementary statements is displayed during the development of the dialogue and left to the authority of **O**. So the formation rule for elementary statements does not really take place at the level of local meaning but at level of global meaning. Since the local reasons for the elementary statements are left to **O**'s authority, what we now need is to describe the process of *synthesis* and *analysis* for local reasons of the logical constants. However, before starting to enrich the language of the standard dialogical framework with local reasons for logical constants let us discuss how to implement a dialogical notion of *formation rules*. The formation rules together with the synthesis and analysis rules settle the local meaning of dialogues for immanent reasoning.

The local meaning of local reasons
Here is an introduction of the formation rules, the synthesis rules, and the analysis rules for local reasons. But we first need to make a clarification on statements and add a piece of notation to the framework:

Statements in dialogues for immanent reasoning
Dialogues are games of giving and asking for reasons; yet in the standard dialogical framework, the reasons for each statement are left implicit and do not appear in the notation of the statement: we have statements of the form **X**! A for instance where A is an elementary proposition. The framework of dialogues for immanent reasoning allows to have explicitly the reason for making a statement, statements then have the form **X**a : A for instance where a is the (local) reason **X** has for stating the proposition A. But even in dialogues for immanent reasoning, all reasons are not always provided, and sometimes statements have only implicit reasons for bringing the proposition forward, taking then the same form as in the standard dialogical framework: **X**! A. Notice that when (local) reasons are not explicit, an exclamation mark is added before the proposition: the statement then has an implicit reason for being made.

A statement is thus both a proposition and its local reason, but this reason may be left implicit, requiring then the use of the exclamation mark.

Adding concessions
In the context of the dialogical conception of CTT we also have statements of the form

$$\mathbf{X}!\, \pi(x_1, \ldots, x_n)\, [x_i : A_i]$$

where "π" stands for some statement in which $(x_1, \ldots, x_n)$ occurs, and where $[x_i : A_i]$ stands for some condition under which the statement $\pi(x_1, \ldots, x_n)$ has been brought forward. Thus, the statement reads:

Table 10.1 Formation rules, condensed presentation

	Connective	Quantifier	Falsum
Move	**X** AKB : *prop*	**X** $(Qx : A)\ B(x)$: *prop*	**X** $\bot$: *prop*
Challenge	**Y** $?_{FK\ 1}$ and/or **Y** $?_{FK2}$	**Y** $?_{FQ1}$ and/or **Y** $?_{FQ2}$	–
Defence	**X** A : *prop* (resp.) **X** B : *prop*	**X** A : *set* (resp.) **X** $B(x)$: *prop* $(x : A)$	–

X states that $\pi(x_1, \ldots, x_n)$ under the condition that the antagonist concedes x_i: A_i. We call *required concessions* the statements of the form [x_i: A_i] that condition a claim. When the statement is challenged, the antagonist is accepting, through his own challenge, to bring such concessions forward. The concessions of the thesis, if any, are called *initial concessions*. Initial concessions can include formation statements such as *A*: *prop*, *B*: *prop*, for the thesis, $A \supset B$: *prop*.

Formation rules for local reasons: an informal overview

It is presupposed in standard dialogical systems that the players use well-formed formulas (*wff*). The well formation can be checked at will, but only with the usual meta reasoning by which one checks that the formula does indeed observe the definition of a *wff*. We want to enrich our CTT-based dialogical framework by allowing players themselves to first enquire on the formation of the components of a statement within a play. We thus start with dialogical rules explaining the formation of statements involving logical constants (the formation of *elementary* propositions is governed by the Socratic rule, see the discussion above on material truth). In this way, the well formation of the thesis can be examined by the Opponent before running the actual dialogue: as soon as she challenges it, she is de facto accepting the thesis to be well formed (the most obvious case being the challenge of the implication, where she has to state the antecedent and thus explicitly endorse it). The Opponent can ask for the formation of the thesis before launching her first challenge; defending the formation of his thesis might for instance bring the Proponent to state that the thesis is a proposition, provided, say, that *A* ***is a set*** is conceded; the Opponent might then concede that *A* ***is a set***, but only after the constitution of *A* has been established, though if this were the case, we would be considering the constitution of an elementary statement, which is a material consideration, not a formal one.

These rules for the *formation* of statements with logical constants are also particle rules which are added to the set of particle rules determining the local meaning of logical constants (called synthesis and analysis of local reasons in the framework of dialogues for immanent reasoning).

These considerations yield the following condensed presentation of the logical constants (plus *falsum*), in which "K" in *AKB* " expresses a connective, and "Q" in "(Q*x* : *A*) *B*(*x*) " expresses a quantifier (Table 10.1).

Because of the *no entity without type* principle, it seems at first glance that we should specify the type of these actions during a dialogue by adding the type "*formation-request*". But as it turns out, we should not: an expression such as "$?_F$*: formation-request*" is a judgement that some action $?_F$ is a formation-request, which

Table 10.2 Synthesis of a local reason for implication

	Implication
Move	**X** ! $A \supset B$
Challenge	**Y** $p_1 : A$
Defence	**X** $p_2 : B$

Table 10.3 General structure for the synthesis of a local reason for a constant

	A constant K
Move	**X** ! φ[K] *X claims that* ϕ
Challenge	*Y asks for the reason backing such a claim*
Defence	**X** $p : \varphi$[K] *X states the local reason p for* φ*[K] according to the rules for the synthesis of local reasons prescribed for K*

should not be confused with the actual act of requesting. We also consider that the force symbol $?_F$ makes the type explicit.

Synthesis of local reasons

The synthesis rules of local reasons determine how to produce a local reason for a statement; they include rules of interaction indicating how to produce the local reason that is required by the proposition (or set) in play, that is, they indicate what kind of dialogic action—what kind of move—must be carried out, by whom (challenger or defender), and what reason must be brought forward.

Implication

For instance, the synthesis rule of a local reason for the implication $A \supset B$ stated by player **X** indicates:

i. that the challenger **Y** must state the antecedent (while providing a local reason for it): **Y** p_1: A.[23]
ii. that the defender **X** must respond to the challenge by stating the consequent (with its corresponding local reason): **X** p_2: B.

In other words, the rules for the synthesis of a local reason for implication are as follows (Table 10.2).

The general structure for the synthesis of local reasons

More generally, the rules for the synthesis of a local reason for a constant K is determined by the following triplet (Table 10.3).

Analysis of local reasons

Apart from the rules for the synthesis of local reasons, we need rules that indicate how to parse a complex local reason into its elements: this is the *analysis* of local reasons. In order to deal with the complexity of these local reasons and formulate

[23]This notation is a variant of the one used by Keiff (2004, 2009).

general rules for the analysis of local reasons (at the play level), we introduce certain operators that we call *instructions*, such as $L^{\vee}(p)$ or $R^{\wedge}(p)$.

Approaching the analysis rules for local reasons

Let us introduce these instructions and the analysis of local reasons with an example: player **X** states the implication $(A \wedge B) \supset A$. According to the rule for the synthesis of local reasons for an implication, we obtain the following:

Move	$\mathbf{X}\,!\,(A \wedge B) \supset B$
Challenge	$\mathbf{Y}\, p_1 : A \wedge B$

Recall that the synthesis rule prescribes that **X** must now provide a local reason for the consequent; but instead of defending his implication (with $\mathbf{X}p_2 : B$ for instance), **X** can choose to parse the reason p_1 provided by **Y** in order to force **Y** to provide a local reason for the right-hand side of the conjunction that **X** will then be able to copy; in other words, **X** can force **Y** to provide the local reason for B out of the local reason p_1 for the antecedent $A \wedge B$ of the initial implication. The analysis rules prescribe how to carry out such a parsing of the statement by using *instructions*. The rule for the analysis of a local reason for the conjunction $p_1 : A \wedge B$ will thus indicate that its defence includes expressions such as

- the left instruction for the conjunction, written $L^{\wedge}(p_1)$, and
- the right instruction for the conjunction, written $R^{\wedge}(p_1)$.

These instructions can be informally understood as carrying out the following step: for the defence of the conjunction $p_1 : A \wedge B$ separate the local reason p_1 in its left (or right) component so that this component can be adduced in defence of the left (or right) side of the conjunction.

The general structure for the analysis rules

	Move	Challenge	Defence
Conjunction	$\mathbf{X}p : A \wedge B$	$\mathbf{Y}?\,L^{\wedge}$ or $\mathbf{Y}?\,R^{\wedge}$	$\mathbf{X}L^{\wedge}(p)^{X} : A$(resp.) $\mathbf{X}R^{\wedge}(p)^{X} : B$
Disjunction	$\mathbf{X}p : A \vee B$	$\mathbf{Y}?_{\vee}$	$\mathbf{X}L^{\vee}(p)^{X} : A$ or $\mathbf{X}R^{\vee}(p)^{X} : B$
Implication	$\mathbf{X}p : A \supset B$	$\mathbf{Y}L^{\supset}(p)^{Y} : A$	$\mathbf{X}R^{\supset}(p)^{X} : B$

The superscripts with the player label indicate which player is entitled to decide how to resolve the instruction, that is, to decide which local reason to bring forward when carrying out the instruction.

Interaction procedures embedded in instructions

Carrying out the prescriptions indicated by instructions require the following three interaction-procedures:

1. *Resolution of instructions*: this procedure determines how to carry out the instructions prescribed by the rules of analysis and thus provide an actual local reason.
2. *Substitution of instructions*: this procedure ensures the following; once a given instruction has been carried out through the choice of a local reason, say b, then every time the same instruction occurs, it will always be substituted by the same local reason b.
3. *Application of the Socratic rule*: the Socratic rule prescribes how to constitute equalities out of the resolution and substitution of instructions, linking synthesis and analysis together.

Let us discuss how these rules interact and how they lead to the main thesis of this study, namely that immanent reasoning is equality in action.

From Reasons to Equality

As we have already discussed to some extent one of the most salient features of dialogical logic is the so-called, Socratic rule (or Copy-cat rule in the standard—that is, non-CTT—context), establishing that the Proponent can play an elementary proposition only if the Opponent has played it previously.

The Socratic rule is a characteristic feature of the dialogical approach: other game-based approaches do not have it. With this rule the dialogical framework comes with an internal account of elementary propositions: an account in terms of interaction only, without depending on metalogical meaning explanations for the non-logical vocabulary. More prominently, this means that the dialogical account does not rely—-contrary to Hintikka's GTS games—on the model-theoretical approach to meaning for elementary propositions.

The rule has a clear Platonist and Aristotelian origin and sets the terms for what it is to carry out a formal argument: see for instance Plato's Gorgias (472b-c). We can sum up the underlying idea with the following statement:

> there is no better grounding of an assertion within an argument than indicating that it has been already conceded by the Opponent or that it follows from these concessions.[24]
>
> What should be stressed here are the following two points:
>
> 1. formality is understood as a kind of *interaction*; and
> 2. formal reasoning *should not* be understood here as devoid of content and reduced to purely syntactic moves.

Both points are important in order to understand the criticism often raised against formal reasoning in general, and in logic in particular. It is only quite late in the history of philosophy that formal reasoning has been reduced to syntactic manipulation—presumably the first explicit occurrence of the syntactic view of logic is Leibniz's "pensée aveugle" (though Leibniz's notion was not a reductive one). Plato and

[24]Recent work (Crubellier 2014, pp. 11–40) and (Rahman et al. 2015) claim that this rule is central to the interpretation of dialectic as the core of Aristotle's logic. Neither Ian Lukasiewicz's (1957) famous reconstruction of Aristotle's syllogistic, nor the Natural Deduction approach of Kurt Ebbinghaus (1964, 2016) and John Corcoran (1974) deploy this rule, but Marion and Rückert (2015) showed that this rule displays Aristotle's view on universal quantification.

Aristotle's notion of formal reasoning is neither "static" nor "empty of meaning"—to use Hegel's words quoted in the introduction. In the Ancient Greek tradition logic emerged from an approach of assertions in which meaning and justification result from what has been brought forward during argumentative interaction. According to this view, dialogical interaction is constitutive of meaning. Some former interpretations of standard dialogical logic did understand formal plays in a purely syntactic manner. The reason for this is that the standard version of the framework is not equipped to express meaning at the object-language level: there is no way of asking and giving reasons for elementary propositions. As a consequence, the standard formulation simply relies on a syntactic understanding of *Copy-cat moves*, that is, moves entitling **P** to copy the elementary propositions brought forward by **O**, regardless of its content. The dialogical approach to CTT (dialogues for immanent reasoning) however provides a fine-grain study of the contextual aspects involved in formal plays, much finer than the one provided by the standard dialogical framework. In dialogues for immanent reasoning which we are now presenting, a statement is constituted both by a proposition and by the (local) reason brought forward in defence of the claim that the proposition holds. In formal plays not only is the Proponent allowed to copy an elementary proposition stated by the Opponent, as in the standard framework, but he is also allowed to adduce in defence of that proposition the *same* local reason brought forward by the Opponent when she defended that same proposition. Thus immanent reasoning and equality in action are intimately linked. In other words, a formal play displays the *roots of the content* of an elementary proposition, and *not* a syntactic manipulation of that proposition. Statements of definitional equality emerge precisely at this point. In particular reflexivity statements such as

$$p = p : A$$

express from the dialogical point of view the fact that if **O** states the elementary proposition A, then **P** can do the same, that is, play the same move and do it on the same grounds which provide the meaning and justification of A, namely p. These remarks provide an insight only on simple forms of equality and barely touch upon the finer-grain distinctions discussed above; we will be moving to these by means of a concrete example in which we show, rather informally, how the combination of the processes of analysis, synthesis, and resolution of instructions lead to equality statements.

The dialogical roots of equality: dialogues for immanent reasoning

In this section we will spell out all the relevant rules for the dialogical framework incorporating features of Constructive Type Theory—that is, a dialogical framework making the players' reasons for asserting a proposition explicit. The rules can be divided, just as in the standard framework, into rules determining local meaning and rules determining global meaning. These include:

1. Concerning *local meaning* (Sect. 0):

    a. formation rules (p. 204);

Table 10.4 Formation rules

	Move	Challenge	Defence
Conjunction	**X** $A \wedge B$: ***prop***	**Y** ? $F_{\wedge 1}$ or **Y** ? $F_{\wedge 2}$	**X** A : ***prop*** (resp.) **X** B : ***prop***
Disjunction	**X** $A \vee B$: ***prop***	**Y** ? $F_{\vee 1}$ or **Y** ? $F_{\vee 2}$	**X** A : ***prop*** (resp.) **X** B : ***prop***
Implication	**X** $A \supset B$: ***prop***	**Y** ? $F_{\supset 1}$ or **Y** ? $F_{\supset 2}$	**X** A : ***prop*** (resp.) **X** B : ***prop***
Universal quantification	**X** $(\forall x : A)B(x)$: ***prop***	**Y** ? $F_{\forall 1}$ or **Y** ? $F_{\forall 2}$	**X** A : ***set*** (resp.) **X** $B(x)$: ***prop***$[x : A]$
Existential quantification	**X** $(\exists x : A)B(x)$: ***prop***	**Y** ? $F_{\exists 1}$ or **Y** ? $F_{\exists 2}$	**X** A : ***set*** (resp.) **X** $B(x)$: ***prop***$[x : A]$
Subset separation	**X**$\{x : A \mid B(x)\}$: ***prop***	**Y** ? F_1 or **Y** ? F_2	**X** A : ***set*** (resp.) **X** $B(x)$: ***prop***$[x : A]$
Falsum	**X** $\bot$: ***prop***	–	–

   b. rules for the synthesis of local reasons; and
   c. rules for the analysis of local reasons.

2. Concerning *global meaning*, we have the following (structural) rules (Sect. 0)

   a. rules for the resolution of instructions;
   b. rules for the substitution of instructions;
   c. equality rules determined by the application of the Socratic rules; and
   d. rules for the transmission of equality.

We will be presenting these rules in this order in the next two subsections, along with the adaptation of the other structural rules to dialogues of immanent reasoning in the second subsection. The following subsection provides a series of exercises and their solution.

Local meaning in dialogues of immanent reasoning
The formation rules
Formation rules for logical constants and falsum
The formation rules for *logical constants* and for *falsum* are given in the following table. Notice that a statement '⊥: prop' cannot be challenged; this is the dialogical account for falsum '⊥' being by definition a proposition (Table 10.4).

The substitution rule within dependent statements
The following rule is not really a formation-rule but is very useful while applying

Table 10.5 Substitution rule within dependent statements (subst-D)

	Move	Challenge	Defence
Subst-D	$\mathbf{X}\pi(x_1, \ldots, x_n)[x_i : A_i]$	$\mathbf{Y}\tau_1 : A_1, \ldots, \tau_n : A_n$	$\mathbf{X}\pi(\tau_1, \ldots, \tau_n)$

formation rules where one statement is dependent upon the other such as $B(x)$: $\boldsymbol{prop}[x : A]$ (Table 10.5).[25]

In the formulation of this rule, "π" is a statement and "τ_i" is a local reason of the form either $a_i : A_i$ or $x_i : A_i$.

A particular case of the application of Subst-D is when the challenger simply chooses the same local reasons as those occurring in the concession of the initial statement. This is particularly useful in the case of formation plays:

Example of a formation-play

Here is an example of a formation play with some explanation. The standard development rules are enough to understand the following plays.

In this example, the Opponent provides initial concession before the Proponent states his thesis. Thus the Proponent's thesis is

$$(\forall x : A)(B(x) \supset C(x)) : \mathbf{prop}$$

given these three provisos that appear as initial concessions by the Opponent:

$$A : \mathbf{set},$$
$$B(x) : \mathbf{prop}[x : A]$$
$$\textit{and } C(x) : \mathbf{prop}[x : A],$$

This yields the following play:

Play 6: formation-play with initial concessions: first decision-option of **O**

	O			P	
0.1	A : **set**				
0.2	$B(x)$: **prop** $[x : A]$				
0.3	$C(x)$: **prop** $[x : A]$			$(\forall x : A)\ B(x) \supset C(x)$: **prop**	0
1	$m := 1$			$n := 2$	2
3	$?F_{\forall 1}$	0		A : **set**	4

P wins.

[25]This rule is an expression at the level of plays of the rule for the substitution of variables in a hypothetical judgement. See (Martin-Löf 1984, pp. 9–11).

Explanation:

- 0.1 to 0.3: **O** concedes that A is a **set** and that $B(x)$ and $C(x)$ are propositions provided x is an element of A.
- Move 0: **P** states that the main sentence, universally quantified, is a proposition (under the concessions made by **O**).
- Moves 1 and 2: the players choose their repetition ranks.
- Move 3: **O** challenges the thesis by asking the left-hand part as specified by the formation rule for universal quantification.
- Move 4: **P** responds by stating that A is a **set**. This has already been granted with the concession 0.1 so even if **O** were to challenge this statement the Proponent could refer to her initial concession.

This dialogue obviously does not cover all the aspects related to the formation of

$$(\forall x : A)B(x) \supset C(x) : \textbf{prop}.$$

Notice however that the formation rules allow an alternative move for the Opponent's move 3,[26] so that **P** has another possible course of action, dealt with in the following play.

Play 7: formation-play with initial concessions: second decision-option of **O**

	O			P	
0.1	A : **set**				
0.2	$B(x)$: **prop** $[x : A]$				
0.3	$C(x)$: **prop** $[x : A]$			$(\forall x : A)\, B(x) \supset C(x)$: **prop**	0
1	$m := 1$			$n := 2$	2
3	$?\, F_{\forall 2}$	0		$B(x) \supset C(x)$: **prop** $[x : A]$	4
5	$x : A$	4		$B(x) \supset C(x)$: **prop**	6
7	$?\, F_{\supset 1}$	6		$B(x)$: **prop**	10
9	$B(x)$: **prop**		0.2	$x : A$	8

P wins.

Explanation:

The second play starts like the first one until move 2. Then:

- Move 3: this time **O** challenges the thesis by asking for the right-hand part.
- Move 4: **P** responds, stating that $B(x) \supset C(x)$ is a proposition, provided that x: A.
- Move 5: **O** challenges the preceding move by granting the proviso and asking **P** to respond (this kind of move is governed by a Subst-D rule).
- Move 6: **P** responds by stating that $B(x) \supset C(x)$ is a proposition.

[26]As a matter of fact, increasing her repetition rank would allow **O** to play the two alternatives for move 3 within a single play. But increasing the Opponent's rank usually yields redundancies (Clerbout 2014a, b, c) making things harder to understand for readers not familiar with the dialogical approach; hence our choice to divide the example into different simple plays.

- Move 7: **O** challenges move 6 by asking the left-hand part, as specified by the formation rule for material implication.

To defend against this challenge, **P** needs to make an elementary move. But since **O** has not played it yet, **P** cannot defend it at this point. Thus:

- Move 8: **P** launches a counterattack against initial concession 0.2 by granting the proviso x: A (that has already been conceded by **O** in move 5), making use of the same kind of statement-substitution (Subst-D) rule deployed in move 5.
- Move 9: **O** answers to move 8 and states that $B(x)$ is a proposition.
- Move 10: **P** can now defend the challenge initiated with move 7 and win this dialogue.

Once again, there is another possible choice for the Opponent because of her move 7: she could ask the right-hand part. This would yield a dialogue similar to the one above except that the last moves would be about $C(x)$ instead of $B(x)$.

Concluding on the formation-play example
By displaying these various possibilities for the Opponent, we have entered the strategic level. This is the level at which the question of the good formation of the thesis gets a definitive answer, depending on whether the Proponent can always win—that is, whether he has a winning strategy. The basic notions related to this level of strategies are to be found in our presentation of standard dialogical logic.

The rules for local reasons: synthesis and analysis
Now that the dialogical account of formation rules has been clarified, we may further develop our analysis of plays by introducing local reasons. Let us do so by providing the rules that prescribe the synthesis and analysis of local reasons. For more details on each rule, see Sect. 0 (Tables 10.6 and 10.7).

Slim instructions: dealing with cases of anaphora
One of the most salient features of the CTT framework is that it contains the means to deal with cases of anaphora.[27]
Notice that in the formalization of traditional syllogistic form *Barbara*, the projection $\mathbf{fst}(z)$ can be seen as the tail of the anaphora whose head is z:

$$\frac{\begin{array}{ll}(\forall z : (\exists x : D)A)B[\mathbf{fst}(z)]\text{true} & \textit{premise}\,1\\ (\forall z : (\exists x : D)B)C[\mathbf{fst}(z)]\text{true} & \textit{premise}\,2\end{array}}{(\forall z : (\exists x : D)A)C[\mathbf{fst}(z)]\text{true} \quad \textit{conclusion}}$$

In dialogues for immanent reasoning, when a local reason has been made explicit, this kind of anaphoric expression is formalized through *instructions*, which provides a further reason for introducing them. For example if a is the local reason for the first premise we have

$$\mathbf{P}p : (\forall z : (\exists x : D)A(x))B(L^{\exists}(L^{\forall}(p)^{\mathbf{O}}))$$

[27] See Sundholm (1986, pp. 501–503) and Ranta (1994, pp. 77–99).

Table 10.6 synthesis rules for local reasons

	Move	Challenge	Defence
Conjunction	$\mathbf{X}!\ A \wedge B$	$\mathbf{Y}?\ L^{\wedge}$ or $\mathbf{Y}?\ R^{\wedge}$	$\mathbf{X}p_1 : A$ (resp.) $\mathbf{X}p_2 : B$
Existential quantification	$\mathbf{X}!\ (\exists x : A)B(x)$	$\mathbf{Y}?\ L^{\exists}$ or $\mathbf{Y}?\ R^{\exists}$	$\mathbf{X}p_1 : A$ (resp.) $\mathbf{X}p_2 : B(p_1)$
Subset separation	$\mathbf{X}!\ \{\ x : A \mid B(x)\}$	$\mathbf{Y}?\ L$ or $\mathbf{Y}?\ R$	$\mathbf{X}p_1 : A$ (resp.) $\mathbf{X}p_2 : B(p_1)$
Disjunction	$\mathbf{X}!\ A \vee B$	$\mathbf{Y}?^{\vee}$	$\mathbf{X}p_1 : A$ or $\mathbf{X}p_2 : B$
Implication	$\mathbf{X}!\ A \supset B$	$\mathbf{Y}p_1 : A$	$\mathbf{X}p_2 : B$
Universal quantification	$\mathbf{X}!\ (\forall x : A)B(x)$	$\mathbf{Y}p_1 : A$	$\mathbf{X}p_2 : B(p_1)$
Negation	$\mathbf{X}!\ \neg A$ Also expressed as $\mathbf{X}!\ A \supset \bot$	$\mathbf{Y}p_1 : A$	$\mathbf{X} : \bot$ (**X** gives up[a])

[a]The reading of stating bottom as giving up stems from (Keiff 2007)

Table 10.7 Analysis rules for local reasons

	Move	Challenge	Defence
Conjunction	$\mathbf{X}p : A \wedge B$	$\mathbf{Y}?\ L^{\wedge}$ or $\mathbf{Y}?\ R^{\wedge}$	$\mathbf{X}L^{\wedge}(p)^X : A$ (resp.) $\mathbf{X}R^{\wedge}(p)^X : B$
Existential quantification	$\mathbf{X}p : (\exists x : A)B(x)$	$\mathbf{Y}?\ L^{\exists}$ or $\mathbf{Y}?\ R^{\exists}$	$\mathbf{X}L^{\exists}(p)^X : A$ (resp.) $\mathbf{X}R^{\exists}(p)^X :$ $B(L^{\exists}(p)^X)$
Subset separation	$\mathbf{X}p : \{\ x : A \mid B(x)\}$	$\mathbf{Y}?\ L$ or $\mathbf{Y}?\ R$	$\mathbf{X}L^{\{\ldots\}}(p)^X : A$ (resp.) $\mathbf{X}R^{\wedge}(p)^X :$ $B(L^{\{\ldots\}}(p)^X)$
Disjunction	$\mathbf{X}p : A \vee B$	$\mathbf{Y}?^{\vee}$	$\mathbf{X}L^{\vee}(p)^X : A$ or $\mathbf{X}R^{\vee}(p)^X : B$
Implication	$\mathbf{X}p : A \supset B$	$\mathbf{Y}L^{\supset}(p)^Y : A$	$\mathbf{X}R^{\supset}(p)^X : B$
Universal quantification	$\mathbf{X}p : (\forall x : A)B(x)$	$\mathbf{Y}L^{\forall}(p)^Y : A$	$\mathbf{X}R^{\forall}(p)^X :$ $B(L^{\forall}(p)^Y)$
Negation	$\mathbf{X}p : \neg A$Also expressed as$\mathbf{X}p : A \supset \bot$	$\mathbf{Y}L^{\neg}(p)^Y : A$ $\mathbf{Y}L^{\supset}(p)^Y : A$	$\mathbf{X}R^{\neg}(p)^X : \bot$ $\mathbf{X}R^{\supset}(p)^X : \bot$

However, since the thesis of a play does not bear an explicit local reason (we use the exclamation mark to indicate there is an implicit one), it is possible for a statement to be bereft of an explicit local reason. When there is no explicit local reason for a statement using anaphora, we cannot bind the instruction $L^{\forall}(p)^{\mathbf{O}}$ to a local reason p. We thus have something like this, with a blank space instead of the anaphoric local reason:

$$\mathbf{P}!\,(\forall z : (\exists x : D)A(x))B(L^{\exists}(L^{\forall}(\ )^{\mathbf{O}}))$$

But this blank stage can be circumvented: the challenge on the universal quantifier will yield the required local reason: **O** will provide $a : (\exists x : D)A(x)$, which is the local reason for z. We can therefore bind the instruction on the missing local reason with the corresponding variable—z in this case—and write

$$\mathbf{P}!\,(\forall z : (\exists x : D)A(x))B(L^{\exists}(L^{\forall}(z)^{\mathbf{O}}))$$

We call this kind of instruction, *slim instructions*. For the substitution of slim instructions the following two cases are to be distinguished:

Substitution of Slim Instructions 1
Given some slim instruction such as $L^{\forall}(z)^{\mathbf{Y}}$, once the quantifier $(\forall z : A)B(\ldots)$ has been challenged by the statement a: A, the occurrence of $L^{\forall}(z)^{\mathbf{Y}}$ can be substituted by a. The same applies to other instructions.
In our example we obtain:

$$\mathbf{P}!\,(\forall z : (\exists x : D)A(x))B(L^{\exists}(L^{\forall}(z)^{\mathbf{O}}))$$
$$\mathbf{O}a : (\exists x : D)A(x)$$
$$\mathbf{P}b : B(L^{\exists}(L^{\forall}(z)^{\mathbf{O}}))$$
$$\mathbf{O}?\,a/L^{\forall}(z)^{\mathbf{O}}$$
$$\mathbf{P}b : B(L^{\forall}(a))$$
$$\ldots$$

Substitution of Slim Instructions 2
Given some slim instruction such as $L^{\forall}(z)^{\mathbf{Y}}$, once the instruction $L^{\forall}(c)$—resulting from an attack on the universal $\forall z$: φ—has been resolved with a: φ, then any occurrence of $L^{\forall}(z)^{\mathbf{Y}}$ can be substituted by a. The same applies to other instructions.

Global Meaning in dialogues for immanent reasoning
We here provide the structural rules for dialogues for immanent reasoning, which determine the global meaning in such a framework. They are for the most part similar in principle to the precedent logical framework for dialogues; the rules concerning instructions are an addition for dialogues for immanent reasoning.

Structural Rules
SR0: Starting rule

The start of a *formal dialogue of immanent reasoning* is a move where **P** states the *thesis*. The thesis can be stated under the condition that **O** commits herself to certain other statements called *initial concessions*; in this case the thesis has the form ! A [B_1, …, B_n], where A is a statement with implicit local reason and $B_1, \ldots, B_n$ are statements with or without implicit local reasons.

A dialogue with a thesis proposed under some conditions starts if and only if **O** accepts these conditions. **O** accepts the conditions by stating the *initial concessions* in moves numbered 0.1, … 0.n before choosing the repetition ranks.

After having stated the thesis (and the initial concessions, if any), each player chooses in turn a positive integer called the *repetition rank* which determines the upper boundary for the number of attacks and of defences each player can make in reaction to each move during the play.

SR1: Development rule

The Development rule depends on what kind of logic is chosen: if the game uses intuitionistic logic, then it is SR1i that should be used; but if classical logic is used, then SR1c must be used.

SR1i: Intuitionistic Development rule, or Last Duty First

Players play one move alternately. Any move after the choice of repetition ranks is either an attack or a defence according to the rules of formation, of synthesis, and of analysis, and in accordance with the rest of the structural rules.

If the logical constant occurring in the thesis is not recorded by the table for local meaning, then either it must be introduced by a nominal definition, or the table for local meaning needs to be enriched with the new expression.[28] Players can answer only against the *last non-answered* challenge by the adversary.

Note: This structural rule is known as the Last Duty First condition, and makes dialogical games suitable for intuitionistic logic, hence the name of this rule.

SR1c: Classical Development rule

Players play one move alternately. Any move after the choice of repetition ranks is either an attack or a defence according to the rules of formation, of synthesis, and of analysis, and in accordance with the rest of the structural rules.

If the logical constant occurring in the thesis is not recorded by the table for local meaning, then either it must be introduced by a nominal definition, or the table for local meaning needs to be enriched with the new expression.

Note: The structural rules with SR1c (and not SR1i) produce strategies for classical logic. The point is that since players can answer to a list of challenges in any order (which is not the case with the intuitionistic rule), it might happen that the two options of a P-defence occur in the same play—this is closely related to the classical development rule in sequent calculus allowing more than one formula at the right of the sequent.

SR2: Formation rules for formal dialogues

[28] If the logical constant occurring in the thesis is not recorded by the table for local meaning, then either it must be introduced by a nominal definition based on some logical constant already present in the local rules, or the table for local meaning needs to be enriched with the new expression.

A formation-play starts by challenging the thesis with the formation request **O** $?_{\mathbf{prop}}$; **P** must answer by stating that his thesis is a proposition. The game then proceeds by applying the formation rules up to the elementary constituents of **prop/set**. After that the Opponent is free to use the other particle rules insofar as the other structural rules allow it.

Note: The constituents of the thesis will therefore not be specified before the play but as a result of the structure of the moves (according to the rules recorded by the rules for local meaning).

SR3: Resolution of instructions

1. A player may ask his adversary to carry out the prescribed instruction and thus bring forward a suitable local reason in defence of the proposition at stake. Once the defender has replaced the instruction with the required local reason we say that the instruction has been resolved.
2. The player index of an instruction determines which of the two players has the right to choose the local reason that will resolve the instruction.

   a. If the instruction I for the logical constant K has the form $I^K(p)^{\mathbf{X}}$ and it is **Y** who requests the resolution, then the request has the form **Y** $?.../I^K(p)^{\mathbf{X}}$, and it is **X** who chooses the local reason.
   b. If the instruction I for the logic constant K has the form $I^K(p)^{\mathbf{Y}}$ and it is player **Y** who requests the resolution, then the request has the form **Y** $p_i\ /I^K(p)^{\mathbf{Y}}$, and it is **Y** who chooses the local reason.

3. In the case of a sequence of instructions of the form $\pi[I_i(\ldots(I_k(p))\ldots)]$, the instructions are resolved from the inside $(I_k(p))$ to the outside (I_i).

This rule also applies to functions.

SR4: Substitution of instructions

Once the local reason b has been used to resolve the instruction $I^K(p)^{\mathbf{X}}$, and if the same instruction occurs again, players have the right to require that the instruction be resolved with b. The substitution request has the form $?\ b/I_k(p)^{\mathbf{X}}$. Players cannot choose a different substitution term (in our example, not even **X**, once the instruction has been resolved).

This rule also applies to functions.

SR5: Socratic rule and definitional equality

The following points are all parts of the Socratic rule, they all apply.

SR5.1: Restriction of P statements

P cannot make an elementary statement if **O** has not stated it before, except in the thesis.

An elementary statement is either an elementary proposition with implicit local reason, or an elementary proposition and its local reason (not an instruction).

SR5.2: Challenging elementary statements in formal dialogues

Challenges of elementary statements with implicit local reasons take the form:

Table 10.8 Non-reflexive cases of the Socratic rule

	Move	Challenge	Defence
SR5.3.1a	$\mathbf{P}a : A$	$\mathbf{O}? = a$	$\mathbf{P}\, \mathrm{I} = a : A$
SR5.3.1b	$\mathbf{P}a : A(b)$	$\mathbf{O}? = b^{A(b)}$	$\mathbf{P}\, \mathrm{I} = b : D$
SR5.3.1c	$\mathbf{P}\, \mathrm{I} = b : D$ (this statement stems from **SR5.3.1b)**	$\mathbf{O}? \ldots = A(b)$	$\mathbf{P}\, A(\mathrm{I}) = A(b) : \boldsymbol{prop}$

$$X!\, A$$
$$Y?_{reason}$$
$$Xa : A$$

where A is an elementary proposition and a is a local reason.

P cannot challenge O's elementary statements, except if O provides an elementary initial concession with implicit local reason, in which case **P** can ask for a local reason, or in the context of transmission of equality.

SR5.3: Definitional equality

O may challenge elementary **P**-statements, challenge answered by stating a definitional equality, expressing the equality between a local reason introduced by **O** and an instruction also introduced by **O**.

These rules do not cover cases of transmission of equality. The Socratic rule also applies to the resolution or substitution of functions, even if the formulation mentions only instructions. We distinguish reflexive and non-reflexive cases.

SR5.3.1: Non-reflexive cases of the Socratic rule

We are in the presence of a *non-reflexive case* of the Socratic rule when **P** responds to the challenge with the indication that **O** gave the same local reason for the same proposition when she had to resolve or substitute instruction I.

Here are the different challenges and defences determining the meaning of the three following moves (Table 10.8):

Presuppositions:

(i) The response prescribed by SR5.3.1a presupposes that **O** has stated A or $a = b : A$ as the result of the resolution or substitution of instruction I occurring in $\mathrm{I} : A$ or in $\mathrm{I} = b : A$.

(ii) The response prescribed by SR5.3.1b presupposes that **O** has stated A and b: D as the result of the resolution or substitution of instruction I occurring in a: A(I).

(iii) SR5.3.1c assumes that $\mathbf{P}\, \mathrm{I} = b : D$ is the result of the application of SR5.3.1b. The further challenge seeks to verify that the replacement of the instruction produces an equality in **prop**, that is, that the replacement of the instruction with a local reason yields an equal proposition to the one in which the instruction was not yet replaced. The answer prescribed by this rule presupposes that **O** has already stated $A(b)$: **prop** (or more trivially $A(I) = A(b)$: **prop**).

The **P**-statements obtained after defending elementary **P**-statements cannot be attacked again with the Socratic rule (with the exception of SR5.3.1c), nor with a rule of resolution or substitution of instructions.

SR5.3.2: Reflexive cases of the Socratic rule

We are in the presence of a *reflexive case* of the Socratic rule when **P** responds to the challenge with the indication that **O** adduced the same local reason for the same proposition, though that local reason in the statement of **O** is not the result of any resolution or substitution.

The attacks have the same form as those prescribed by SR5.3.1. Responses that yield reflexivity presuppose that **O** has previously stated the same statement or even the same equality.

The response obtained cannot be attacked again with the Socratic rule.

SR6: Transmission of definitional equality

Transmission of definitional equality I: Substitution within dependent or independent statements. The expression "type" refers to either **prop** or **set**. For more explanations on this structural rule, see next section.

Move	Challenge	Defence
X $b(x)$: $B(x)$ $[x: A]$	**Y** $a = c$: A	**X** $b(a) = b(c)$: $B(a)$
X $b(x) = d(x)$: $B(x)$ $[x : A]$	**Y** a: A	**X** $b(a) = d(a)$: $B(a)$
X $B(x)$: *type* $[x: A]$	**Y** $a = c$: A	**X** $B(a) = B(c)$: *type*
X $B(x) = D(x)$: *type* $[x : A]$	**Y** $?_{B(x)=D(x)}$ a: A or **Y** $?_{B(x)=D(x)}$ $a = c : A$	**X** $B(a) = D(a)$: *type* or **X** $B(a) = D(c)$: *type*
X $A = B$: *type*	**Y** $?_{A=D}$ a: A or **Y** $?_{A=D}$ $a = c : A$	**X** a: B or **X** $a = c$: B

Transmission of definitional equality II:

Move	Challenge	Defence
X A: *type*	**Y** $?_{type}$- *refl*	**X** $A = A$: *type*
X $A = B$: *type*	**Y** $?_B$- *symm*	**X** $B = A$: *type*
X $A = B$: *type* **X** $B = C$: *type*	**Y** $?_A$- *trans*	**X** $A = C$: *type*
X a: A	**Y** $?_A$- *refl*	**X** $a = a$: A
X $a = b$: A	**Y** $?_b$- *symm*	**X** $b = a$: A
X $a = b$: A **X** $b = c$: A	**Y** $?_a$- *trans*	**X** $a = c$: A

Table 10.9 Transmission of definitional equality I: Substitution within dependent or independent statements

Move	Challenge	Defence
X $b(x) : B(x)$ $[x : A]$	**Y** $a = c : A$	**X** $b(a) = b(c) : B(a)$
X $b(x) = d(x) : B(x)$ $[x : A]$	**Y** $a : A$	**X** $b(a) = d(a) : B(a)$
X $B(x)$: *type* $[x : A]$	**Y** $a = c : A$	**X** $B(a) = B(c)$: *type*
X $B(x) = D(x)$: *type* $[x : A]$	**Y** $?_{B(x)=D(x)}$ $a : A$ or **Y** $?_{B(x)=D(x)}$ $a = c : A$	**X** $B(a) = D(a)$: *type* or **X** $B(a) = D(c)$: *type*
X $A = B$: *type*	**Y** $?_{A=D}$ $a : A$ or **Y** $?_{A=D}$ $a = c : A$	**X** $a : B$ or **X** $a = c : B$

SR7: Winning rule for plays

The player who makes the last move in a dialogue wins the dialogue. If **O** stated $\bot$ (or $p : \bot$), at move n then **P** wins with the move **O**-*gives up*(n). **P** can also adduce **O**-*gives up*(n) as local reason in support for any statement that he has not defended before **O** stated $\bot$ at move n.

Rules for the transmission of definitional equality

As can be expected, definitional equality is transmitted by reflexivity, symmetry,[29] and transitivity. Definitional equalities however can also be used in order to carry out a substitution within dependent statements—they can in fact be seen as a special form of application of the substitution rule for dependent statement Subst-D presented in the first section for local meaning, with the formation rules (0, p. 205). We use the expression "type" as encompassing **prop** and **set** (Table 10.9).

Reading adjuvant for the fourth rule (dependent statements):

If **X** stated that $B(x)$ and $D(x)$ are equal propositional functions, provided that x is an element of the set A—that is, **X** $B(x) = D(x)$: **prop** $[x: A]$—, then **Y** can carry out two kinds of attacks:

1. Stating himself that some local reason, say a, can be adduced for A—$Y a : A$—, and request at the same time of **X** that he replaces x with a in $B(x) = D(x)$, that is stating $B(a) = D(a)$: **prop**.
2. Stating himself an equality such as $a = c$: A, and request at the same time **X** to carry out the corresponding substitutions in $B(x) = D(x)$, that is to state **X** $B(a) = D(c)$: **prop** (Table 10.10).

Reading adjuvant:

In order to trigger reflexivity, transitivity, and symmetry from some equality statements the challenger can attack an equality by asking for each of these properties. For example, if **X** stated $A = B$: **prop/set**, **Y** can ask **X** to state the commutated equality $B = A$: **prop/set** by calling on *symmetry*. The notation of such an attack is as follows: **Y** $?_{B\text{-}symm}$. Similarly, **Y** $?_{A\text{-}refl}$ and **Y** $?_{A\text{-}trans}$ respectively request reflexivity and transitivity.

[29] Symmetry used here is not the same notion as the symmetry of section 0.

Table 10.10 Transmission of definitional equality II

Move	Challenge	Defence
X $A : type$	**Y** $?_{type}$- *refl*	**X** $A=A : type$
X $A=B : type$	**Y** $?_B$- *symm*	**X** $B=A : type$
X $A=B : type$ **X** $B=C : type$	**Y** $?_A$- *trans*	**X** $A=C : type$
X $a : A$	**Y** $?_A$- *refl*	**X** $a=a : A$
X $a=b : A$	**Y** $?_b$- *symm*	**X** $b=a : A$
X $a=b : A$ **X** $b=c : A$	**Y** $?_a$- *trans*	**X** $a=c : A$

Appendix 3: The identity-predicate Id

The dialogical meaning explanation of the identity predicate **Id**(*x*, *y*, *z*)—where *x* is a **set** (or a **prop**) and *y* and *z* are local reasons in support of *A*—is based on the following: X's statement Id(A, a, b) presupposes that a: A and b: A, and expresses the claim that "a and b are identical reasons for supporting *A*. The presupposition yields already its formation rule, the second requires a formulation of the Socratic Rule specific to the identity predicate. Let us start with the formation:

Formation of Id

Statement	Challenge	Defence
X ! Id(A, a_i, a_j) : prop	**Y** $?_{F1}$ **Id**	**X** ! A : **set**
	Y$?_{F2}$ **Id**	**X** ! $ai : A$
	Y$?_{F3}$ **Id**	**X** ! $aj : A$

Socratic Rules for Id

Opponent's statements of identity can only be challenged by means of the rule of global analysis or by Leibniz-substitution rule

The following rules apply to statements of the form **Id**(*A*, *a*, *a*) and the more general statement of identity **Id**(*A*, *a*, *b*). Let us start with the reflexive case.

SR-Id.1 Socratic Rules for Id(*A*, *a*, *a*)

If the Proponent states **P** !**Id**(*A*, *a*, *a*), then he must bring forward the definitional equality that conditions statements of propositional intensional identity (see chapter II.8). Furthermore, the statement **P** !**Id**(*A*, *a*, *a*) commits the proponent to make explicit the local reason behind his statement, namely, the local reason **refl**(*A*, *a*) **specific** of **Id**-statements, the only internal structure of which is its dependence on *a*. Thus; the dialogical meaning of the instruction **refl**(*A*, *a*) amounts to prescribing the definitional equality $a =$ **refl**$(A, a) : A$ as defence to the challenge **O** ?= **refl**(*A*, *a*). The following two tables display the rules that implement those prescriptions.

Socratic Rule for the Global Synthesis of the local reason for P ! Id(*A*, *a*, *a*)

Statement	Challenge	Defence
P ! Id(A, a, a)	**O** ? $_{reasonId}$	**P refl**(A, a) : **Id**(A, a, a)
P refl(A, a) : **Id**(A, a, a)	**O** ?=**refl**(A, a)	**P** $a=a:A$

(This rule presupposes that the well-formation of **Id(*A*,** *a*, *a*) has been established) The following rule is just applying the general Socratic Rule for local reasons to the specific case of **refl**(A, a) and shows that the local reason **refl**(A, a) is in fact equal to a.

Socratic Rule for the challenge upon P's use of refl(A, a)

Statement	Challenge	Defence
P refl(A, a) : Id(A, a, a)	**O** ?=**refl**(A, a)	**P** a =**refl**(A, a) : A

Since in the dialogues of immanent reasoning it is the Opponent who is given the authority to set the local reasons for the relevant sets, **P** can always trigger from **O** the identity statement **O** p : **Id**(A, a, a) for any statement **O** $a:A$ has brought forward during a play. This leads to the next table that *constitutes one of the exceptions* to the interdiction on challenges on **O**'s elementary statements

Socratic Rule for triggering the reflexivity move O ! Id(*A*, *a*, *a*)

Statement	Challenge	Defence
O $a:A$	**P** $?^{\mathbf{Id}-a}$	**O refl**(A, a) : **Id**(A, a, a)

Remarks

Notice that it looks as if **P** will not need to use this rule since according to the rule for the synthesis of the local reason for an Identify statement by **P**, he can always state **Id**(A, a, a), provided **O** stated $a:A$. However, in some case, such as when carrying out a substitution based on identity, **P** might need **O** to make an explicit statement of identity suitable for applying that substitution-law.

This rule

The next rule prescribes how to analyse some local reason p brought forward by **O** in order to support the statement **Id(*A*, *a*, *a*)**

Analysis I The Global Analysis of O *p*: Id(*A*, *a*, *a*)

Statement	Challenge	Defence
O p : Id(A, a, a)	**P** ? $_{\mathbf{Id}=p}$	**O** p=**refl**(A, a) : **Id**(A, a, a)

The second rule for analysis involves statements of the form Id(A, a, b), so we need to general rules for statements that are not restricted to reflexivity. In fact the rules for Id(A, a, b) can be obtained by re-writing the precedent rules – with the exception of the rule that triggers statements of reflexivity by O.

We will not write the rules for Id(A, a, b) down but let us stress two important points

(1) the unicity of the local reason **refl**(A, a).
(2) the non-inversibily of the intensional predicate of identity in relation to judgmental equality.

(1) In relation to the first remark, the point is that the local reason produced by a process of synthesis for any identity statement is always **refl**(A, a). In other words, the local reason prescribed by the procedures of synthesis involving the statement ! **Id**(A, a, a) and the statement **Id**(A, a, a), is the *same one*, namely **refl**(A, a).
(2) In relation to our second point, It is important to remember that the **global synthesis** rule refers to the commitments undertaken by **P** when he affirms the identity between a and b. Such commitment amount to i) providing a local-reason for such identity ii) stating $a = b : A$.
On the contrary the rule of **global analysis of an identity statement by O** prescribes what **P** may require from O's statement. In that case, P cannot force **O** to state $a = b : A$ only because she stated **Id**(A, a, b). This is only possible with the so-called extensional version of propositional identity (see II.8 above and thorough discussion in Nordström et al. 1990, pp. 57–61). The dialogical view of non-reversibility here is that the rule of synthesis set the conditions P must fulfil when he states and identity, not *what follows from his statement of identity*:

Id is transmitted by the rules of reflexivity, symmetry, transitivity and by the **substitution of identicals**.

References

Bishop, E. (1967). *Foundations of constructive mathematics*. New York, London: McGraw-Hill.
Clerbout, N. (2014a). First-order dialogical games and tableaux. *Journal of Philosophical Logic, 43*(4), 785–801.
Clerbout, N. (2014b). *Étude sur quelques sémantiques dialogiques: Concepts fondamentaux et éléments de métathéorie*. London: College Publications.
Clerbout, N. (2014c). Finiteness of plays and the dialogical problem of decidability. *IfCoLog Journal of Logics and their Applications, 1*(1), 115–140.
Clerbout, N., & Rahman, S. (2015). *Linking game-theoretical approaches with constructive type theory: dialogical strategies as CTT-demonstrations*. Dordrecht: Springer.
Crubellier, M. (2014). Aristote, premiers analytiques. Traduction, introduction et commentaire. Garnier-Flammarion.

Ebbinghaus, K. (1964). *Ein formales Modell der Syllogistik des Aristoteles*. Göttingen: Vandenhoeck & Ruprecht GmbH.
Ebbinghaus, K. (2016). *Un modèle formel de la syllogistique d'Aristote*. (C. Lion, Trans.) College publication.
Fraenkel, A., Bar-Hillel, Y., & Levy, A. (1973). *Foundations of set theory* (2nd ed.). Dordrecht: North-Holland.
Goodman, N. D., & Myhill, J. (1978). Choice implies excluded middle. *Zeitschrigt für mathematische Logik und Grundlagen der Mathematik, 24,* 461.
Hintikka, J. (1973). *Logic, language-games and information: Kantian themes in the philosophy of logic*. Oxford: Clarendon Press.
Hintikka, J. (1996). *The principles of mathematics revisited*. Cambridge: Cambridge University Press.
Hintikka, J. (2001). Intuitionistic logic as epistemic logic. *Synthese, 127,* 7–19.
Jovanovic, R. (2013). Hintikka's take on the axiom of choice and the constructive challenge. *Revista de Humanidades de Valparaíso, 2,* 135–152.
Jovanovic, R. (2015). *Hintikka's Take on realism and the constructive challenge*. London: College Publications.
Keiff, L. (2007). *Le Pluralisme dialogique: Approches dynamiques de l'argumentation formelle*. Lille: PhD.
Keiff, L. (2009). Dialogical logic. In E. N. Zalta (Ed.) *The stanford encyclopedia of philosophy*. Retrieved from: http://plato.stanford.edu/entries/logic-dialogical.
Krabbe, E. C. (1982). *Studies in dialogical logic*. Rijksuniversiteit, Gröningen: PhD.
Krabbe, E. C. (1985). Formal systems of dialogue rules. *Synthese, 63,* 295–328.
Krabbe, E. C. (2006). Dialogue logic. In D. Gabbay & J. Woods (Eds.), *Handbook of the history of logic* (Vol. 7, pp. 665–704). Amsterdam: Elsevier.
Lorenz, K. (1970). *Elemente der Sprachkritik. Eine Alternative zum Dogmatismus und Skeptizismus in der Analytischen Philosophie*. Frankfurt: Suhrkamp.
Lorenz, K. (2001). Basic objectives of dialogical logic in historical perspective. *Synthese, 127,* 255–263.
Lorenz, K. (2010a). *Logic, language and method: On polarities in human experiences*. Berlin/New York: De Gruyter.
Lorenz, K. (2010b). *Philosophische Variationen: Gesammelte Aufsätze unter Einschluss gemeinsam mit Jürgen Mittelstrass geschriebener Arbeiten zu Platon und Leibniz*. Berlin/New York: De Gruyter.
Lorenz, K., & Lorenzen, P. (1978). *Dialogische Logik*. Damstadt: Wissenschaftliche Buchgesellschaft.
Marion, M. (2006). Hintikka on Wittgenstein: From language games to game semantics. *Acta Philosophica Fennica, 78,* 223–242.
Marion, M., & Rückert, H. (2015). Aristotle on universal quantification: A study from the perspective of game semantics. *History and Philosophy of Logic, 37*(3), 201–209.
Martin-Löf, P. (1984). *Intuitionistic type theory. Notes by Giovanni Sambin of a series of Lectures given in Padua, June 1980*. Naples: Bibliopolis.
Martin-Löf, P. (2006). 100 Years of Zermelo's axiom of choice: What was the problem with it? *The Computer Journal, 49*(3), 345–350.
Poincaré, H. (1905). *La Valeur de la Science*. Paris: Flammarion.
Poincaré, H. (2014). The value of science, science and method. Online edition, Adelaide: University of Adelaide.
Rahman, S., & Clerbout, N. (2013). Constructive type theory and the dialogical approach to meaning. *The Baltic International Yearbook of Cognition, Logic and Communication: Games, Game Theory and Game Semantics, 11,* 1–72.
Rahman, S., & Clerbout, N. (2015). Constructive type theory and the dialogical turn: A new approach to Erlangen constructivism. In J. Mittelstrass & C. von Bülow (Eds.), *Dialogische Logik* (pp. 91–148). Münster: Mentis.

Rahman, S., & Keiff, L. (2005). On How to be a Dialogician. In D. Vanderveken (Ed.), *Logic, thought and action* (pp. 359–408). Dordrecht: Kluwer.
Rahman, S., & Keiff, L. (2010). La Dialectique entre logique et rhétorique. *Revue de métaphysique et de morale, 66*(2), 149–178.
Rahman, S., & Redmond, J. (2016). Armonía Dialógica: tonk, Teoría Constructiva de Tipos y Reglas para Jugadores Anónimos. (Dialogical Harmony: Tonk, constructive type theory and rules for anonymous players). Theoria. *An International Journal for Theory, History and Foundations of Science, 31*(1), 27–53.
Rahman, S., Clerbout, N., & Keiff, L. (2009). On dialogues and natural deduction. In G. Primiero & S. Rahman (Eds.), *Acts of knowledge: History, philosophy and logic: essays dedicated to Göran Sundholm* (pp. 301–336). London: College Publications.
Rahman, S., McConaughey, Z., & Crubellier, M. (2015). *A dialogical framework for Aristotle's syllogism*. Work in progress.
Rahman, S., McConaughey, Z., Klev, A., & Clerbout, N. (2018). *Immanent reasoning. A Plaidoyer for the play level*. Dordrecht: Springer.
Ranta, A. (1988). Propositions as games as types. *Syntese, 76*, 377–395.
Ranta, A. (1994). *Type-theoretical grammar*. Oxford: Clarendon Press.
Redmond, J., & Rahman, S. (2016). Armonía Dialógica: tonk Teoría Constructiva de Tipos y Reglas para Jugadores Anónimos. *Theoria, 31*(1), 27–53.
Sundholm, G. (1986). Proof theory and meaning. In D. Gabbay, & F. Guenthner. (Ed.), *Handbook of Philosophical Logic, 3*, 471–506. Dordrecht: Reidel.
Sundholm, G. (1997). Implicit epistemic aspects of constructive logic. *Journal of Logic, Language and Information, 6*(2), 191–212.
Sundholm, G. (1998). Inference versus consequence. In T. Childers (Ed.), *The logica yearbook 1997* (pp. 26–36). Prague: Filosofía.
Sundholm, G. (2001). A Plea for logical Atavism. In O. Majer (Ed.), *The Logica yearbook 2000* (pp. 151–162). Prague: Filosofía.
Sundholm, G. (2009). A Century of judgement and inference, 1837–1936. Some strands in the development of logic. In L. Haaparanta (Ed.), *The development of modern logic* (pp. 263–317). Oxford: Oxford University Press.
Sundholm, G. (2012). Inference versus consequence revisited: Inference, conditional, implication. *Syntese, 187*, 943–956.
Sundholm, G. (2013). Containment and variation. Two strands in the development of analyticity from Aristotle to Martin-Löf. In M. van der Schaar (Ed.), *Judgment and epistemic foundation of logic* (pp. 23–35). Dordrecht: Springer Netherlands.
Sundholm, G. (2016). Independence friendly language is first order after all? Logique et Analyse, forthcoming.
Zermelo, E. (1904). Neuer Beweis, dass jede Menge Wohlordnung werden kan (Aus einem an Hern Hilbert gerichteten Briefe). *Mathematische Annalen, 59*, 514–516.
Zermelo, E. (1908). Untersuchungen über die Grundlagen der Mengenlehre, I. *Mathematische Annalen, 65*, 261–281.

Chapter 11
Avicenna: Mathematics and Philosophy

Roshdi Rashed

Abstract Like his Greek and Arab predecessors, Avicenna's research in mathematics concerned the development of methods of exposition, proof procedures and analytical tools. But Avicenna belonged to a new era of mathematics, and the question that this paper seeks to examine is how Avicenna applied this new mathematical knowledge in his philosophy. A treatment of the subject of Avicenna and mathematics cannot, then, confine itself to generalities, important as these are, but must start from the degree of knowledge he had of the various branches and the use he actually made of them. I shall restrict myself here to a number of examples which seem to me to possess evidential value and which all reveal a new conception of the connections between philosophy and mathematics. It will turn out that what Avicenna in fact did was to develop an analytical philosophy of mathematical concepts.

Like his Greek and Arab predecessors, Avicenna's research in mathematics concerned the development of methods of exposition, proof procedures and analytical tools. Like them, he occupied himself in clarifying mathematical knowledge and determining its place in the encyclopaedia of rational sciences that he wanted to build up. But Avicenna belonged to a new era of mathematics, beyond the experience even of his nearest predecessors, like al-Kindī and al-Fārābī: mathematics and mathematical sciences had undergone two centuries of uninterrupted expansion in which they had been enriched by new disciplines and acquired greater depth. This new age is adorned by a constellation of glorious names: dynasties of scholars such as the Banū

R. Rashed (✉)
CNRS SPHERE, Université Paris 7, Paris, France
e-mail: rashed@paris7.jussieu.fr

H. Tahiri (ed.), *The Philosophers and Mathematics*, Logic, Epistemology, and the Unity of Science 43, https://doi.org/10.1007/978-3-319-93733-5_11

Mūsā, the Banū Qurra, the Banū al-Ṣabāḥ, the Banū al-Karnīb; and illustrious mathematicians such as al-Khāzin, al-Qūhī, al-Sijzī, al-Būzjānī among others. Avicenna is himself a contemporary of Ibn al-Haytham and a member of the court at Khwārizm[1] which also included the famous Ibn Irāq and al-Bīrūnī. This is the period in which mathematicians were pushing back the boundaries of Hellenistic mathematics and creating new disciplines such as algebra, elementary algebraic geometry, indeterminate analysis, the geometry of the projections of the sphere, spherical geometry and so on. There also took place in this period the systematic mathematization of the traditional disciplines such as astronomy and optics, and likewise the development of new social and juridical disciplines. Avicenna was then indeed a man of a new age, an age in which mathematical disciplines had multiplied and an age also of enforced specialisation; but this specialisation was different in extension and in understanding from that which had been practised in Alexandria in the third and second centuries before our era, and from that which is to be seen in Baghdad over the second half of the eighth century and at the beginning of the ninth. Henceforth there is a distinction to be made between algebraists and geometers, between these and the astronomers, the astrolabists, and the rest. This new specialisation led the mathematicians to take up questions that had, up until then, been the prerogative of philosophers, relating to the foundations of mathematics and the theory of demonstration, as the works of Thābit ibn Qurra, Ibrāhīm ibn Sinān, al-Qūhī, Ibn al-Haytham, among many others, attest.

This new environment must be kept in mind in any discussion of 'Avicenna and mathematics'. It is to be expected that, given the extent to which mathematical sciences had now been added to and transformed, the connections between the disciplines would no longer be the same (in particular those between mathematics and philosophy) and that the dividing lines between the professions would be called into question. It was by this time much less easy than in al-Kindī's time to be both a philosopher and a creative mathematician, and the philosopher had to maintain connections with the new mathematics that took account of the numerous distinct disciplines it embraced. A treatment of the subject of Avicenna and mathematics cannot, then, confine itself to generalities, important as these are, but must start from the degree of knowledge he had of the various branches and the use he actually made of them. I shall restrict myself here to a number of examples which seem to me to possess evidential value and which all reveal a new conception of the connections between philosophy and mathematics, while taking care not to omit an examination of mathematical knowledge and its acquisition. What Avicenna in fact did was to develop an analytical philosophy of mathematical concepts.

[1] Al-Nizāmī al-'Arūdī al-Samarqandī, *Jihār Maqāla*, translated into Arabic by 'A. 'Azām and Y. al-Khashshāb (Cairo, 1949), pp. 81f.:

كان لأبي العباس مأمون خوارزمماشاه وزير اسمه أبو الحسين محمد السهلي، كان حليم الطبع كريم النفس فاضلاً، وكذلك كان خوارزمشاه حليم الطبع صديقًا لأهل الفضل، وبفضلهما اجتمع كثير من الحكماء وأهل الفضل في هذه الحضرة مثل أبي علي بن سينا وأبي سهل المسيحي وأبي الخير الخمار وأبي الريحان البيروني وأبي نصر بن عراق.

See R. Rashed, 'Abū Nasr ibn 'Irāq: 'indamā kāna al-Amīr āliman (When the Prince was a scientist)', *al-Tafahom*, 40 (2013), pp. 145–170.

In *al-Shifā'* as in many other writings, Avicenna devotes substantial space to the mathematical sciences. In *al-Shifā'* alone, no fewer than four books are given over to mathematics. These 'contain all the sciences of the ancients, including music'.[2] Thus, unlike his predecessors such as al-Kindī, Avicenna no longer thought of mathematical research as constituting original research of a new kind, separate from the body of philosophical knowledge as a whole, but rather as an integral part of it. What is involved is quite evidently the sum total of the mathematics regarded as propaedeutic to the study of physics and metaphysics, and so indispensable to the education of a philosopher and to the construction of *al-Shifā'* as an encyclopaedia of rational disciplines. The sum total concerned is essentially traditional, that is, Hellenistic, even if it includes some new acquisitions and presents some new features. The astronomy, for example, is Ptolemy's, but in the first part of his edition Avicenna eschews the use of the term 'chord of double arc' in favour of the quite modern language of the sines and cosines used by al-Būzjānī, Ibn 'Irāq and al-Bīrūnī in their spherical geometry.

The *Arithmetic* of *al-Shifā'* is roughly that of Neo-Pythagoreans like Nicomachus of Gerasa. Avicenna however incorporates Thābit ibn Qurra's theorem on amicable numbers, and also introduces the language and some of the notions of algebra; and to this must be added some of the results achieved in the study of congruencies developed by his predecessors and contemporaries. In *Demonstration* in *al-Shifā'*, Avicenna also brings in al-Khāzin's theorem on the impossibility of the equation $x^3 + y^3 = z^3$ in rational numbers, but without taking notice of a completely new logical question, that of negative propositions, where the proof does not rapidly boil down to a *reductio ad absurdum*.[3] In the same way, in the *Metaphysics*,[4] he refers to the matter of the derivation of all figures from the circle, a question dealt with by al-Sijzī, without, however, paying attention to the mathematical and methodological consequences that the latter draws from this. These two examples point to what may have been the limits of Avicenna's knowledge of the advanced mathematics of his time.

This knowledge presents a double aspect, as it emerges from the reading of his various works. It is wide-ranging and multiform, and encompasses the disciplines treated in Euclid's *Elements*, and in Euclid's *Optics*; in Ptolemy's *Almagest* and *Harmonics*, and also, perhaps, his *Book of Hypotheses*; in the *Arithmetic* of Nicomachus of Gerasa. We note, however, the absence of the treatises from the Hellenistic age on advanced mathematics, those by Archimedes, by Apollonius, by Menelaus and by Diophantus, for example, even though these had been translated into Arabic and worked at and developed by mathematicians before Avicenna. This choice is deliberate, for it is also to be observed that Avicenna was, without any doubt, *au fait*

[2]Letter to Kiyā, in 'A. Badawī, *Aristū 'inda al-'Arab* (Kuwait, 1978), p. 121:
وقد قضيتُ الحاجة في ذلك فيما صنفته من كتاب **الشفاء** العظيم المشتمل على جميع علوم الأوائل، حتى الموسيقى،
بالشرح والتفصيل والتفريع على الأصول.

[3]Ibn Sīnā, *al-Shifā'*, *al-Mantiq*, 5. *al-Burhān*, ed. A. 'Afifi (Cairo, 1956), pp. 194–195:
لو أن إنسانًا سأل في الهندسة عن الأضداد هل عِلْمها واحد، فقد سأل مسألة من حق الفلسفة الأولى. أو عن عددين
مكعبين هل يجتمع منهما مكعب كما يجتمع من عددين مربعين مربع، فقد سأل مسألة حسابية.

[4]*Al-Shifā'*. *al-Ilāhiyyāt* (1) (The *Metaphysics*), Text established and edited by G. C. Anawati and S. Zayed, revised and furnished with an introduction by I. Madkour (Cairo, 1960), p. 145.

with the algebra of al-Khwārizmī, and also with Indian arithmetic and the theory of numbers of Thābit ibn Qurra and al-Khāzin.

The question that this paper seeks to examine is how Avicenna applied this mathematical knowledge in his philosophy. This is a question I have often asked myself, particularly in connection with the study of his ontology. I should like here to look at two other contexts in his work in which mathematics and philosophy meet. These are the main contexts and they fall into two categories:

1. The first covers cases where the philosopher thinks he can resolve a mathematical difficulty by means of an analysis of the concepts. In these, Avicenna's procedure is to embark on a philosophical elucidation of the difficulty met with by the ancient as well as the contemporary mathematicians. That was how he thought he could surmount the difficulties presented by the study of the angle and the angle of contact. The second example, which involves the use of the method of exhaustion to solve the problem of squaring the circle, is no less important.
2. The second comprises cases where the philosopher develops a mathematical (geometrical or kinematic) model in pursuing his research in metaphysics, in physics and in logic, for example when he is engaged in criticism of infinitist theories, the theory of geometrical and physical magnitudes or the arithmetical theory of congruencies.

11.1 When the Philosopher Thinks He Can Solve a Mathematical Difficulty by Means of an Analysis of the Concepts: The Concept of the Angle

How are the concepts of the angle, the mixtilinear angle, the angle of contact etc., as they are presented in Euclidean geometry, to be defined? This is a question that was continually asked, fully or partially, from the time of the ancient Greeks and into late antiquity.

In their enquiry into the angle, the mixtilinear angle and especially the angle of contact, mathematicians found themselves confronted by additional questions: are these two angles of the same kind? Are they magnitudes, and, if so, of what kind? Are they comparable and what are the scales of comparison? and so on. For at least two millennia mathematicians lacked any real means of providing answers—to do that, they needed to be able to study the notion of the length of a curve, and that of contact of curves, to measure curvature and to put in order infinitely small entities; and even if they managed to take a few tentative steps in this direction, it was not until Newton and his successors, in particular Euler, in his work of 1778, that a first solution was sketched out, one that is today but one element in an entire chapter devoted to the asymptotic study of functions, later developed by P. D. G. du Bois-Reymond, G. H. Hardy, and many others. Throughout this period, all that mathematicians could do was to combine their mathematical solution with philosophical reflexions on the infinite, magnitude, continuity, the nature of the angle as a mathematical object, and the like. Philosophers, for their part, were faced with a paradox: here was an object, the angle, something that might have belonged in the category of quantity, but which, however, did not verify the definition of quantity. Must they admit that it fell under several categories at once—quantity, quality, relation, position—or that it corresponded to none of the categories defined by Aristotle? And here was another object—the angle of contact—which did not verify the notion of magnitude as defined in V.4 and X.1 of the *Elements*. Whatever sense was given to Aristotelian classification by categories, the question remained unresolved. Aristotle himself touches on the mixtilinear angle, and not the particular case of the angle of contact, in at least two places: the *Posterior Analytics* and the *Meteorologica*. Aristotle's commentators and his successors ceaselessly puzzled over the angle as a mathematical object, and their thoughts deserve all the more attention because the angle and the angle of contact played a part in philosophical argumentation.[5] The angle and the angle of contact together constituted a combination of two complex notions, which had not yet been mastered: the infinitely small and non-Archimedean magnitude, something not yet conceived of which would have to await discovery until the foundation of non-Archimedean geometry (D. Hilbert, G. Veronese, etc.). To this difficulty must be added another supplementary problem, which has to do with classification. Because of its form, the angle of contact is neither rectilinear nor curvilinear: it is mixtilinear; but as a result of the way it is generated, it differs from the other mixtilinear angles, since it is not formed by a chord, but by a tangent. The research on these objects seems in fact to have occasioned a movement towards the clarification of the notion of magnitude: the difficulties encountered by mathematicians and philosophers in their pursuit of this research undermined and jeopardised a widely held idea to the effect that the notion of magnitude as defined in the *Elements*—Books V and X—is the only one.

We know of two essays by Avicenna on the angle and a fragment on the angle of contact. The first essay is a paper entirely and uniquely devoted to the notion of angle in general. This piece had, therefore, been composed before Avicenna had gone on to compile his *magnum opus*, *al-Shifā'*.

Al-Shifā' is the work in which Avicenna develops his doctrine on the angle. Moreover, he returns to it more than once, in order to discuss the category under which it falls, its definition, the knowledge we can have of it and so on. This time his intention is plain: to take advantage of the philosophical research carried out in that masterly book to provide a philosophical elucidation of this impenetrable mathematical concept.

First of all he had to try to define the concepts of angle, mixtilinear angle, etc. as they are presented in Euclid's geometry. This was, in any case, a matter that had, as a whole or in part, been a constant puzzle since the time of the ancient Greeks and into late antiquity, as Proclus bears witness. Mathematicians and philosophers in their turn took over the search for an answer, always within the framework of

[5]This is found particularly in the discussion of Atomism by the theologian-philosophers (al-Mutakallimūn); cf. M. Rashed, 'Kalām e filosofia naturale', in *Storia della scienza*, vol. III: *La civiltà islamica, Enciclopedia Italiana* (Rome, 2002), pp. 49–72.

the theory of Euclidian magnitudes as set out in the fifth and the tenth books of the *Elements*. Avicenna and his contemporaries at the end of the tenth century were familiar with the gist of this literature. In fact while he was making his effort to answer the question in Iran, the mathematician Ibn al-Haytham was trying to do the same in Egypt. The situation is replete with both historical and logical significance: while the mathematician was at work on the first theoretical model, which would not be improved upon until several centuries later, by Newton, the philosopher was engaged upon an avowedly interpretative task: an attempt to penetrate the darkness of the notion of angle. As one who had read Simplicius and John Philoponus, among others, Avicenna could not but be aware of the debates and controversies in which ancients and moderns alike had expressed opposing views on the subject of the angle. Besides, if he opted for the interpretative approach, it was not because he lacked the means, as a mathematician, to discuss the notion of angle using geometry, as is borne out by his edition of Euclid's *Elements*. If further evidence is required, Avicenna addresses the question in the books of *al-Shifā'* which deal with logic, physics and metaphysics, and not in the sections devoted to *Geometry*—particularly the third book—which is where we would expect it.

It was, then, as a philosopher that he meant to study the angle, intending to settle a debate that had gone on for more than a thousand years. In the *Categories* in *al-Shifā'*, he begins by looking for a definition of the angle sufficiently general to accommodate its two species, the plane and the solid. He therefore raises the matter of the category or categories the angle falls under, a question already argued over by his predecessors and which he had himself taken up in the *Epistle*. Moreover, everything points to the fact that, right from the start, Avicenna intends to show that the angle does not belong to one category only, but to several. He is brought to this position by the difficulty encountered in seeking to define the angle in the way the other figures are defined, by the category of quantity alone.

By 'angle', then, in the primary sense, is meant a magnitude with several limits that end in a common limit; on this account, it comes under quantity. It is, as Avicenna says, a 'figured magnitude' (*miqdār mushakkal*), that is, a magnitude that has a certain geometrical figure. In that case, the angle as surface or solid must essentially possess the three properties of any quantity: it has one part; it has a magnitude; and it admits of equality and inequality.

Now it is precisely with this property—the equality relation—that the difficulty starts: Avicenna could not be unaware that angles are not always comparable. At any rate he tries to resolve the difficulty, at least for rectilinear angles, by basing equality on superposition.

But if Avicenna puts forward the notion of superposition to resolve the difficulty raised by defining the angle as magnitude, he also brings in another idea for the same purpose. For continuous magnitudes, he distinguishes between a magnitude that is continuous by itself, like a line, a surface, and a solid, and a magnitude that is continuous by another, like an angle.

We are not here concerned with Avicenna's expositions on the subjects of continuity and contact. It is however worth mentioning that when it comes to the angle, the Euclidean definition is Avicenna's, often implicit, reference. Thus the plane angle is a surface bounded by two limits, two lines that have the same point as their extremity, but without one line's prolonging the other; and the solid angle is a solid bounded by two plans which have a line as their common extremity.

But Avicenna had learned from the *Elements* that the angle does not verify the definition of magnitude given in the fifth book. He himself declares:

> In general, the definition of quantity is that there exists in it a thing of it that has the value of a one for counting, and which is in itself, whether the value is existential or hypothetical.[6]

Now the idea of a unit of measurement, which exists for continuous magnitudes and allows setting in order and comparison, does not exist for the angle.

To say that the angle falls under quantity is therefore unsatisfactory, unless the other categories that are involved in its definition are explicitly mentioned. Now Avicenna had always given the impression that the category of quality had an important part to play. But here he could not side-step the question of whether the same thing can fall under several categories at once. There is no doubt that this was a question Avicenna asked himself, for he writes:

> We know that the categories are different and that it is not valid that two categories at once can predicate one and the same thing, a predication <by> genus, so that the same thing comes by its essence under two categories, even if it is possible for the thing to fall under one category per se and under the other by way of accident.[7]

Further on he writes:

> There is nothing to stop something falling under two categories, in two ways: in one way 'by itself' (*i.e.* per se), in that it is a species for it; in the other, 'by accident', in that it is the object to which it (*sc.* the accidental property) happens.[8]

The difficulty in the case of the angle is in determining which is the category it falls under essentially, and which are the ones it falls under by accident, a difficulty made the more formidable by the diversity of the angles that there are: rectilinear, curvilinear, of contact, solid, etc.

The strategy adopted by Avicenna, consists, apparently, in circumventing the obstacle by formally modifying the set of components that were, in his view, needed to form the concept, 'angle'.

He begins by making explicit the part played by the category of quality. The quality in question is what happens *per accidens* to a quantity, and of which there are

---

[6]*Al-Shifā'. Al-Ilāhiyyāt* (1) (*Metaphysics*), ed. G. C. Anawati and S. Zayed, revised and preceded by an introduction by I. Madkour (Cairo, 1960), p. 118, 14–15:

فالكمية بالجملة حدها هي أنها التي يمكن أن يوجد فيها شيء منها يصح أن يكون واحدًا عادًا، وبكون ذلك لذاته سواء كانت الصحة وجودية أو فرضية.

[7]*Al-Shifā'. Al-Mantiq*—1. *al-Madkhal*, *al-Maqūlāt*, ed. G. C. Anawati, M. Khudayrī and F. al-Ahwānī, p. 156, 1–3:

إنا نعلم أن المقولات متباينة، وأنه لا يصلح أن تحمل مقولتان معًا على شيء واحد حمل الجنس حتى يكون الشيء الواحد يدخل من جهة ماهيته في مقولتين، وإن كان قد يدخل الشيء في مقولة بذاته، وفي الآخر على سبيل العرض.

[8]Ibid., p. 225, 7–8:

ولا بأس أن يدخل الشيء في مقولتين، على وجهين؛ أما في أحدهما، فبالذات، على أنه نوع له. وأما في الآخر، فبالعرض، على أنه موضوع لعروضه له.

three species: figure, non-figure and a combination of the two. Figure is an accident of magnitude *qua* magnitude. Avicenna then distinguishes two senses of figure: the geometers' sense and that of figure as a quality. Initially, 'figure' is understood in the sense of 'what is surrounded by a limit or limits', like a circle or a square. According to Avicenna, geometers understand the term in the sense of 'figured magnitude'. What he himself understands by 'figure' is, however, 'configuration', *al-hay'a*.

So, along with quality, Avicenna introduces a key notion, both for him and also for his successors: configuration, *al-hay'a*. At the time, the term was in current use in expressions like *hay'at al-'ālam*, 'the configuration of the universe', *'ilm al-hay'a*, 'astronomy', etc. The term can also be rendered as 'structure', or 'model'.

The notion of configuration plays a major part in Avicenna's doctrine of the angle: it orders all the other notions involved, and, ultimately, it is what determines their sense. It also implicitly contains other categories besides quantity and quality.

Numerous examples could be provided which illustrate that Avicenna saw the angle as falling under two categories at once, and for specific reasons: under quantity as a surface or solid; under quality, as 'configuration'. Viewed in this way, it would be, as it were, a qualitative organization of quantity, or a quality with the particular capacity to endow a quantity with a certain form, as quadrature does for a square, rectilinearity for a straight line, circularity for a circle, convexity and concavity for curves, etc.

Thus, the notion of the angle embraces several elements:

1. A surface or a solid. 2. The limits—lines or surfaces—that bound it. 3. The points or lines in which these boundaries meet. 4. The form of their meeting. 5. The relationships of the different parts to each other. 6. These elements are put together according to a certain configuration which, in the end, determines the angle. 'Configuration' here covers both a geometrical figure and an appropriately arranged combination of all these elements.

But this conception of an angle depends on two other categories. The first is the category of position, since an angle is, in a sense, a magnitude, and a magnitude has a position; or, as Avicenna puts it, 'Quantity with a position is magnitude'.[9] The second is the category of relation, since when we contemplate an angle we are looking at the relationships between its different elements. If the involvement of this last category is only implicit in the notion of angle in general, it does, on the other hand, become absolutely necessary when Avicenna raises the question of the definition of a particular angle, for example an acute angle.

But this position as such in itself brings us back to the category of relation. In fact such aspects as the inclination of one side to the other, and their proximity or distance apart, clearly fall under this category. Would it be necessary to bring in this category explicitly to define an acute angle, for example? Avicenna's reply is that,

[9] *Al-Shifā'. Al-Mantiq*—1. *al-Madkhal*, *al-Maqūlāt*, ed. G. C. Anawati, M. Khudayrī and F. al-Ahwānī, p. 129, 2:

والكم ذو الوضع هو المقدار.

even if that 'does not actually point to a relation by reason of its difficulty, it does point to it potentially, by actually introducing a relation'.[10]

This brief account of the gist of the angle doctrine as it is presented in *al-Shifā'* seems to show that Avicenna believes that the angle is an object belonging to several categories. He has with delicacy and subtlety endeavoured to qualify these category memberships so as not to place them on the same level, in particular by appealing to the doctrine of actuality and potentiality. The notion of configuration was introduced in order to integrate the different elements he had identified, in a structure that, in the end, defines an angle. In so doing, it endowed the object 'angle' with, so to speak, a more formal mode of existence than that enjoyed by an object belonging to only one category. Furthermore it relieves Avicenna of the need to discuss the taxonomic and ontological consequences of one and the same object's belonging to several categories, essentially and by accident, something which would lead to endless complications. The configuration notion also enables him to steer clear of the particularly obscure 'inclination' idea put forward by Euclid. Lastly, it enables Avicenna to avoid offering an account of an angle as a measurable magnitude, something he could not begin to explain: the configuration is a locus formed by two lines which have different positions and a common limiting point.

In order to explain the notion 'angle', Avicenna, in the knowledge that this does not verify the Euclidean definition of magnitude, brings in the pair 'by itself/by another' (on one occasion) and the pair 'by essence/by accident' (on another occasion) to obtain a philosophical solution for this mathematical problem. But the difficulty already encountered in studying the angle—plain, curvilinear, mixtilinear, solid—is still greater when it comes to the angle of contact, which, as a magnitude, is non-Archimedean.

The other example where Avicenna thinks he can resolve a mathematical difficulty using an analysis of concepts is that of squaring the circle, which once again involves the thorny question of comparing non-homogeneous magnitudes.

11.2 When the Philosopher Develops a Mathematical Model (Geometrical or Kinematic) in Order to Further His Research in Metaphysics, Physics and Logic

I now turn to another meeting point of mathematics and philosophy in Avicenna, when the philosopher develops a mathematical model to further his research in metaphysics, physics and logic.

The employment of mathematical arguments in the investigation of philosophical problems is, as everyone knows, a practice as old as philosophy itself. We have only to leaf through Avicenna's books, on logic, and likewise those on physics or

[10]Ibid., p. 250, 15–16:
لم يدل على هذه الإضافة بالفعل لصعوبتها فقد دل عليها بالقوة في إدخال إضافة بالفعل.

metaphysics, to observe that arguments like this appear in different guises and at several levels of thought. They range from a straightforward reference to a familiar theorem to a calculation of congruencies, for example, or again to borrowing or constructing a mathematical model to shore up a philosophical thesis. This last move is what interests me here and, to preserve the unity of my account, I shall confine myself to considering a single theme: the theory of magnitudes and the criticism of infinitist theories. But, before going on to examine some individual examples of these, let us remind ourselves of three features that distinguish them and clarify their range as well as their limits.

The models, whether borrowed or constructed, were expressed in terms of Euclidean geometry, and so assume the parallel postulate and the common notion that 'the whole is greater than the part'. But, unlike Aristotle and Euclid himself, and following the geometers of his time such as al-Qūhī and al-Sijzī, among others, Avicenna has no hesitation in introducing movement into his geometrical models.

A second feature is shared by the models he uses in dealing with cosmological problems. He begins by reducing these to geometrical problems involving infinity, and then goes on to develop a kinematic model to refute the infinitists.

Finally, in the case of infinity, Avicenna considers only whether it exists and never the question of its size—or its measure—whereas the latter question was broached, for the first time in history, by his predecessor, the mathematician Thābit ibn Qurra. This is the third feature to which I should like to call attention.

In the course of these analytical and critical exercises, Avicenna succeeded in advancing proofs that were irrefutable for anyone confined to the framework of Euclidean geometry. It would not be until the early years of the nineteenth century, and under other skies, that it would finally be possible to show what was wrong with them. We understand that Avicenna's successors, and by no means insignificant scholars, since Naṣīr al-Dīn al-Ṭūsī is counted among them, returned to these models without finding anything the matter with them; and when those hostile to Avicenna, such as Abū al-Barakāt al-Baghdādī or Ibn al-Ghaylān, offered some criticisms in regard to them, these rebounded on them.[11]

I shall confine myself to two examples, one geometric and the other kinematic.

I. The first example, in which Avicenna's purpose is to refute infinitist theories in geometry, is as follows: Let AB and AC be two straight lines which represent the two magnitudes, with origin A, produced to infinity, and let the points B_i and C_i ($i = 1, 2, \ldots$) be such that $BB_1 = B_1\,B_2 = \ldots = B_{n-1}\,B_n = \ldots$ and $CC_1 = C_1\,C_2 = \ldots = C_{n-1}\,C_n = \ldots$; then the parallel straight lines BC, $B_1\,C_1$, ..., B_nC_n, ... have equal increments $B_i\,D_i$ ($i = 1, 2, \ldots$). Avicenna states that, if infinite magnitudes exist, then there exists an interval $B_\infty C_\infty$ composed of an infinite number of equal increments, limited by two points B_∞ and C_∞ respectively on AB and BC produced. Thus we would have an infinite number of increments on a finite interval, which is absurd.

Avicenna reasons like this:

We have

[11] Abū al-Barakāt al-Baghdādī, *Kitāb al-Mu'tabar* (Hyderabad 1358H), pp. 60 ff.; Ibn Ghaylān, *Ḥudūth al-'ālam*, ed. M. Mohaghegh (Teheran 1998).

$$B_1C_1 = BC + B_1D_1$$

$$B_nC_n = BC + nB_1D_1;$$

since the intervals B_iD_i are equal,

$$B_\infty C_\infty = BC + \sum_{k=1}^{\infty} B_kD_k.$$

For him, $B_\infty C_\infty$ is a finite interval, since it is closed, whereas the sum of BC and intervals B_iC_i is infinite.

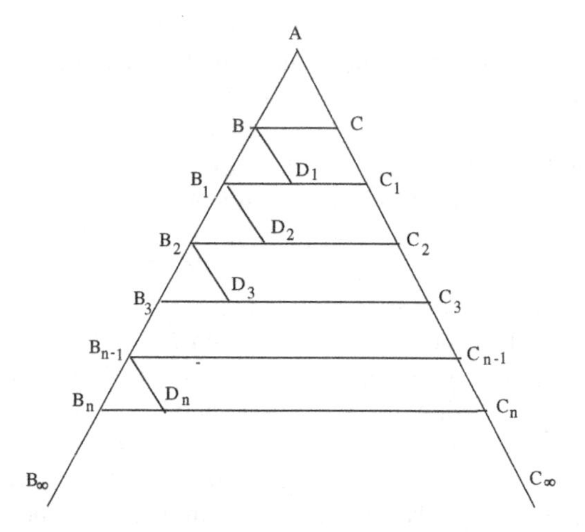

Avicenna achieves this contradiction because he admits the existence of the endpoints B_∞ and C_∞ on AB and BC produced to infinity. He therefore thinks of an infinite straight line as a closed interval with two endpoints.

II. Avicenna develops several models in order to refute infinitist theories in physics. In the *Physics* of *al-Shifā'*, after reducing a cosmological problem to one of geometry, among other questions relating to infinity, he raises the question of the finitude of physical magnitudes. I shall reproduce here the only model he develops to demonstrate the impossibility of circular motion in a void.

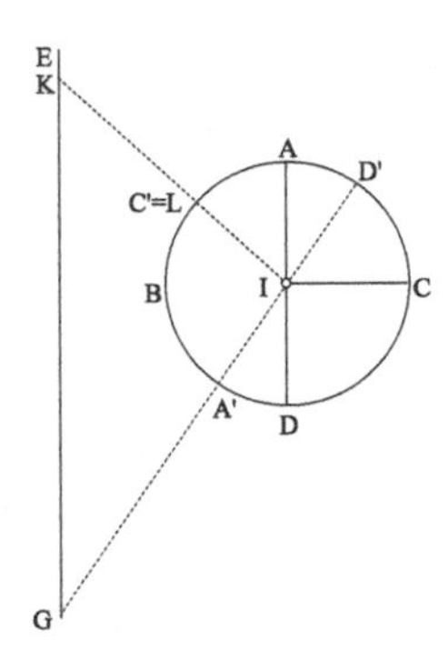

This is how he proceeds:

Let there be a circle *ABCD*, with centre *I*; it is assumed that *I* is fixed and that the circle turns around *I*, carrying with it point *C* and diameter *AD* perpendicular to *IC*.

A straight line *EG* is drawn parallel to the initial direction of *AD* in the half-plan that does not contain *C*; it is assumed that *EG* does not meet the circle and that it is fixed.

Initially the ray *IC* produced does not meet *EG*, according to the hypothesis, for, if any radius of the circle met *EG*, this line would completely envelop the circle. The situation remains the same at the start of the movement as long as the straight line *AD* has not reached a position perpendicular to *EG*. But when *AD* has passed that position, the half-line *IC* necessarily meets *EG*.

Avicenna assumes that there exists a direction *IH*, the point *H* on *EG*, which separates the directions of *IC* that do not meet *EG* from those that meet *EG*. From this, he deduces a contradiction; for, if *K* is a point on *EG* beyond point *H*, direction *IK* is reached by *IC* before *IH*; and *IC does* meet *EG* at *K*.

The model rests on the assumed existence of point *H* at a finite distance, whereas the point in question is actually pushed out to infinity. The kind of geometry that could supply this knowledge still lay in the future.

In *al-Najāt*, Avicenna returns to this kinematic argument in the following, slightly longer version:

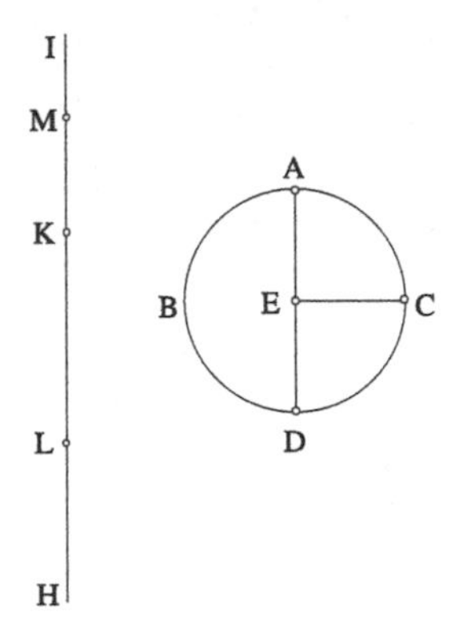

> We say: let there be a circular movement in an infinite void, if it were possible for an infinite void to exist. Let the body in motion be as the sphere *ABCD* which moves around its centre<the point *E*>. Let us imagine the straight line *IH* in the infinite void. Let the straight line *EC* – from the centre to one side of the circumference – not meet the straight line *IH* in the direction of *H*, even if it is produced to infinity. But, if the sphere turns, this line [*EC* produced] will be such as to cut it [*IH*], move along it and separate from it. The meeting and the separation are then necessarily in the direction of two points, let them be *K* or *L*. But point *M* is in its direction before point *K*. But point *K* is the first direction point<of *EC*>; which is absurd. But circular movement exists; therefore the void is not infinite.[12]

Thus, when the sphere turns, it carries the straight line *EC* produced with it in its rotation. This line will be parallel to *IH* before ceasing to be so and meeting it at some point and then going on to cut it and pass along it. On this occasion, what Avicenna considers is not a circle, as in the *al-Shifā'* text, but a sphere that turns around its centre *E*, and an infinite straight line. He in fact envisages two points on *IH*, call them H_1 and H_2, such that the directions EH_1 and EH_2 separate the parts of *EC* meeting *IH* from those that do not meet it (in both of its directions of travel along *IH*, towards *I* and towards *H*). Avicenna assumes that the two points H_1 and H_2 are at a finite distance. Now this is not the case. The founders of non-Euclidean geometry in the first third of the nineteenth century (Gauss, Lobatchevski and Bolyai) employed a line of argument perfectly analogous to Avicenna's but understood that the points H_1 and H_2 are necessarily pushed out to infinity. In the eleventh century, Avicenna's remarkable intuition had brought him surprisingly close to the truth!

The themes that I have just been looking at illustrate both the importance and the diversity of the connections between theoretical philosophy and mathematics in Avicenna's work. The philosopher-mathematicians who were his heirs, like Naṣīr al-Dīn al-Ṭūsī, Quṭb al-Dīn al-Shīrāzī, Athīr al-Dīn al-Abharī among others, followed in his footsteps on the path he had opened up and built on the foundations he had laid in the philosophy of mathematics.

We have seen that the use of philosophy to clarify mathematical concepts that had remained impenetrable is not to be confused with another more traditional approach. It is no longer a matter of discoursing in general terms on the foundations of mathematics and the nature of their apodeicticity. In the era of specialization and of the transformation of mathematics, indeed, the latter task was taken up again by mathematicians like Thābit ibn Qurra, Ibn Sinān, al-Qūhī, al-Sijzī and Ibn al-Haytham, who took the process further and in much greater depth. The approach that Avicenna takes as a philosopher is, then, a new one, which may be described as both epistemic and aetiological: clarifying mathematical concepts, and especially those which pose most problems, in an effort to determine the cause of the difficulty simply by analysing them. In discussion of the angle or the method of exhaustion, for exam-

[12] *Al-Najāt*, ed. M. Kurdī (1938), p. 123:

فنقول: لتكن حركة مستديرة في خلاء غير متناه، إن أمكن أن يكون خلاء غير متناهٍ. وليكن الجسم المتحرك مثل كرة ا ب جـ د المتحركة على مركزها <نقطة ه>. ولنتوهم في الخلاء الغير المتناهي خط ط ح. وليكن ه جـ من المركز إلى جهة من المحيط لا يلاقي خط ط ح من جهة ح، وإن أخرج بغير نهاية. لكن الكرة إذا دارت صار هذا الخط بحيث يقاطعه ويجري عليه وينفصل عنه. فيكون الالتقاء والانفصال بمسامتة نقطتين لا محالة وليكونا ك ول. لكن نقطة م تسامته قبل نقطة ك ونقطة ك أول نقطة تسامت، هذا خلف لكن الحركة المستديرة موجودة، فالخلاء ليس بلا نهاية.

ple, diagnosing the difficulty depends on a 'clinical' approach, which, put simply, involves an analytical philosophy of mathematical concepts.

But this 'analytical' outlook is more marked and is combined with a 'proto-formal' tendency when Avicenna examines mathematical concepts that are more general than the angle concept. Thus, when he tackles a concept like 'magnitude', he cannot himself avoid bringing some mathematical tools to bear; he likewise uses mathematical models in studying the concept of infinite magnitude. The term 'proto-formal' is used here to describe reflexion on objects outside the empirical world of sense, that can be represented by means of mathematics, or even by a combinatorics of letters of the alphabet, as Avicenna did in his work entitled *al-Nayrūziya*. I believe that I showed not long ago that this 'proto-formal' tendency implicitly underpins Avicenna's ontology of emanation. In this ontology, it is in fact possible to designate all existents by letters of the alphabet, and thereafter proceed by combination. Now it is precisely this combinatorial step that Naṣīr al-Dīn al-Ṭūsī initiated and successfully carried through, followed by others, like Ibrāhīm al-Ḥalabī.

References

Abū al-Barakāt al-Baghdādī. (1358). *Kitāb al-Mu'tabar*, Hyderabad.

Al-Nizāmī al-'Arūdī al-Samarqandī. (1949). *Jihār Maqāla*, translated into Arabic by 'A. 'Azām and Y. al-Khashshāb, Cairo.

Badawī, 'A. (1978). *Aristū 'inda al-'Arab*, Kuwait.

Ibn Ghaylān. (1998). *Ḥudūth al-'ālam*, ed. M. Mohaghegh, Teheran.

Ibn Sīnā. (1956). *al-Shifā', al-Mantiq*, 5. *al-Burhān*, ed. A. 'Afifi, Cairo.

Ibn Sīnā. (1960). *Al-Shifā'. al-Ilāhiyyāt* (1) (The *Metaphysics*). Text established and edited by G. C. Anawati and S. Zayed, revised and furnished with an introduction by I. Madkour, Cairo.

Ibn Sīnā. (1938). *Al-Najāt*, ed. M. Kurdī.

Ibn Sīnā. (1959). *Al-Shifā', Al-Mantiq* – 1. *al-Madkhal, al-Maqūlāt*, ed. G. C. Anawati, M. Khudayrī and F. al-Ahwānī, Cairo.

Rashed, M. (2002). 'Kalām e filosofia naturale'. In: *Storia della scienza*, vol. III. *La civiltà islamica, Enciclopedia Italiana*, Rome.

Rashed, R. (2013). 'Abū Nasr ibn 'Irāq: 'indamā kāna al-Amīr āliman (When the Prince was a scientist)', *al-Tafahom*, 40.

Chapter 12
For a Continued Revival of the Philosophy of Mathematics

Jean-Jacques Szczeciniarz

Envoi. This essay is a friendly and grateful tribute to Roshdi Rashed. Needless to say, this article will deal with the philosophy of mathematics and particularly what we call recent mathematics. As a historian of mathematics, Roshdi Rashed (like the great Neugebauer) is a tireless reader of contemporary mathematics. He knows how to draw, for example from category theory examples and ways of thinking that serve as benchmarks for exploring the conceptual history of mathematics.

Abstract This paper argues in favor of a nonreductionist and nonlocal approach to the philosophy of mathematics. Understanding of mathematics can be achieved neither by studying each of its parts separately, nor by trying to reduce them to a unique common ground which would flatten their own specificities. Different parts are inextricably interwined, as emerges in particular from the practice of working mathematicians. The paper has two topics. The first one concerns the conundrum of the unity of mathematics. We present six concepts of unity. The second topic focuses on the question of reflexivity in mathematics. The thesis we want to defend is that an essential motor of the unity of the mathematical body is this notion of reflexivity we are promoting. We propose four kinds of reflexivity. Our last argument deals with the unity of both of the above topics, unity and reflexivity. We try to show that the concept of topos is a very powerful expression of reflexivity, and therefore of unity.

J.-J. Szczeciniarz (✉)
CNRS SPhERE University of Paris Diderot, Paris, France
e-mail: jean-jacques.szczeciniarz@paris7.jussieu.fr

H. Tahiri (ed.), *The Philosophers and Mathematics*, Logic, Epistemology, and the Unity of Science 43, https://doi.org/10.1007/978-3-319-93733-5_12

12.1 Introduction

12.1.1 Mathematics

In his paper *A view of Mathematics* Alain Connes comments on the role of mathematics:

> Mathematics is the backbone of modern science and a remarkably efficient source of new concepts and tools to understand the "reality" in which we participate. It plays a basic role in the great new theories of physics of the XXth century such as General Relativity and Quantum Mechanics. The nature and inner workings of this mental activity are often misunderstood or simply ignored, even among scientists from other disciplines. They usually only make use of rudimentary mathematical tools that were already known in the XIXth century and miss completely the strength and depth of the constant evolution of our mathematical concepts.[1]

This is even more true of philosophy. Of course there are some exceptions like Cavaillès, Lautman, and some historians of mathematics, nevertheless one can say that the living heart of the activity of mathematics in action is generally ignored.

Our aim here is to provide some elements to change this situation. We can only propose a modest contribution in the face of the immense task that should be undertaken. The main point is to insist on the fact that an essential reason for this situation lies in the neglect or ignorance of the unity of mathematics. This unity accounts for the remarkable efficiency of new concepts and their ability to understand the reality in which we participate.

12.1.2 Some Essential Feature of the Mathematical Landscape

At first glance the mathematical landscape seems immense and diverse: it appears to be a union of separate parts such as geometry, algebra, analysis, number theory etc. Some parts are dominated by (various aspects of) our understanding of the concept of "space", others by the art of manipulating "symbols", and others by the problems occurring in our thinking about "infinity" and "the continuum".

This first view is not completely false, but this breaking down of mathematics into different regions of inquiry also misses much—it has a superficial aspect and needs to be rectified and re-elaborated by bringing together different elements; to go through the surface and the depth of this landscape amounts to understanding its unity. And to understand the unity is also to understand the reasons for it.

---

[1] Alian Connes, 2008 *A View of Mathematics*: Concepts and Foundations vol. 1 www.colss.net/ or Eolss. http://www.eolss.net/sample-chapters/c02/E6-01-01-00.pdf.

12.2 The Unity of Mathematics

The most essential feature of the mathematical world is the following: it is virtually impossible to isolate any of the above parts from the others without depriving them of their essence according to Alain Connes in the same article (Connes 2008). In order to describe this profound unity of mathematics we must take into account the nature of mathematical abstraction, and the manner in which it is related to the unity of mathematics.

According to this view the essence of mathematics is linked to its unity. The first way to think of this unity is to compare the mathematical body with a biological entity: it can function and flourish only as a whole and would perish if separated into disjoint pieces. There are many ways to think of this organic metaphor for the unity of mathematics. I would like to emphasize four aspects.

12.2.1 Four Features of Unity of Mathematics

Firstly: this first unity comes from a very old view in the history of science. It is a conception whose scope is universal and which serves to order our understanding not only of mathematics but also of the manner in which the whole physical or biological world is to be thought of as a unity. Plato, for instance, builds an organic unity, hierarchical set of entities that form the universe. The Forms (Ideas) that preside over this hierarchical unity—from the intelligible world to the sensible world—are geometric. This geometry is translated into a unit having two faces. The one is that of the universe (theory of proportions for the cosmology or for the politics, analogy of the Line[2] the other is that of the geometry itself which takes advantages for its own unity, of the intelligible world.

Penrose is fascinated by the crucial role that complex numbers play, both in quantization and in the geometry of spinors. He has always been motivated by the idea that complex structures provide an important link between these two objects. The physical universe can be explored by means of complex numbers. Moreover, complex geometry contributes to understanding the unity of mathematics. This first unity would be the unity of mathematics as the intellectual unity of science and at the same time a deepening of unification of the mathematics.

Secondly: as an organic unity, it develops from inside, just like a living being. I will say more about this feature below. It is the unity of mathematics as unity of his extension and expansion movement.

Third: this growth can take a variety of directions which carry simultaneous and multiple meanings and exhibit, so to speak, different rhythms of development. Grothendieck becomes the contemporary of Galois, Riemann of Archimedes.

---

[2]Plato, Rep. VI, 508–509, Platon, *Œuvres complètes* Texte établi par Auguste Diès, Paris, Les Belles Lettres, Budé T. 7-1, *Platonis Opera* John Burnet, Oxford Classical Texts, Clarendon Press 1900–1907, G. Leroux, Garnier Flammarion, Paris 2002, nelle édition de la *République* 2003.

Fourth: the "topologies" of these different kinds of increase can take very different forms, including—for instance, metaphorically speaking—non-classical topological spaces of representation. We will see that layers of mathematics can be cut several times, (topology and algebra, integers and real numbers) or that different domains are non-separable (group theory, topological group theory, (topological) vector spaces).

12.2.2 Internal Endogenic Growth for the Second Kind of Unity

An essential feature of this organic development of mathematics is the extension of the body by new elements emerging from within, as in a living being. A theory may typically provide the resources for the expression of a further theory which develops by means of a *reflection* on the first. I will add some elements of analysis to develop this topic below. It will be our second topic.

Consider for example the calculus that Newton and Leibniz in different ways invented. It is only when it became a question for the mathematical body, when it adapted itself to a host structure produced by the body, that one arrives at "the" calculus. Differential calculus was invented in the deepening of geometrical research of the Archimedean tradition. He could then give a conceptual framework to solve physical problems. The production of a purely mathematical concept is the result of the elaboration of a notion, arising from geometrical reflection, by the mathematical body and after this process physical issues may have started to be dealt with. Consider the case of Leibniz's contribution, which provided a clear set of rules for working with infinitesimal quantities, allowing the computation of second and higher derivatives, and providing the product rule and chain rule, in their differential and integral forms. Unlike Newton, Leibniz paid a lot of attention to the formalism, often spending days determining appropriate symbols for concepts. And this is purely mathematical working in the sense of an internal development.

This slow process of transformation of Euclidean concepts of motion has been studied by Panza (2005).[3]

Any apparently external element, object, idea, image must be integrated and reconstructed in a mathematical manner and form. It is not certain that the calculus could have appeared without the intervention of physics, but the physical question had to be entirely transformed, mathematized, conceived as a mathematical problem in the passage from its initial Euclidean setting to the analytic one. This is the case for the concept of force and acceleration. What is important is the mathematical development of conceptual tools, whose different steps we can describe as an internalizing of external elements.

Moreover there are some difficulties with the organic metaphor. It misses a central aspect that characterizes mathematics: the fact that different disciplines have appeared that are essential for all existing mathematics. For example, topology, or

---

[3]Marco Panza, *Newton et les origines de l'analyse: 1664–1666,* Blanchard, Paris, 2005.

algebraic geometry. . . . Topology impacts on all mathematics and has helped to renew old theories and approach them in a new light. Each discipline has effects on others in various ways. Thus a supplementary body appears to be essential for another. It is possible to follow the various ways in which a discipline (such as algebra, topology, geometry. . .) leaves the marks of its growth within the body. They are able to go through very different stages of growth and roles.

12.2.3 Difficulties with the Organicist Concept of Unity, Particularly for the Third Conception: Two Opposite Concepts of the Mathematical Body

The Coming-to-be of mathematics appears as autonomous, unpredictable, and endogenous, and in accordance with a temporality such that the overall structure is out of reach. It typically involves bifurcations, branches, breaks, continuity, recovery, neighborhood relations, and moments of partial unification. We can try to propose two "optimal" forms of such development.

(a) Labyrinthine

There are many underground networks: Archimedes is related to Lebesgue and Riemann, but Archimedes is also related to Pascal and Leibniz, Lagrange to Galois and Galois to Grothendieck. There are profound underground paths, sometimes surprising. It also happens that new proofs of the same theorem come as secondary benefit of a new theory. Reintroduction of the Pythagorean theorem in infinitesimal geometry renewed its sense. Multiple timeframes are sometimes involved in this. Surprise and multiplicity of different temporalities disturb the coherence of the organic metaphor. We can nevertheless retain the affirmation of endogenous growth.

These necessarily succinct remarks on this dispersed diachronic of mathematics go hand in hand with the synchronic dimension of the mathematical body.

(b) Architectonic

There is an underground network of connections between various trajectories, whose reality or forms we do not appreciate. When these connections appear frequently and unexpectedly we can reconstruct a new region of the already known territory. We then join the architectonic organization of the mathematical body.

> Where things get really interesting is when unexpected bridges emerge between parts of the mathematical world that were previously believed to be very far removed from each other in the natrural mental picture that a generation had elaborated. At that point one gets the feeling that a sudden wind has blown out the fog that was hiding parts of a beautiful landscape.[4]

[4]Alain Connes, A view of mathematics. *ibid.*

I recall some of the principal new ideas Grothendieck considered as essential to his work [R and S 1985][5]

1. Topological tensor products and nuclear spaces.
2. Continuous and discrete duality (derived categories, "six operations").
3. Riemann-Roch-Grothendieck Yoga (K-theory, relation with intersection theory).
4. Schemes.
5. Topos.
6. Etale and l-adic Cohomology.
7. Motives and motivic Galois group (Grothendieck categories).
8. Crystal and crystalline cohomology, yoga of "de Rham coefficients ", "Hodge coefficients"....
9. "Topological Algebra": 1-stacks, derivators; cohomological topos formalism, as inspiration for a new homotopic algebra.
10. Tame Topology.
11. Algebraic anabelian geometry Yoga, Galois-Teichmüller theory.
12. Schematic or arithmetic point of view for regular polyhedras and regular configurations in all genera.

Each of these "new ideas" plunges deeply into the mathematical body and imposes on it a new systematic unity, or at least re-shapes our perspective on the different forms of unity it exhibits and enables us to trace new connections between them. The fact that we can distinguish these two opposite conceptions is as such significant. They are two forms of the creative productivity of mathematics. The first is that form in which it escapes us. The second is the form in which it gives ways of exercising control over its forms of expansion. We are able to recognize new trajectories and detect new relations, for example, the program of derived algebraic geometry, that consider polynomial equations up to homotopy. This is a new trajectory within a program. More precisely, it is a combination of schema theory and homotopy theory. Schema theory is re-worked from a homotopical perspective. The synthesis of both theories retains the power of each within a further unity. This allows a higher level viewpoint, permitting us to reinterpret both theories, and at the same time provides them with greater power. As a matter of fact, the unity as a synthesis of different elements, or different disciplines. Among the examples given above that would require immense development. We will be interested (only partially) in the theory of schemes (Hartshorne 1977).[6] We will proceed in four steps in order to explain the elementary concept of a scheme.

[5]Alexandre Grothendieck, *Reaping and Sowing* 1985 Récoltes et Semailles Part 1. The life of a mathematician. Reflections and Bearing Witness. Alexander Grothendieck 1980, English Translation by Roy Lisker, Begun December 13, 2002.

[6]Robin Hartshorne, *Algebraic Geometry*, Springer, New York, 1977.

12.2.4 Example of Synthetic Unity: The Concept of Scheme

(a) We construct the space $SpecA$ associated to a ring A. As a set we define $SpecA$ to be the set of all prime ideals of A. We assume known the concept of ring and ideal and prime ideal. If $\mathfrak{a}$ is an ideal of A, we define the subset $V(\mathfrak{a}) \subseteq SpecA$ to be the set of all prime ideals which contain $\mathfrak{a}$. These concepts are purely algebraic concepts. They refer to an important part of commutative algebra: the theory of ideals.[7]

(b) Now we define a topology on $SpecA$ by taking the subsets of the form $V(\mathfrak{a})$ to be the closed subsets.[8] We show that finite unions and arbitrary intersections of set of the form $V(\mathfrak{a})$ are again of that form. $V(\emptyset) = SpecA$, and $V(0) = SpecA$. You can see how algebra and topology form a specific unity. But this synthesis is not yet complete.

(c) The concept of a sheaf provides a systematic way of (Hartshorne 1977) [Grothendieck EGA I][9] discerning and taking account of local data. Sheaves are essential in the study of schemes. The concept of sheaf is another synthesis between algebra and topology. We give the definition.

Let X be a topological space; A *presheaf* $\mathcal{F}$ of abelian groups on X consists of the data

(i) for every open subset $U \subseteq X$, an abelian group $\mathcal{F}(U)$ and

(ii) for every inclusion $V \subseteq U$ of open subsets of X, a morphism of abelian groups $\rho_{UV} : \mathcal{F}(U) \to \mathcal{F}(V)$ subject to the conditions

(1) $\mathcal{F}(\emptyset) = 0$ where $\emptyset$ is the empty set

(2) ρ_{UU} is the identity map $\mathcal{F}(U) \to \mathcal{F}(U)$

(3) if $W \subseteq V \subseteq U$ are three open subsets then $\rho_{UW} = \rho_{VW} \circ \rho_{UV}$.

A presheaf is a concept that is easy to express in the language of the categories that makes this unity of domains or disciplines appear: a presheaf is just a contravariant functor from the category $\mathfrak{Top}$[10] of open subsets of a topological space X to the category $\mathfrak{Ab}$ of abelian groups.

If $\mathcal{F}$ is a presheaf on X we refer to $\mathcal{F}(U)$ as the section of the presheaf $\mathcal{F}$ over the open set U. Indeed we have to understand that we dispose a map s from U to $\mathcal{F}$, denoted as sections of the presheaf.

---

[7]Actually he concept of $SpecA$ is based on two types of earlier constructions: the correspondence between the points of an affine algebraic variety $X \subset kn$, on an algebraically closed field k and the *maximal* ideals of $k(X) = k[t_1 \cdots t_n]/I$ I being the ideal of null polynomials on X: Zariski's topology on X 2) correspondence between the points of a compact topological space K and the maximal ideals of *C(K)* algebra of continuous functions on K with complex values and reconstruction of the topology from this algebra (Gel'fand).

[8]The great innovation of Grothendieck (he proceeds in this way very often) consists in considering a geometrical object associated with any commutative ring A; to have a spectrum that depends functorially on A it is necessary to replace the maximal ideals by the primes ideals.

[9]EGA I, Le langage des schémas. *Publ. Math. IHES* 4, 1960.

[10]It could be noticed : $Ouv(X)$

A sheaf is a presheaf satisfying some extra conditions. We will give only one condition in mathematical form.

If U is an open set, if $\{V_i\}$ is an open covering of U, and if $s \in \mathcal{F}(U)$ is an element such that $s_{|V_i} = 0$ for all i then $s = 0$.

The second condition is the condition that says that sections that coincide in the intersection of both open sets glue together in an unique section. It is the essential property of gluing that makes one pass from the local to the global. The different syntheses above syntheses are able to give a philosophical synthesis: the reflexive synthesis that allows us to know if a property can be globalized. This unity is beyond the unity between concepts or between different disciplines, it is a synthetic unity that "constructs" the globality of a property. There are many examples of sheaves, such as the sheaf of continuous functions on a topological space, the sheaf of distributions etc. In this construction algebra and topology play a role at different levels.

(d) Let A be a ring. The *spectrum* of A is the pair consisting of the topological space $Spec A$ together with the sheaf of rings $\mathcal{O}$ defined above. To each ring A we have associated its spectrum $Spec A, \mathcal{O}$. This association is not complete. We would like that this correspondence/association is really synthetic unity which could be constructed as a conceptual, mathematical unity. If the ring A can be seen as a category, and it is also the case for the Spectrum we require this correspondence to be fonctorial.[11] The appropriate notion is the category of locally ringed spaces. So a *ringed space* is a pair $(X, \mathcal{O}_X)$ consisting of a topological space X and a sheaf of rings $\mathcal{O}_X$ on X. And next we must define what a morphism of ringed spaces consists of. A varies in the category of commutative rings and $(Spec A; \mathcal{O})$ varies in the category *locally* ringed spaces, i.e., ringed spaces $(X, \mathcal{O}_X)$ such that for any section f of $\mathcal{O}_X X$ is a union of the two open in which f and $1 - f$ are respectively invertible. The structural sheaf $\mathcal{O}$ is defined on the spectrum $Spec A$, its value in the open where a section f is not zero is the ring $A[f^1]$.[12]

We get our last definition. An *affine scheme* is a locally ringed space $\mathcal{O}_X$ which is isomorphic (as a locally ringed space) to the spectrum of some ring. A scheme is a locally ringed space $\mathcal{O}_X$ in which every point has an open neighborhood U such that the topological space together with the restricted sheaf $\mathcal{O}_{X|U}$ is an affine scheme. We call X the *the underlying topological space* of the scheme $(X, \mathcal{O}_X)$ and $\mathcal{O}_X$ its *structure sheaf*.

This is a complicated unity, about which we shall make some remarks. The synthesis we have carried out makes it possible to make the various elements applies,

---

[11] For the concept of functor see Sect. 5.3.3(ii).

[12] More deeply, $(Spec A, \mathcal{O})$ solves a universal problem: the ring A defines a contravariant functor F_A of the category of locally ringed spaces in the category of sets such that $F_A(X, \mathcal{O}_X)$ is the set of ringed space morphisms of $(X\mathcal{O}_X$ in $(\bullet, A)$ ($\bullet$, is the punctual space); an element of $F_A(X, \mathcal{O}_X)$ is a rings morphism $A \to \Gamma(X, \mathcal{O}_X)$. For any locally ringed space $(X, \mathcal{O}_X)$ any ϕ: $A \to \Gamma(X, \mathcal{O}_X)$ defines a unique morphism $\Phi : \Gamma(X, \mathcal{O}_X) \to (Spec A, \mathcal{O})$ of locally ringed spaces such as $\Phi^* (O) \to (\mathcal{O}_X)$ defines an homomorphism $A \simeq \Gamma(Spec A, \mathcal{O}) \to \Gamma(X, \mathcal{O}_X)$ that identifies itself to ϕ.

in particular the topological element. But this one in turn plays within the algebraic control of the topological structure.

This description can be repeated for every theme developed by Grothendieck.

> To speak frankly these innumerable questions, notions, and formulations of which I've just spoken, indeed, the countless questions, concepts, statements I just mentioned, only make sense to me from the vantage of a certain "point of view" - to be more precise, they arise spontaneously through the force of a context in which they appear self evident: in much the same way as a powerful light (though diffuse) which invades the blackness of night, seems to give birth to the contours, vague or definite, of the shapes that now surround us. Without this light uniting all in a coherent bundle, these 10 or 100 or 1000 questions, notions or formulations look like a heterogeneous yet amorphous heap of "mental gadgets", each isolated from the other - and not like parts of a totality of which, though much of it remains invisible, still shrouded in the folds of night, we now have a clear presentiment. The fertile point of view is nothing less than the "eye", which recognizes the simple unity behind the multiplicity of the thing discovered. And this unity is, veritably, the very breath of life that relates and animates all this multiplicity.[13]

12.3 This Architectonic Unity Takes Different Forms in the History of Mathematics. Three Philosophical Forms

12.3.1 Unity as Logical Unity or Operational Concept

We will build on Lautman to develop this philosophical conception and eventually to criticize the opposite conceptions. The first conception we appeal to here is founded on logical developments, like those proposed by Russell and Carnap. The second on Wittgenstein. As a matter of fact, this second conception explains that mathematical statements should be explained in terms of logical *operations*. Nevertheless, this approach is divorced from mathematical reality (which we also want to promote and analyze), and for this reason it was rejected by Lautman. Above all, he refuses the reduction of philosophy to the syntactic study of scientific utterances, and he rejects the reduction of philosophy to a role of clarification of propositions which intervene in "what is generally called the theory of knowledge". For example, propositions on space and time must be subject to criticism from the syntactic point of view.[14]

> Mathematical philosophy is often confused with the study of different logical formalisms. This attitude generally results in the affirmation of the tautological character of mathematics. The mathematical edifices which appear to the philosopher so difficult to explore, so rich in results and so harmonious in their structures, would in fact contain nothing more than the principle of identity. We would like to show how it is possible for the philosopher to

[13]Récoltes et Semailles Part I, The life of a mathematician. Reflection and Bearing Witness, Alexander Grothendieck, English translation by Roy Lisker, Begun December 13, 2002.

[14]Lautman (2006, pp. 52–53) *Les mathématiques, les idées et le réel physique* Vrin, Paris. Introduction and biography by Jacques Lautman; introductory essay by Fernando Zalamea. Preface to the 1977 edition by Jean Dieudonné. Translated in Brandon Larvor *Dialectics in Mathematics*. Foundations of the Formal Science, 2010.

> dismiss such poor conceptions and to find within mathematics a reality which fully satisfies the expectation that he has of it.[15]

Lautman talks about the fading away of mathematical reality, and his judgement holds for Russell, Carnap and Wittgenstein. Alongside this logicist philosophical unity he excludes, Lautman retains two other conceptions of unity which are close to his own. From one side, mathematical reality can be characterized by the way one apprehends and analyzes its organization. From the other side, it can also be characterized in a more intrinsic fashion, from the point of view of its own structure. The first case was illustrated by Hilbert's position where he stressed the dominant role of metamathematical notions compared to those of the mathematical notions they serve to formalize. On this view, a mathematical theory receives its value from the mathematical properties that embody its structure in some generic sense. We recognize in this approach one (very influential) structural conception of mathematics. Indeed Hilbert substitutes for the method of genetic definitions the method of axiomatic definitions. He introduces new variables and new axioms, from logic to arithmetic and from arithmetic to analysis, which each time enlarge the area of consequences. For example, in order to formalize the analysis, it is necessary to be able to apply the axiom of choice, not only to numerical variables, but to a higher category of variables, those in which the variables are functions of numbers.[16] The mathematics is thus presented as successive syntheses in which each step is irreducible to the previous stage.

Moreover, it is necessary to superimpose a metamathematical approach on this formalized mathematical theory which takes it as an object of analysis from the point of view of non-contradiction and its completion.[17,18,19] When we recall the Hilbertian point of view, we see that a duality of plans between formalized mathematics and the metamathematical study of this formalism entail the dominant role of metamathematical notions in relation to formalized mathematics.

Lauman quotes Hilbert *Gesammelte Abhandlungen* t. III, p. 196 sq.[20] and Paul Bernays, *Hilberts Untersuchungen über die Grundlagen der Arithmetik.*[21]

---

[15]Lautman (2006) *ibid.*

[16]*ibid.*, (Lautman 2006, p. 130).

[17]Lautman means completeness in the sense of completion. The system is said to be completed if any proposition of the theory is either demonstrable or refutable by the demonstration of its negation. The property of completion is said to be structural because its attribution to a system or a proposal requires an internal study of all the consequences of the considered system.

[18]Recall that I am analyzing the philosophical architectonic unity of mathematics. This was illustrated by Hilbert's position. He stressed the dominant role of metamathematical notions compared to those of the mathematical notions they serve to formalize. On this view, a mathematical theory receives its value from the metamathematical properties that embody its structure in some generic sense. We recognize in this approach one (very influential) structural conception of mathematics.

[19]Lautman (2006, p. 30).

[20]David Hilbert, *Gesammelte Abhandlungen*, Verlag Julius Springer Berlin, 1932.

[21]Paul Bernays, *Hilberts Untersuchungen über die Grundlagen der Arithmetik*, Springer, 1934.

In this structural, synthetic conception,[22] mathematics is seen—if not as a completed whole—then at least as a whole within which theories are to be regarded as qualitatively distinct and stable entities whose interrelationships can in principle be thought of as completely specifiable,

What is the meaning of Hilbert's structuralism? Lautman[23] takes the example of the Hilbert space. The consideration of a purely formal mathematics leaves the place in Hilbert to the dualism of a topological structure and functional properties in relation to this structure. The object studied is not the set of propositions derived from axioms, but complete organized beings having their own anatomy and physiology. For Lautman the Hilbert space is "defined by axioms which give it a structure appropriate to the resolution of integral equations. The point of view that prevails here is that of the synthesis of the necessary conditions and not that of the analysis of the first conditions"[24]

As the second conception we recognize a more dynamic diachronic picture of the interrelationships, which sees each theory as coming with an indefinite power of expansion beyond its limits bringing connections with others, of a kind which confirms the unity of mathematics, especially from the standpoint of mathematical epistemology.

In Hilbert's metamathematics one aims to examine mathematical notions in terms of notions of non-contradiction and completion. This ideal turned out to be unattainable. Metamathematics can consider the idea of certain perfect structures, possibly realized by effective mathematical theories. Lautman wanted to develop a framework that combines the fixity of logical concepts and the development that gives life theories.

12.3.2 Dialectics

Lautman (in a third conception, the one he defends) wanted to consider other logical notions that may also be connected to each other in a mathematical theory such that solutions to the problems they pose can have an infinite number of degrees. On this picture mathematics set out partial results, reconciliations stop halfway, theories are explored in a manner that looks like trial and error, which is organized thematically and which allows us to see the kind of emergent linkage between abstract ideas that Lautman calls *dialectical*.

Contemporary mathematics, in particular the development of relations between algebra group theory and topology appeared to Lautman to illustrate this second—our words—"labyrinthine" model of the dynamic evolving unity of mathematics, struc-

[22]Lautman, *Essai sur les notions de structure et d'existence*, Hermann, Paris 1937: the structural point of view to which we must also refer is that of Hilbert's metamathematics etc.

[23]Lautman (2006, pp. 48–49).

[24]Lautman, *ibid.*

tured around oppositions such as local/global, intrinsic/extrinsic, essence/existence. It is at the level of such oppositions that philosophy intervenes in an essential way.

> It is insofar as mathematical theory supplies an answer to a dialectical problem that is definable but not resolvable independently of mathematics that the theory seems to me to participate, in the Platonic sense, in the Idea with regard to which it stands as an Answer to a Question.

(Lautman 2006, p. 250)[25]

12.3.3 Philosophical Choices

Lautman seeks to study specific mathematical structures in the light of oppositions such as continuous/discontinuous, global/local, finite/infinite, symmetric/ antisymmetric. Brendan Larvor[26] remarks that in *New Research on the Dialectical Structure of Mathematics* Lautman offers a slightly different list of dialectical poles: "whole and parts situational and intrinsic properties, basic domains and objects defined on these domains, formal systems and their models etc."[27] In his book (Lautman 2006), there is a chapter about "Local/global". He studies the almost organic way that the parts are constrained to organize themselves into a whole and the whole to organize the parts. Lautman says "almost organic": one thinks of this expression in the following way. There exists an "organic" unity within mathematics, reminiscent of biological systems, as Alain Connes amongst many others has noted (see above). We want to develop this recognition further below by making use of the notion of reflexivity. In his chapter on extrinsic and intrinsic properties with title "Intrinsic properties and induced properties", Lautman examines whether it is possible to reduce the relationships that some system maintains with an ambient medium to properties inherent to this system. In this case he appeals to classical theorems of algebraic topology. More well-known is his text on "an ascent to the absolute", in which an analysis of Galois theory, class field theory, and the uniformization of algebraic functions on a Riemannian surface is presented. Lautman wanted to show how opposite philosophical categories are incarnated in mathematical theories. Mathematical theories are data for the exploration of ideal realities in which this material is involved.

[25]Brendon Larvor, *Albert Lautman: Dialectics in Mathematics*, Foundations of formal Science, 2010.

[26]Brendan Larvor, *Albert Lautman, ibid.*

[27]Lautman (2006).

12.3.4 Concerning On the Unity of the Mathematical Sciences

This is the first of Lautman's two theses. It takes as its starting point a distinction that Hermann Weyl made in his 1928 work on group theory and quantum mechanics.[28] Weyl distinguished between "classical" mathematics, which found its highest flowering in the theory of functions of complex variables, and the "new" mathematics represented[29] by the theory of groups and topology (Lautman 2006, p. 83). For Lautman, the classical mathematics of Weyl's distinction is essentially analysis,[30] that is, the mathematics that depends on some variables tending toward zero, convergent series, limits, continuity, differentiation and integration. It is the mathematics of arbitrary small neighborhoods, and it reached maturity in the nineteenth century. And, Brendan Larvor continues, the 'new mathematics of Weyl's distinction is global': it studies structures of "wholes".[31] Algebraic topology, for example, considers the properties of an entire surface (how many holes) rather than aggregations of neighborhoods.

Having illustrated Weyl's distinction, Lautman re-draws it.[32]

> In contrast to the analysis of the continuous and the infinite, algebraic structures clearly have a finite and a discontinuous aspect. Through the elements of a group, a field or an algebra (in the restricted sense of the word) may be infinite, the methods of modern algebra usually consists in dividing these elements into equivalence classes, the number of which is, in most applications, finite.[33]

The chief part of Lautman's "unity" thesis is taken up with four examples in which theories of modern analysis [see Brendan Larvor] depends in their most intimate details on results and techniques drawn from the "new" (Larvor 2010) algebraic side of Weyl's distinction. Algebra comes to the aid of analysis. That is, dimensional decomposition in function theory; non-Euclidian metrics in analytic function theory; non commutative algebras in the equivalence of differential equations; and the use of finite, discontinuous algebraic structures to determine the existence of the function of continuous variables.[34]

Lautman transforms a broad historical distinction (between the local, analytic, continuous and infinitistic mathematics of the nineteenth century, and the 'new' global, synthetic, discrete and finitistic style) into a family of dialectical dyads (local/global, analytic/synthetic, continuous/discrete, infinitistic/finitistic. These pairs find their content in the details of mathematical theories (Larvor 2010), that,

---

[28] Hermann Weyl, *Gruppentheorie and Quantenmechanik*, Hirzel, Leipzig, 1928.

[29] Larvor (2010).

[30] Brendan Larvor, *ibid.*

[31] Lautman (2006, p. 84).

[32] Larvor (2010), Lautman, 2005, p. 196.

[33] Lautman (2006), pp. 86–87.

[34] Lautman (2006, p. 87).

though they belong to analysis, sometimes employ a characteristically algebraic point of view.

12.4 Mathematical Reflexivity

The topic of this section is the study of the development of forms of *reflexivity* in mathematics, which imply the history of the concept of space and the history of several disciplines. Mathematical activity involves, as an essential aspect, examining concepts, theories or structures through (the lens of) other concepts theories and structures, which we recognize as reflecting them in some way.

There are many ways to understand the notion of reflexivity in mathematical practice. One such way is illustrated by the stacking of algebraic structures, groups, rings, fields, vector space, modules, etc. Each level is the extension of the previous one—a vector space for example is a certain kind of module. The extension here consists in adding a property or a law. This imposition of additional structure brings a new perspective on the initial structure. A second way involves the addition of some property coming from another domain altogether, as seen in the notions of topological group, Lie group, differential or topological field. This synthesis also yields a new view of the initial structure. This is reflexivity in the weak sense. The effect of such new syntheses makes up much of the history of mathematics. But one has synthesis also between structures or concepts.

This other kind of the reflexivity includes the case where one discipline, for example algebra, reflects some concepts or some structures from another discipline. For example in algebraic topology, algebraic concepts and methods are used to translate and to control some topological properties. The same holds for algebraic geometry. And it can happen that algebraic topology and geometry themselves cross-fertilize by means of such reflexive interactions. All the phenomena of translation of one discipline in to another also illustrate such forms of reflexivity. This is a local manifestation of the fact that mathematics is permeated with such "reflecting surfaces".

It is possible also to construct the history of one concept, for example, the concept of number or of space viewed from this standpoint. Gilles-Gaston Granger, a French philosopher of mathematics, says that these concepts are "natural". But they are also the most opaque.[35]

Notice that the history of the concept of space through the concept of a manifold involves the intersection of multiple disciplines and the development of multiple forms of *reflexivity*.

[35]Gilles-Gaston Granger, *Formes opérations, objets* Paris, Vrin, 1994, pp. 290–292.

12.4.1 The Concept of a Manifold

In the case of space, there was a long process whereby a deepening reflection on the concept of surface was produced in mathematics. Along the way, such a concept as that of variety was revealed. The concept of variety arose as a geometrical *reflection* on the concept of surface: First came the notion of an abstract surface parameterized by coordinates, then that of abstract place covered by topological opens (maps, atlas) in relation to an ambient space. These notions were extracted from such "reflexive" contexts as autonomous concepts which could be seen as defining a new kind of mathematical entity.

This extraction involved abstraction from the concept of surface, an abstraction which at the same time brought a change of point of view on the earlier concept: one passed from a concept defined via coordinates to one resting on parameters. That passage was effected by a reflection on the sense of using coordinates. One can understand that a surface is nothing but the different forms of the variation of its coordinates. And when one speaks in terms of maps and atlases the concept is further deeply reworked as was achieved by Hermann Weyl in his *Concept ofRiemann surface* 1912.[36] This new entity now acts as the carrier of topological properties, and manifolds come to be seen as autonomous entities and indeed as a fundamental concept. The act whereby we obtained a surface is geometrically displaced, so to speak, and in this act of displacement the entity to which it is related is re-defined. The notion of a variety is likewise designated in functional terms: it is the range of variation of the values of certain functions. Functions can now reflect their nature by means of this new entity. A manifold becomes the support and mirror for the properties of functions that are defined on it. These functions with their properties constitute the new objects we should consider as new basis and point of departure for a further stage of geometrization.

We would like to give a particularly striking example. We remind that a functor **F** from a category C to another category D is a structure-preserving function from C to D. Intuitively, if C is een as a network of arrows beteween objects, then **F** maps that network onto network of arrows of D. Every category C has an identity functor $1_A : C \to C$ which leaves the objects and arrows of C unchanged, and given functors, $\mathbf{F} : C \to D$ and $\mathbf{G} : D \to E$ there is a composite $\mathbf{G} \circ \mathbf{F} : C \to E$. So it is natural to speak of a category of all categories, which we call **CAT**, the objects of which are all the categories and the arrows of which are all functors. And Colin McLarty asks whether **CAT** is a category in itself. His answer is to treat **CAT** as a regulative idea; an inevitable way of thinking about categories and functors, but not a strictly legitimate entity.[37,38] In a not so formal sense we can get a notion of common

[36] Hermann Weyl, *The Concept of a Riemann Surface*, Addison and Wesley, 1964, First Edition *die Idee der Riemanschen Fläche*,Teubner, Berlin, 1912.

[37] Immanuel Kant, *Kritik der reinenVernunft*, Hartnoch Transl. N. Kemp Smith (1929) as *Critique of Pure Reason*, Mcmillan.

[38] Colin McLarty, *Elementary Categories, Elementary Toposes*, Clarendon Press Oxford, 1992, p. 5 "Compare the self, the universe and God in Kant 1781".

foundation for mathematics in the elementary notions that constitute categories. The author believes, in fact, that the most reasonable way to arrive at a foundation meeting these requirements is simply to write down axioms descriptive of properties which the intuitively-conceived category of all categories has until an intuitively adequate list is attained; that is essentially how the theory described below was arrived at.[39] Thus our notion of space changes status, it becomes an intelligible object in itself, and that is why it can provide a reflexive context in which to reconceptualize the previous notion of a surface. At the same time, the act of measuring can be considered as such and made the object of study as a structure within the mathematical body. The concept of a metric on a manifold makes possible this new reflexion. Any such expression of magnitude can be reduced to a quadratic form, and thereby expresses the most general law that defines the distance between two infinitely near points of a variety.

This entity in turn enables us to construct new spaces: we can now define the notions of algebraic manifold, topological manifold, differentiable manifold, analytic manifold, arithmetic manifold. In this way we are given the means to pass from one discipline to another. This passage between formerly separated disciplines involved both an upward movement (in the formation of the concept of manifold) and horizontal and synthetic extension of concepts (across several domains and disciplines).

12.5 Reflexivity and Unity

One of the most powerful tools we use to explain the reflexivity is the concept of topos, and moreover the concept of Grothendieck's topos.

12.5.1 Prerequisites for Understanding the Search for the Unity of Mathematics According to Grothendieck

I distinguish three prerogatives that underlie the arguments of Grothendieck.

(a) The unity of mathematics according to Grothendieck is that of the discrete and the continuous, and the structure of mathematics must be able to account for it.
(b) Three aspects of mathematical reality are traditionally distinguished. Number, or the arithmetical aspect; size, or the analytical aspect; form, or the analytical aspect.[40] Grothendieck took an interest in form as embodied in structures.

[39]William Lawvere, The category of categories as a foundation of mathematics by, *Proceedings of the Conference on Categorical Algebra, La Jolla Calif.* 1965, pp. 1–20, Springer Verlag, New York, 1966.

[40]Mathieu Belanger, *La vision unificatrice de Grothendieck: au-delà de l'unité (méthodologique?) de Lautman* Philosophiques vol 37 Numéro 1–2010.

> This means that if there is one thing in mathematics that has always fascinated me more than any other, it is neither number nor size, but always form. And among the thousand and one faces that form chooses to reveal itself to us, the one that has fascinated me more than any other and continues to fascinate me is the hidden structure in mathematical things.[41]

(c) Grothendieck adopts a resolutely realistic attitude.

> The structure of a thing is by no means something we can invent. We can only patiently update, humbly get to know it, "**discover**" it. If there is inventiveness in this work, if we happen to be a blacksmith or indefatigable builder, it is not to "shape" or to build structures ... It is to **express** as faithfully as we can these things that we are discovering and probing, this reluctant structure to indulge ...

The sequel of the quotation describes both tasks

> Inventing language capable of expressing more and more finely the intimate structure of the mathematical thing and ... constructing, with the aid of this language, progressively and from scratch, the theories which are supposed to account for what has been seen and apprehended.

(RS 1985)[42]

> One might say that Numbers are what is appropriate for grasping the structure of discontinuous or discrete aggregates. These systems, often finite, are formed from "elements" or "objects" conceived as isolated with respect to one another. "Magnitude" on the other hand is the quality, above all, susceptible to "continuous variation", and is most appropriate for grasping continuous structure and phenomena: motion, space, varieties in all their forms, force, field, etc. Therefore arithmetic appears to be (over-all) the science of discrete structures while analysis is the science of continuous structures.[43]

It is therefore necessary to understand that the point of view of number is used for the discrete structure, whereas the point of view of magnitude is used to grasp the structure of the continuum. According to Grothendieck "arithmetic is the science of discrete structures and analysis is the science of continuous structures",[44] and for him, geometry intersects both the discrete structures and the continuous structures. The study of geometrical figures could be done from two distinct points of view. First, the combinatorial topology in Euler's sequence was linked to the discrete properties of the figures. Second, geometry (synthetic or analytic) examined the continuous properties of the same figures. It was based in particular on the idea of size expressed in terms of distances. Geometry studied both the discrete and the continuous, but distinctly.[45] The development of abstract algebraic geometry in the 20th century inaugurated a renewal of the aspect of form by imposing a single point of view that directly participates in both the study of discrete structures and the study of continuous structures.

---

[41] Grothendieck, *Récoltes et Semailles,* 1985, *Reaping and Sowing* my translation.

[42] *ibid.*, my translation.

[43] A. Grothendieck, Récoltes et Semailles, traduction Roy Lisker p. 66.

[44] *Ibid.*

[45] Mathieu Belanger [Belanger p. 15 online].

12.5.2 *Prerequisites to the Search for Unity as Implementation of Reflexivity*

The search for unity through the creation of a new discipline consists in seeing how analysis can be reflected in arithmetic, and how arithmetic can be reflected in analysis. Whenever a concept of one discipline is enlightened by another, it is analyzed by the other: by associating a concept of the concerned discipline and a form of abstraction. It refers this form to the first discipline. This reflexivity has taken place in the new algebraic geometry of Grothendieck in a double manner, or in a mirror with two faces: one face for arithmetic and the other for analysis, Grothendieck called it "arithmetical geometry".

12.5.3 *The rôle Played by Weil's Conjectures*

Working on abstract algebraic geometry, the great French mathematician André Weil formulated four conjectures concerning the zeta function of algebraic manifolds on finite fields. We cannot expose the very great complexity of these conjectures. It is sufficient to know for our purposes that the very great generality of these conjectures and their difficulty was due to the fact that they required the application of topological invariants to algebraic varieties. According to Grothendieck, Weil's conjectures required the construction of a bridge between continuous structures and discrete structures. The Weil conjectures served as a guide to the elaboration of the new geometry. We can see unity and reflexivity in the context of the new arithmetical geometry.

> It may be considered that the new geometry is above all else a synthesis between these two adjoining and closely connected but nevertheless separate parts: the arithmetical world in which the so-called spaces live without the principle of continuity, and the world of continuous quantity, that is, space in the proper sense of the term, accessible through analysis (and for that very reason) accepted by him as worthy to live in the mathematical city. In the new vision, these worlds, formerly separated, form but one.[46]

12.5.4 *Reflecting Space to Produce a New Topological Concept*

The traditional concept of space does not have the flexibility required by the topological invariants of arithmetic geometry. However, no concept of space was more general than that which prevailed in the 1950s.

[46]Reaping and Sowing, 1985, my translation.

12.5.5 The Generality of the Topological Space

A topology is considered as the most stripped-down and therefore the most general structure available to a space. Let E be any set. Constructed from a family of sets, topology $\mathcal{P}(\mathcal{P}(E))$ chooses those that respect a stability for any union set operations (for the definition of topological open sets) and for finite intersection. This is a first level of reflexivity. Indeed, the operation of taking the parts of a set is redoubled on itself, and makes it possible to choose, according to a rule, certain subsets. It is also a way of analyzing the subsets of a set. The iteration is identified with a form of reflexivity. The concept of topological space encompasses all other space concepts.

We see here the intervention of set concepts to give the notion of space a form that goes beyond its static bases thanks to the set operations which possess a structure of algebra. But the most flexible spatial structure available to mathematicians was not sufficient for the problem raised by Weil's conjectures. The cornerstone of the new geometry therefore had to be a concept of space allowing one to go beyond the maximum generality of the concept of traditional topological space.

12.5.6 The Concept of Topos According to Grothendieck

The concept of topos provides maximum generality. It allows us to form a unit and realizes a form of reflexivity.

12.5.7 Back to Concept of Sheaf

Grothendieck replaces the lattice of open subsets, which defines the structure of a topological space in the traditional sense. He uses the notion of sheaf, which we have defined above. A sheaf is a mathematical concept allowing one to define a mathematical structure defined locally on a space X by a process of restriction and gluing. Some definitions are in order

(i) A nonempty subset Y of a topological space X is irreducible if it cannot be expressed as the union $Y = Y_1 \cup Y_2$ of two proper subsets, each one of which is closed in Y. The empty set is not considered to be irreducible.
(ii) Let k be a fixed algebraically closed field. We define affine n-space over k denoted $\mathbf{A}^n$ to be the set of all n-uples of elements of k. An *affine algebraic variety*, (or simply *affine variety*) is an irreducible closed subset of $\mathbf{A}^n$ with the induced topology from the topology of $\mathbf{A}^n$ which is the Zariski topology. An open subset of an affine variety is a *quasi-affine variety*.
(iii) A function $f : Y \to k$ is regular at a point $P \in Y$ if there is an open neighborhood U with $U \subseteq Y$ and polynomials $g, h \in A = k[x_1, \ldots x_n]$ such that h

is nowhere zero on U and $f = g/h$ on U. $A = k[x_1, \dots x_n]$ is the polynomial ring on a field k.

(iv) Let X be a variety over the field k. For each open set $U \subseteq X$ let $\mathcal{O}(U)$ the ring of regular functions from U to k. $\mathcal{O}$ verify the conditions of presheaf and of sheaf. These functions form a ring because they verify the ring's operations.

As we remarked we can consider the unity that ring structure gives functions and the way these functions are reflected by means of algebraic structure.

One can define the sheaf of continuous real-valued functions, the sheaf of differentiable functions on a differentiable manifold, or the sheaf of holomorphic functions on a complex manifold.

If we consider the lattice of open subsets of a topological space X, (we can also denote with $\mathcal{O}(X)$), the real-valued functions $f : U \in \mathbb{R}$, the restrictions of $f_{|V}$ on the open subsets $V \subset U$. By means of correct choice of V_i it is possible to reconstruct the function from its restrictions.

12.5.8 *The Language of Category Theory*

(i) Grothendieck considers the totality of sheafs on a topological space; It is the remarkable generative effect produced by this approach. All the sheafs on a topological space X form a category, denoted $Sh(X)$. This category is essential because it makes it possible to find the topological structure of space, that is to say the lattice of the open spaces $Ouv(X)$.

The topological structure is in fact determined by the category of the sheafs, which is much more flexible than the topological structure. It possesses the flexibility sought to transcend the apparent generality of the concept of traditional topological space.

12.5.9 *Grothendieck Topology*

In the usual definition of a topological space and of a sheaf on that space, one uses the open neighborhoods U of a point in a space X. Such neighborhoods are topological maps: $U \to X$ which are injective. For algebraic geometry it turned out that it was important to replace these injections (inclusions) by more general maps $Y \to X$ which are no necessarily injective. It extends the application by relieving constraints and preserving only the application whose source is any object of the category. The idea of replacing inclusions $U \to X$ by more general maps $U' \to X$ led Grothendieck to define "the open covers" of X.[47] We see that Grothendieck systematically considers the point of view of applications (morphisms) instead of objects (sets).

---

[47] These morphisms are *étales* morphisms in *étales* topology. Grothendieck has built other topologies in his theory of descent

(i) covering families

Let **C** be a category and let C be an object in **C**. Consider the indexed families

$$S = \{f_i : C_i \to C | i \in I\}$$

and suppose that for each object C of **C** we have a set

$$K(C)\{S, S', S'', \dots, \}$$

of certain such families called the *coverings* of C under the rule K. Thus for these coverings we can repeat the usual topological definition of a sheaf.

(ii) category equipped with covering families

To introduce a general notion of a category equipped with *covering families* we first use a functor. A functor—as we know—is a map (morphism) from a category to another category. If we are given a category C we are interested in the category Ĉ of all contravariant functors $\mathbf{C}^{op} \to$ **Set**. (We take the opposite category denoted $\mathbf{C}^{op}$). It is a category for which the maps (morphisms) are reversed with respect to the morphisms of the starting category. There is a functor from the category $\mathbf{C}^{op}$ to the category **Set**.

Thus we dispose the functor category denoted $\mathbf{Set}^{\mathbf{C}^{op}}$. Let us note the rise in abstraction, first the categories (objects and arrows) then the opposite categories, then the functors, passage from one category to another, and finally the category whose objects are the functors. We do not define "natural" applications that are not necessary for us. The use of a functor is necessary to see at the same time the passage from one category to another, which thus forges a possible unity and reflection reflected from one category to another. Each time this unity and this reflection take on a different meaning.

(iii) sieve

Grothendieck defines a notion of topology anchored in the category theory. It is also more general and more flexible than the traditional set of concepts. It uses the concept of a sieve. A sieve S may be given as a family of morphisms in **C** all with codomain C, such that

$$f \in S \Rightarrow f \circ g \in S$$

whenever this composition makes sense; in other words S is a right ideal. If S is a sieve on C and $h : D \to C$ is any arrow to C then

$$h^*(S) = \{g | cod(g) = D, h \circ g \in S\}$$

is a sieve on D. A sieve is a conceptual tool that makes it possible to gather arrows that are composed. Intuitively it is a collection of arrows that is "to hang" one to the other.

(iv) A Grothendieck topology on a category **C** is a function J which assigns to each object C of **C** a collection $J(C)$ of sieves on C, in such a way that

(a) the maximal sieve $t_C = \{f | cod(f) = C\}$ is in $J(C)$;
(b) (stability axiom) if $S \in J(C)$ then $h^*(S) \in J(D)$ for any arrow $h : D \to C$;
(c) (transitivity axiom) if $S \in J(C)$ and R is any sieve on C such that $h^*(R) \in J(D)$ for all $h : D \to C$ in S then $R \in J(C)$.

A Grothendieck topology is at a higher level a reflexion of topology. Here is a quotation by F. William Lawvere.

> A Grothendieck topology appears most naturally as a modal operator, of the nature "it is locally the case of".[48]

Grothendieck topology chooses some sieves. It is first and foremost a way of making the covering families respecting a stability of operative composition on the objects that it targets. If $S \in J(C)$, one says that S is a *covering sieve* or that S *covers* C (or, if necessary, that S J- covers C). Reflexivity here takes the form of the dynamic establishment of the conditions under which one can construct a topology.

In the case of an ordinary topological space, one usually describes an open cover U as just a family, $\{U_i, i \in I\}$ of open subsets of U with union $\bigcup U_i = U$. Such a family is not necessarily a sieve, but it does generate a sieve-namely, the collection of all those open $V \subseteq U$ with $V \subseteq U_i$ for some U_i. [Saunders Mac Lane, Ieke Moerdijk 1992].[49] (Informally V goes through the sieve if it fits through one of holes U_i of the sieve).[50]

(v) Site

A Grothendieck topology has a basis from which we give elements and properties.

A basis for a Grothendieck topology on a category C with pullbacks (that is with some sort of inverse image) is a function K which assigns to each object C a collection $K(C)$ consisting of families of morphisms with codomain C such that

(i') if $f : C' \to C$ is an isomorphism, then $\{f : C' \to C\} \in K(C)$;
(ii') if $\{f_i : C_i \to C | i \in I\} \in K(C)$ then for any morphism $g : D \to C$ the family of pullbacks $\{\pi_2; C_i \times_C D \to D\}$ is in $K(D)$;
(iii') if $\{f_i : C_i \to C | i \in I\} \in K(C)$, and if for each $i \in I$ one has a family $\{g_{ij} : D_{ij} \to C_i | j \in I_i\} \in K(C)$ then the family of composites $\{f_i \circ g_{ij} : D_{ij} \to C | i \in I, j \in I_i\}$ is in K(C) .

Condition (ii') is again called the stability axiom, and (iii') the transitivity axiom. The pair (**C**, K)is also called a *site* and the elements of the set $K(C)$ are called *covering families* or *covers* for this site. Covering families (we can denote *Cov*) are also called a *pretopology*. The pair (C, *Cov*) is a site.

The definition of the base pushes the notion of stability very far.

---

[48]See below Sect. 6.1.7.

[49]Saunders Mac lane, Jeke Moerbijk, *Sheaves and Geometry*, Springer, New York, 1992, pp. 110, 111.

[50]Mac Lane, Moerbijk, *ibid.*

12.5.10 Grothendieck Topos

A Grothendieck topos is a category which is equivalent to the category $Sh(\mathbf{C}, J)$ of sheaves on some site $\mathbf{C}, J$. Or in other words let be a *stack* or with another name a *presheaf* of sets over a category $\mathcal{C}$: it is a (contravariant) functor $F : \mathcal{C} \to \mathbf{Set}$.

We need to add the following remarks. The category $\mathbf{St}(\mathcal{C})$ of all stacks over $(\mathcal{C})$ is equivalent to the topos $\mathbf{Set}^{\mathcal{C}^{op}}$. This is an elementary topos like **Set** or **Finset** and other simple topoi. They are, generally speaking, some domain where mathematics can be developed, roughly speaking, without problems. We can do mathematics without thinking about it. Grothendieck topoi are more complicated.

We need to consider the subcategory of $\mathbf{St}(\mathcal{C})$ generated by those objects that are sheaves over the site $(\mathcal{C}, Cov$. It will denoted $\mathbf{Sh}(Cov)$. A Grothendieck topos is, by definition, any category that is equivalent to one of the form $\mathbf{Sh}(Cov)$.

12.6 Return to the Question of the General Unity of Mathematics and of the Reflexivity

Topoi theory is the way that Grothendieck constructed in order to find the solution of the problem of unity of the discrete and the continuous to resolve Weil's conjectures.

> The idea of topos encompasses, in a common topological intuition, the traditional (topological) spaces embodying the world of continuous magnitude, the (supposed) "spaces" or "varieties" of impenitent algebraic geometricians, as well as innumerable other types of structures, which until then had seemed irrevocably bound to the "arithmetical world" of "discontinuous" or "discrete" aggregates.[51]

The topoi tool allowing us to apply the topological invariants to an algebraic variety on a finite field makes the geometry a bridge between the arithmetical and analytical aspects of the mathematics, that is to say between the discrete and the continuous.

> It is the theme of the topos... which is this "bed" or "deep river" where geometry, algebra, topology and arithmetic, mathematical logic and the theory of categories come together, the world of continuous and that of 'discontinuous' or discrete structures.[52,53]

[51]Grothendieck, 1985, *Reaping and Sowing*, my translation.

[52]Grothendieck, 1985 *ibid.*

[53]Olivia Caramello developed a deep and powerful work on the "topos-theoretic background, and on the concept of a bridge" see "the bridge-building technique" in Olivia Caramello, 'Topos-theoretic background" IHES, September, 2014.

12.6.1 *Brief Considerations on Topoi, Apropos of Reflexivity and Unity*

12.6.2 *From Grothendieck Topos to "Elementary" Topos*

We have recalled the more general notion of coverings in a category (Grothendieck topology), the resulting "sites" as well as the topos formed as the category of all sheaves of sets on such a site. Then, [Mac Lane and Morbijk] said, in 1963, Lawvere embarked on the daring project of a purely categorical foundation of all mathematics, beginning with an appropriate axiomatization of the category of sets, thus replacing set membership by the composition of functions. This replacement is an essential movement of reflexivity, a transition to dynamic operations that transform any static basic link in set theory. Lawvere soon observed that a Grothendieck topos admits basic operations of set theory as the formation of sets Y^X of functions (all functions from X to Y) and of power sets $P(X)$ (all subsets of X). Lawvere and Tierney discovered an effective axiomatization of categories of sheaves of sets (and in particular, of the category of sets) via an appropriate formulation of set-theoretic properties. They defined, in an elementary way, free of all set-theoretic assumptions, the notion of an "elementary topos". They yield a final axiomatization of "beautiful and amazing simplicity" [Mac Lane and Moerdijk, p. 3]. An elementary topos is a category with finite limits, function objects Y^X for any two object Y and X, and a power object $P(X)$ for each object X; they are required to satisfy some simple basic axioms, like first-order properties of ordinary function sets and power sets in naive set theory. A limit is defined by means of a diagram (consisting of objects c in a category $\mathcal{C}$ together with arrows $f_i c \to d_i$), called a *cone*, that makes arrows to commute. A limit is a cone $\{f_i : c \to d_i\}$ with the property that for any other cone $\{f_i' : c' \to d\}$ there exists exactly one arrow $f' : c' \to c$ that makes both cones commute when composed.

Every Grothendieck topos is an elementary topos but not conversely. Lawvere's basic idea was that a topos is a "universe of sets". Intuitionistic logic, and the mathematics based on it, originated with Brouwer's work on the foundations of mathematics at the beginning of the twentieth century. He insisted that all proofs be constructive. That means that he did not allow proof by contradiction and hence that he excluded the classical *tertium non datur*. Heyting and others introduced formal system of intuitionistic logic, weaker than classical logic.

To understand this point let us make the following remark. In a topological space the complement of an open set U is closed but not usually open, so among the open sets the "negation" of U should be the interior of its complement. This has the consequence that the double negation of U is not necessarily equal to U. Thus, as observed first by Stone and Tarski, the algebra of open sets is not Boolean, but instead follows the rules of the intuitionistic propositional calculus. Since these rules were first formulated by A. Heyting, such an algebra was called a Heyting algebra.

Subobjects (defined below) in a category of sheaves have a negation operator which belongs to a Heyting algebra. Moreover—we follow [Mac Lane and and Moerbijk]—there are quantifier operations on sheaves, which have exactly the properties of corresponding quantifiers in intuitionistic logic. This leads to the remarkable result, that the "intrinsic" logic of a topos is in general intuitionistic. There can be particular sheaf categories, where the intuitionistic logic becomes ordinary (classical) logic. An arbitrary topos can be viewed as an *intuitionitistic universe of sets*.[54]

12.6.3 Some Brief Remarks on Benabou-Mitchel and Kripke-Joyal Languages

Mathematical statements and theorems can be formulated with precision in the symbolism of the standard firs-order logic. As Mac Lane and Moerdijk remind.[55] There are at least four objectives for occasional such formulations (say, theorems of interest) as follows:

(1) They provide a precise way of stating theorems.
(2) They allow for a meticulous formulation of the rules of proof of that domain, by stating all "the rules of inference" which allow in succession the deduction of (true) theorems from the axioms in the domain.
(3) They may serve to describe an object of the domain—a set, an integer, a real number-as the set of all so and so's, thus in the language of natural numbers.
(4) They make possible a "semantics" which provides a description of when a formula is "true" (that is universally valid). Such a semantics (in terms of Mac Lane and Moerdijk) in terms of some domains of objects assumed to be at hand.

As in point (3), it is showed that formulas $\phi(x)$ in a variable x of the Mitchell-Benabou can be used to specify objects of $\mathcal{E}$ ($\mathcal{E}$ any topos) in expression of the form

$$\{x|\phi(x)\}$$

-in the fashion common in set theory. This shows how a topos behaves like a "universe of sets". One can for example, mimic the usual set-theoretic constructions of the integers, rationals, reals, and complex numbers and so construct in any topos with a natural numbers object, the object of integers, rationals, reals, ... Mac Lane and Moerdijk also show how the work of Beth and Kripke, in constructing a semantics for intuitionistic and modal logics, can also provide a semantics for the Mitchell Benabou language of a topos $\mathcal{E}$. In practice this means that one can perform many set-theoretic constructions in a topos and define objects of $\mathcal{E}$ as in (1); however- this

[54]This does not implies a revision of mathematics but the following position. There are structural principles of demonstration that most mathematicians use when demonstrating. These principles, if used alone, define a constructive or intuitionistic mathematics. It has structural rules, models, an essential notion of context, soundness, etc. It can be shown that the excluded middle can not be deduced from it etc. the job is to show if this logic is sufficient or to specify what of the classical logic should be available to do some demonstrations.

[55]Saunders Mac Lane and Ieke Moerdijk, *Sheaves in Geometry and Logic A first introduction to Topos Theory*, p. 296 sq.

is important—in establishing properties of these objects within the language of the topos *one should use only constructive and explicit arguments.*

12.6.4 An example, Preliminaries for Its Explanation

I cannot specify the language. I limit myself to giving essential features of this language. It can conveniently be used to describe various objects of $\mathcal{E}$.

I need firstly to define what a classifier consists of. In a category **C** with finite limits a subobject classifier is a monic (monomorphism) ($\equiv$ an injection), $1 \to \Omega$ such that every monic $S \to X$ in **C** there is a unique arrow ϕ which, with the given monic, forms a pullback square

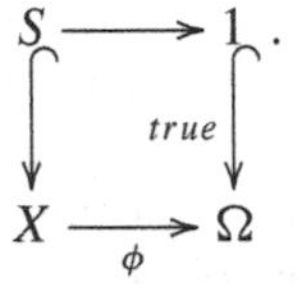

$$\begin{array}{ccc} S & \longrightarrow & 1 \\ \downarrow & & \downarrow \mathit{true} \\ X & \xrightarrow[\phi]{} & \Omega \end{array}.$$

This amounts to saying that the subobject functor is representable. In detail by simplifying in Moerdijk's fashion a subobject of an object X in any category **C** is an equivalence class of monics $m : S \to X$ to X. By a familiar abuse of language we say that the subobject *is* S or m meaning always the equivalence class of m.

A *pullback* or *fibered product* is the following. Given two functions $f : B \to A$ and $g : C \to A$ between sets, one may construct their fibred product as the set

$$B \times_A C = \{(b, c) \in B \times C | f(b) = g(c)\}.$$

Thus $B \times_A C$ is a subset of the product, and comes equipped with two *projections* $\pi_1 : B \times_A C \to B$ and $\pi_2 : B \times_A C \to C$ which fit into a commutative diagram

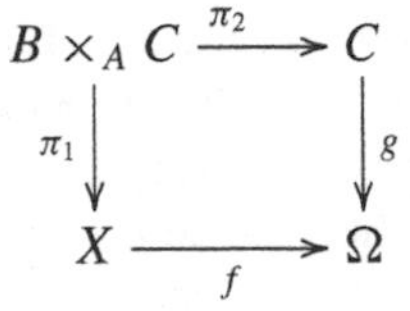

$$\begin{array}{ccc} B \times_A C & \xrightarrow{\pi_2} & C \\ \pi_1 \downarrow & & \downarrow g \\ X & \xrightarrow[f]{} & \Omega \end{array}$$

i.e. $f\pi_1 = g\pi_2$, plus a universal property I do not give here.

I present an important property before coming back to Benabou-Mitchel (BM) language. A category **C** with finite limits and small Hom-sets (small means to be a set) has a subobject classifier if and only if there is an object Ω and an isomorphism

$$\theta_X : Sub_{\mathbf{C}}(X) \cong Hom_{\mathbf{C}}(X, \Omega)$$

natural for $X \in \mathbf{C}$. It is not important to knowing the definition of natural.

12.6.5 The Example as Such

Now let us come back to BM language. It can be used to describe various objects of $\mathcal{E}$. For example the object of epimorpisms.

A morphism $f : C \to D$ is called an epimorphism if for any object E and any two parallel morphisms, g, h

$$D \rightrightarrows E$$

in $\mathbf{C}$ $gf = hf$ implies $g = h$. One writes $f : C \rightarrowtail D$. One defines, for any two objects X and Y in a topos $\mathcal{E}$, an object $Epi(X, Y) \subseteq Y^X$ called the "object of epimorphisms" from X to Y. This object has the property that $Epi(X, Y) \cong 0$ implies that there is no epimorphism: $X \to Y$.

The BM language can describe various objects. For example, "the object of epimorphisms"

$$Epi(X, Y) \rightarrowtail Y^X$$

constructed for giving objects X and Y of a topos $\mathcal{E}$, can be described by the expected formulas, involving variables x, y, f of types X, Y, Y^X

$$Epi(X, Y) = \{f \in Y^X | \forall y \in Y \; \exists x \in X \; f(x) = y\}.$$

More explicitly, we (Moerdijk and Mac Lane) state that the subobjects of Y^X defined above in the language of $\mathcal{E}$ coincides with the subobject $Epi(X, Y)$ defined in purely categorical terms.

12.6.6 Some General Remarks

Deriving new valid formulas from given ones can be carried out as for "ordinary", mathematical proofs using variables as if they were ordinary elements, *provided* that the derivation is explicitly constructive. For a general topos, one cannot use indirect proofs (*reductio ad absurdum*) since the law of excluded middle ($\phi \vee \neg\phi$) need *not* be valid, nor can one use the axiom of choice. More technically, this means that the derivation is to follow the rules of the intuitionistic predicate calculus. Kripke's semantics for intuitionistic logic can also be viewed as a description of truth for the language of a suitable topos.

As we saw the existence of a classifier of the functor of subobjects make possible many developments. It is essential to remember that each topos possesses its own logics. That means that the notion of a statement and tools, i.e., logical connectives, are present in any topos. Each topos contains arrows that represent mathematical statements and all logical statements operate on these arrows. The BM language or internal language is a high level language that make possible the manipulation of arrows. The semantics of this internal language is the Kripke-Joyal semantics. One needs to introduce news expressions as a kind of abbreviation for the terms we dispose until now by means of BM language. The introduction of the internal language is a way of giving meaning to the mathematical statements transposed into a topos. The topos becomes in this way a reconstructor of mathematical statements.

12.6.7 *Reflexivity*

One must more generally consider that the concept of topos, from this point of view, is an in-depth reflection on what a set is. It is, as we have said, a return to oneself of the concept of the whole by extracting from the dynamics that one finds in it a form of self-recovery. But this in-depth reflection on the theory of sets has the consequence of transferring this concept by dynamically recasting it. According to John L. Bell[56] gradually arose the view that the essence of mathematical structure is to be sought not in its internal structure as a set-theoretical entity, but rather in the form of its relationship with other structures through the network of morphisms. John Bell says that the uncritical employment of (axiomatic) set theory in their formulation of the concept of mathematical structure prevented Bourbaki from achieving the structuralist objective of treating structures as autonomous forms with no specified substance.

We will pass on this line of analysis from the concept of category to that of topos. Category theory transcends particular structure not by doing away with it, but by taking it as given and generalizing it. And category theory suggests that the interpretation of a mathematical concept may vary with the choice of "category of discourse".[57] And the category theoretic meaning of a mathematical concept is determined only in relation to a "category of discourse" which can *vary*. John Bell states that the effect of casting a mathematical concept in category-theoretic terms is to confer a *degree of ambiguity of reference* on the concept.

It becomes mandatory,[58] to seek a formulation for the set concept that takes into account its underdetermined character, that is, one that does not bind it so tightly to the absolute universe of sets with its rigid hierarchical structure. Category theory furnishes such a formulation through the concept of *topos*, and its formal counterpart *local set theory*. A local set theory is a generalization of the system of classical set theory, within which the construction of a corresponding category of sets can still be carried out, and shown to be a topos. Any topos can be obtained as the category of sets

[56]John L. Bell, *Toposes and Local Set Theories*, Clarendon Press, Oxford, 1988, p. 236 sq.

[57]John L Bell, *ibid.*, p. 23.

[58]John Bell, *ibid.*

within some local set theory. Topoi are in a natural sense the models or interpretations of local set theories.

12.6.8 Geometric Modalities

Goldblatt, argues (1977)[59] following Lawvere, in favor of a modal interpretation of Grothendieck topology. Modal logic is concerned with the study of one-place connective on sentences that has a variety of meanings, including "it is necessarily the true that", (alethic modality), "it is known that" (epistemic modality), "it is believed that" (doxastic modality), and "it ought to be the case that" (deontic). What we obtained with Grothendieck's topology is what we might call *geometric* modality. Semantically the modal connective corrresponds to an arrow. Lawvere suggests that when the arrow is a topology $j : \Omega \to \Omega$ on a topos, the modal connective has the "natural reading" "it is locally the case that".[60] It is remarkable that the topology becomes thus a way of understanding mathematical reasoning.[61]

12.6.9 Some Analogies with the Theory of Relativity

We need to introduce the notion of geometric morphism, that is $\mathbf{E} \to \mathbf{E}'$. We may think of this morphism as a "nexus between the mathematical worlds represented

---

[59]Robert Goldblatt, *Topoi, the categorial analysis of logic*, North-Holland publishing company, Amsterdam New York Oxford, 1977.

[60]Goldblatt, 1977, p. 382.

[61]These brief remarks are borrowed from the course by Alain Prouté in Paris **VII** *Introduction to categorical logic* and Robert Glodblatt (vs). Let **C** be a small category with a Grothendieck J topology seen as a sub-object $\Omega \in \mathbf{Ob}\,(\hat{\mathcal{O}})$. Ω is the "object of truth values", the characteristic arrow of J, $j : \Omega \to \Omega$ has the following properties

- $j \circ T = T$
- $j \circ j = j$
- $\wedge \circ (j \times j) = j \circ \wedge$

This Grothendieck topology corresponds to Lawvere Tierney topology on a presheaf category. Moreover a topos has an internal language that allows to manipulate the logical or mathematical objects already present in it. This language is given to us by the work by Bénabou Mitchel. And in this language we get a translation of Lawvere-Tiernay topology. The translation by the adverb *locally* corresponds intuitively to the way in which the Lawvere Tierney topologies were constructed as generalizations of Grothendieck topologies. Take the presheaf ζ that associates to the open set U all functions $f : U \to \mathbf{R}$ and the subpresheaf η that associates to the open set U to the functions $f : U \to \mathbf{R}$ which are the functions that are bounded on U. Let $\phi : \zeta \to \Omega$ the characteristic arrow of inclusion $\eta \subset \zeta$. Above the open $U\phi$ sends a function f defined on U on the set of open sets included in U, on which f is bounded, while $j \circ \phi$ sends f on all open which are covered by open on where f is bounded, in other words, on all open sets on which f is locally bounded. The effect of the modality j on the internal predicate ϕ is to attach the adverb*locally to it.*

by **E** and **E**′", or, John Bell adds, "as a method of shifting from **E** to **E**′ and vice versa". There is an analogy here with the physical geometric notion of *change of coordinate system*. In astronomy one effects a change of coordinate system to simplify the description of motions. It also proves possible to simplify the formulation of a mathematical concept by effecting a shift of mathematical framework. Like Bell, we might give as example the topos $\mathbf{Sh}(X)$ of sheaves on X. Here *everything* is varying continuously, so shifting from **Set** to $\mathbf{Sh}(X)$ essentially amounts to placing oneself in a framework which is, according to Bell, so to speak, itself co-moving with the variation over X of any given variable real number. This causes its variation not to be "noticed" in $\mathbf{Sh}(X)$.

Bell notes another analogy. In relativistic physics, invariant physical laws are statements of mathematical physics that, suitably formulated, hold universally, i. e., in every mathematical framework. Analogously, invariant mathematical laws are mathematical assertions that hold universally, i. e., in every mathematical framework. The invariant mathematical laws are those provable *constructively*. Notice in this connection that a theorem of classical logic that is not constructively provable will not hold universally until it has been transformed into its intuitionistic correlate. The procedure of translating classical into intuitionistic logic, Bell said, is thus the logical counterpart of casting a physical law in invariant form.

12.6.10 Brief Complement on Higher Order Logic

We will mention briefly a study that has been made of the relationship between higher-order logic and topoi.[62] Higher order logic[63] has formulae of the form $(\forall X)\phi X$ and $(\exists X)\phi\ X$ may stand for set, a relation, a set of sets, a set of relations, a set of sets of sets..., etc. So for a classical model $\mathfrak{A} = <A, \dots>$ the range of X may be any of $\mathcal{P}(A), \mathcal{P}(A^n), \mathcal{P}(\mathcal{P}(A^n))$. And as Goldblatt mentions, analogues of these exist in any topos, in the form of Ω^a, Ω'^a etc. and so higher logic is interpretable in $\mathcal{E}$ (a topos). In fact the whole topos becomes a model for a manysorted language, having one sort of individual variables for each $\mathcal{E}$-object. Goldblatt[64] mentions some ancient results by Michael Fourman[65] or by Boileau.[66] This provide a full explication of Lawvere 's statement that "the notion of topos summarizes in objective categorical form the essence of 'higher order logic'".[67]

[62] William Goldblatt, *Topoi The categorical analysis of Logic* North–Holland, Amsterdam-New York-Oxford, 1979, p. 286 sq.

[63] We refer not only to second order logics but also to other logics.

[64] Robert Goldblatt, *ibid.* p. 287.

[65] Michael P. Fourman, *Connections between category theory and logic* D. Phil. Thesis Oxford University, 1974.

[66] André Boileau, *Types versus Topos, Thèse de Philosophie Doctor* Université de Montréal, 1975.

[67] William Lawvere, Introduction and ed. for *Toposes, Algebraic Geometry and Logic*, Lectures Notes in Mathematics, Vol. 274, Springer Verlag, 1972.

12.7 Conclusion

Topos theory involves both geometry, especially sheaf theory, and logic, especially set theory. Nevertheless J. Bell in the book we used for the above remarks, provides a systematic presentation of topos theory from the point of view of formal logic. Does this mean that logic is the discipline that can produce a unification of mathematics? We have seen that logic has in fact transformed itself to become categorical logic. As such, it produces forms of unity and non-unity. The path of this unity is linked to modes of reflection on self reference of mathematics, of which we have shown only certain forms. What is striking is that entire disciplines can be reflected in each other.

We have tried to show that the theory of topos can fulfill this dual unifying and reflective function. We must briefly respond to an objection that might affect our attempt.

When we produce any form of unification, each of the unified disciplines loses much of its substance. And if they are reflected in each other it is in a form that is often very reduced. Hence the claim of "true mathematics" against these theories considered speculative and formal. A factual answer is to say that true mathematics uses these theories more and more precisely because they bring forms of unity and reflection.

It should be added that this unity is not only the result of a formal extraction that "crushes" the information. On the contrary, it is a conceptual element which structures and fills a synthesis. From a Platonic unity, explicitly philosophical, to a topos-unity first in Grothendieck's work, then in conceptual reflection on it, like that of Lawvere, real mathematics is instead installed on a more synthetic terrain on which they are energized.

The reflexive syntheses proposed by Lautman presented difficulties. He had tried to fill them with an appeal to Heidegger's philosophy. But this induced other difficulties, notably that of the distinction between dialectics and mathematics. Lautman uses dialectical terms in the Platonic sense and also in the sense of a kind of contradiction theory.

> If dialectics tries to find its own solutions to the problems it expresses, it will "mimic" mathematics with such a collection of subtle distinctions and logical tricks that it will be mistaken for mathematics itself.[68]

This is the fate of the logicism of Frege and Russell. Nevertheless the line between dialectics and mathematics is neither clear nor stable. And, more difficult for the arguments, mathematics itself can provide dialectical answer to a mathematical question.[69] But this criticism of logicism on behalf of mathematics as such uses the dialectical position, for example, definitions by "abstraction" of equivalence, measurement, operators, etc., characterize not a "genre" in extension but possibilities of structuring, integrations, operations, closure, conceived in a dynamic and organizing

---

[68]Lautman (2006, p. 228).

[69]Brendan Larvor, p. 201.

way. The distinction within the same structure between the intrinsic properties of a being and its possibilities of action... seems to be similar to the Platonic distinction between the Same and the Other".[70] Using the concept of a topos as a vector of unity and reflexivity, we have made a choice which first has, for him, to have representatives among the important mathematicians of our century. But it appeared to us to express a movement which characterizes mathematics as indefinite movement of the search for this reflexive unity. But this movement of reflexion possesses a mathematical form indissociable from its philosophical questioning. No doubt in a more intrinsic and immanent way than had been proposed by Lautman, who might have opted for this position if he had lived longer. It seems more relevant to search for a mathematical unity of mathematics by means of the concept of a topos, taking into account that this concept possesses philosophical significance. It nevertheless brings us closer to Cavaillès than to Lautman. Lautman in his letter to the mathematician Maurice Fréchet says: "Cavaillès seems to me in what he calls mathematical experience to assign a considerable role to the activity of the mind. There would therefore be no general characters that constitute mathematical reality.... I think in experience there is more than experience."[71] And then he quotes Cavaillès, and we are closer to his position: "Personally I'm reluctant to ask anything else that would dominate the mathematician's actual thinking, I see the requirement in the problems. . . and if dialectics is only that, we only arrive at very general proposition".[72] This has been our position on the issue of the unity reflexivity of mathematics. I can only add that *within* the mathematical experience, philosophical questions arise.

References

Belanger, B. (2010). La vision unificatrice de Grothendieck : au-delà de l'unité (méthodologique?) de Lautman. *Philosophiques, 37*(1).

Bell, J. L. (1988). *Toposes and local sets theories*. Oxford: Oxford Science Publications.

Bernays, P. (1934). *Hilberts Untersuchungen über die Grundlagen der Arithmetik*. Berlin: Springer.

Boileau, A. (1975). *Types versus topos* (Thèse de Philosophie Doctor Université de Montréal).

Caramello, O. (2014). *Topos-theoretic background*. France: IHES.

Connes, A. (2008). *A view of mathematics Mathematics: Concepts and Foundations* (Vol. 1). www.colss.net/Eolss.

Fourman, M. P. (1974). *Connections between category theory and logic*. D. Phil: Thesis, Oxford University.

Goldblatt, W. (1979). *Topoi the categorical analysis of logic*. Amsterdam, New York, Oxford: North-Holland.

Granger, G. G. (1994). *Formes opérations, objets*. Paris: Vrin.

Grothendieck, A. *et al.Théorie des topos et cohomologie étale des schémas*. Lectures Notes, Vols. (269, 270, 305).

Grothendieck, A. (1985). *Reaping and sowing 1985*. Récoltes et Semailles.

[70] Albert Lautman, *Les mathématiques les idées et le réel physique*, p. 79 my translation.

[71] A. Lautman, *ibid.* p. 263, my translation.

[72] A. Lautman, *ibid.* p. 263, my translation.

Grothendieck, A. (1960). *EGA I Le langage des schémas*, (Vol. 4). France: Publications Mathamatiques de IHES.
Hartshorne, R. (1977). *Algebraic geometry* (p. 1977). New York: Springer.
Hilbert, D. (1932). *Gesammelte Abhandlungen*. Berlin: Springer.
Kant, I. (1929). *Kritik der reinen Vernunft* In J. F. Hartnoch (transl.) N. Kemp Smith as Critique of Pure Reason.
Larvor, B. (2010). Albert Lautman: Dialectics in mathematics. In Foundations of the Formal Science.
Lautman, A. (2006). Les mathématiques, les idées et le réel physique. In J. Lautman, J. Dieudonné, & F. Zalamea (Eds.), *Introduction and biography; introductory essay; Preface to the* (1977th ed.). Paris: Vrin.
Lawvere, W. (1966). The category of categories as a foundation of mathematics. In *Proceedings of the Conference on Categorical Algebra La Jollia Californa, 1965* (pp. 1–20). New York: Springer.
Mac Lane, S., & Moerbijk, I. J. (1992). *Shaeves and geometry*. New York: Springer.
Panza, M. (2005). *Newton et les origines de l'analyse: 1664–1666* Paris: Blanchard.
Proutè, A. (2010). *Introduction à la Logique Catégorique*. Paris-Diderot: Cours Université.
Szczeciniarz, J.-J. (2013). *COED*. Fabio Maia Bertato, Jose Carlos Cifuentes: In the steps of Galois. Hermann Cle Paris Campinas.
Weyl, H. (1912). *The concept of a Riemann Surface Addison and Wesley 1964 First edition die Idee der Riemannschen Fläche*. Berlin: Teubner.
Weyl, H. (1928). *Gruppentheorie and Quantenmechnik*. Leipzig: Hirzel.

Chapter 13
The Foundations of Arithmetic in Ibn Sīnā

Hassan Tahiri

مَنْ عَدِمَ هذه العلوم الاربعة المخصوصة باسم الرياضيات عدم علم الفلسفة... فقد ينبغي
لمن أراد علم الفلسفة أن يُقَدِّم استعمال كتب الرياضيات على مراتبها التي حددتها.
الكندي

There is no knowledge of philosophy without knowledge of the mathematical sciences… hence he who wants to acquire knowledge of philosophy should first begin studying the books of mathematics according to the order I have established.

Al-Kindī in Rasā'il al-Kindī al-falsafiyya, p. 378.

Abstract Ibn Sīnā (980–1037), one of the most influential philosopher-scientists who was known to the West by the Latinised name Avicenna, has introduced a major shift in the philosophy of mathematics. His conception of number is structured into 5 main conceptual developments: (1) the recognition of mathematical objects as intentional entities and the acknowledgment that this amounts to provide an intentional notion of existence; (2) the link between the intentional act of apprehending unity and the generation of numbers by means of a specific act of repetition made possible by memory; (3) the identification of a specific intentional act that explains how the repetition operator can be performed by an epistemic agent; (4) the development of a notion of aggregate (or constructive set) that assumes an inductive operation for the generation of its elements and an underlying notion of equivalence relation; (5) the claim that plurality and unity should be understood interdependently (we grasp plurality by grasping it as instantiating an invariant).

This paper was part of a wider investigation that was recently published under the title *Mathematics and the Mind. An introduction into Ibn Sīnā's Theory of Knowledge*, Springer Brief Series, 2016.

H. Tahiri (✉)
CFCUL—Centre for Philosophy of Science, University of Lisbon, Lisbon, Portugal
e-mail: hassan.tahiri@yahoo.fr

H. Tahiri (ed.), *The Philosophers and Mathematics*, Logic, Epistemology, and the Unity of Science 43, https://doi.org/10.1007/978-3-319-93733-5_13

The second half of the 19th century was dominated by problems akin to the foundations of mathematics and in particular the concept of number which captured the imagination of modern mathematicians and philosophers, are they new? By no means. What is little known is that those questions which have yielded so many lively and heated discussions in the development of contemporary logic and mathematics have been equally subject of research by the ancients who acknowledged it both as an important and difficult. Unfortunately, in most of the contemporary analytic philosophy of mathematics the historical roots of this approach have been mostly overlooked. Mathematicians, logicians, philosophers and historians of mathematics would be surprised to find out that what seems to be a central topic of the 19th century philosophy of science was a topic that reinvented epistemology ten centuries earlier. The purpose of this talk is thus to narrow the major historical gap in our understanding of the evolution of some basic mathematical concepts by examining the concept of number in the classical Arabic-Islamic period.[1] We shall particularly see why and how Ibn Sīnā has introduced a major shift in the philosophy of mathematics which is constituted by 5 main conceptual developments:

1. The recognition of mathematical entities as intentional objects and the acknowledgment that this amounts to provide an intentional notion of existence.
2. The link between the intentional act of apprehending unity and the generation of numbers by means of specific act of repetition made possible by memory.
3. The identification of a specific intentional act that explains how the repetition operator can be performed by an epistemic agent.
4. The development of a notion of aggregate (or constructive set) that assumes an inductive operation for the generation of its elements and an underlying notion of equivalence.
5. The claim that plurality and unity should be understood interdependently (we grasp plurality by grasping it as instantiating an invariant).

Let's finally emphasise that Ibn Sīnā's philosophical enterprise is exclusively motivated by questions of understanding i.e. questions like what are numbers, what is their meaning and where are they? The answers to which should be both rooted in mathematical practice, motivate the better understanding of this practice and contribute to its further development. From this point of view Ibn Sīnā's inquiry is also intended to be a contribution to explain the actual genesis and evolution of the conception of number by the human mind.

[1]The Arabic-Islamic tradition is heavily under-researched because of the structure of modern scientific research that reflects a peculiar perception of the development of the history of science. Little efforts have been made to update the prevailing historical knowledge despite, as Rashed points out, "the accomplishments of the last five decades that outweigh everything we owe to the last two centuries." (Rashed p. 3, 2015) As a a result, many modern scholars continue to skip over the Arabic-Islamic period as if it has never existed! For more on this topic, see Tahiri (2018).

13.1 Numbers as Intentional Objects and Intentionality as Mental Existence

In his monumental *al-Shifā'* (or *The Healing*), Ibn Sīnā has conducted an unprecedented investigation into the concept of number whose status was left in disarray since the discovery of the irrationals and motivated our author to explore new paths despite the difficulty of the task as he seldom admits:

> How difficult it is for us to say something that is reliable on this topic.[2] (p. 80, §3)
>
> فما أعسر علينا أن نقول في هذا الباب شيئا يعتد به.

One decisive point in Ibn Sīnā's strategy is to acknowledge that numbers are objects, if they are objects they possess (some kind of) reality since to say of something that is an object amounts to say:

> If you were to say: 'The reality of such a thing is the reality of such a thing,' or 'The reality of such a thing is a reality' this would be superfluous, useless statement. And if you were to say 'The reality of such a thing is a thing,' this, too, would not be a statement imparting knowledge of what is not known. Even less useful than this is for you to say 'Reality is a thing' unless by thing you mean "the existent"; for then it is as though you have said "The reality of such a thing is an existing reality. (p. 24, §11)
>
> The meaning of existence is permanently concomitant with it because the thing exists either in the concrete or in the estimative faculty and the intellect, if this were not the case, it would not be a thing.[3] (ibid.)

Now the question is what kind of reality or existence is involved in the notion of number?

> Number has an existence in things and an existence in the soul. The statement of one who says that number exists only in the soul is not reliable. But should he say that number stripped from what is counted (مجردا من المعدودات) that are in concrete existence has no existence except in the soul, this would be true.[4] (p. 91, §2)

Numbers are not only an abstraction from some properties; but they have a separate existence in the mind. Thus the necessary condition for something, which does not exist in the external world, to be an object is to exist in the mind. Ibn Sīnā draws our attention to the specificity of our mind: its content is an object whose presence is a specific form of existence. It is in the nature of our mind to have some content as object for to think is to think about something, what Ibn Sīnā calls then existence in

[2] Unless indicated otherwise, all Ibn Sīnā's quotations are taken from his *al-Ilāhiyāt*, the number of pages and paragraphs refers to the bilingual edition of the book which was translated by Marmura under the title "The Metaphysics of *The Healing*".

[3] لو قلت: إن حقيقة كذا حقيقة كذا، أو أن حقيقة كذا حقيقة، لكان حشوا من الكلام غير مفيد. ولو قلت: إن حقيقة كذا شيء، لكان أيضا قولا غير مفيد ما يجهل. وأقل إفادة منه أن تقول: إن الحقيقة شيء إلا أن يعني بالشيء الموجود، كأنك قلت: إن حقيقة كذا حقيقة موجودة.

معنى الموجود يلزمه [الشيء] دائما لأنه يكون إما موجودا في الأعيان أو موجودا في الوهم والعقل، فإن لم يكن كذا لم يكن شيئا.

[4] العدد له وجود في الأشياء ووجود في النفس. وليس قول من قال: إن العدد لا وجود له إلا في النفس بشيء يعتد به، أما إن قال: إن العدد لا وجود له مجردا عن المعدودات التي في الأعيان إلا في النفس، فهو حق.

the mind or mental existence (الوجود الذهني) is the simple presence of an object in the mind, "this is the meaning intended by the thing" he concludes. (p. 25, §11). This is a major ontological extension that provides a general solution to the mystery of how we can meaningfully talk about a very large class of strange objects that have no real existence like fictions. Ibn Sīnā seems to have discovered that the real relation between the mind and its content is simply intentional: the objects of our intentional acts need not be physical, spatiotemporal, or ideal entities, and they need not exist independently of our intentional acts. An idea that he clarifies further by rejecting the claim defended by some faction of *al-Mutakallimūn*'s philosophical school that non-existent is a thing:

> If by the nonexistent is meant the nonexistent in external reality, this would be possible; for it is possible for a thing that does not exist in external things to exist in the mind. But if something other than this is meant, this would be false and there would be no information about it at all. It would not be known except only as something conceived in the soul. To the notion that the nonexistent would be conceived in the soul as a concept that refers to something external [to the mind], we say certainly not.[5] (p. 25, §12)

This significant passage clarifies what seems to be the mysterious relation established earlier between "existence of number in things" and "existence of number in the soul." For Ibn Sīnā, numbers should be conceived as intentional objects for they refer to some extramental entities. Numbers like 10 e.g. exist in our mind as an intentional object as a result of being stripped from what is counted but also exists in the outside world for they refer to particular objects for we can at any time present things that serve as referents, like fingers for example, whose total is ten. Ibn Sīnā's intentionality seems to play then a surprisingly double function: the first is mental existence by stripping numbers from what is counted, and the second is what seems to be the reverse function by referring (back) to particular objects in the outside world. However, as already mentioned, the back reference is not always a simple inverse one to one relation, particularly not in relation to generation of infinite numbers that though it might be launched by a sensible experience it will trigger a repetition process beyond that sensible experience. Moreover, this double mental operation, which explains the powerful creativity of the human mind, raises nevertheless more questions: for how can numbers be stripped from what is counted to become intentional objects? How many numbers can be formed in this way i.e. by abstraction? And once abstracted, how can that what is counted and separated form nevertheless a single entity in the mind? And how can such an abstracted entity refer back to particular objects of the outside world? To tackle these complex issues, Ibn Sīnā reinvents epistemology in which intentionality acts as an interaction between the mind and the world and then distinguishes an act of intentionality specific to the generation of numbers. It is out the scope of this talk to present ibn sina's epistemology. For our purpose, I will present just one significant passage in which he explains the basic idea of his conception of numbers and of his overall theory of knowledge

---

5 إن عني بالمعدوم المعدوم في الأعيان، جاز أن يكون كذلك، فيجوز أن يكون الشيء ثابتا في الذهن معدوما في الأشياء الخارجية. وإن عني غير ذلك كان باطلا، ولم يكن عنه خبر ألبتة ولا كان معلوما إلا على أنه متصور في النفس فقط. فأما أن يكون متصورا في النفس صورة تشير إلى شيء خارج فكلا.

i.e. how the intellect or the mind comes to essences such as a humanity from the perception of individuals like *Zayd* and *Omar*, the key word is invariance:

> Conceptualizing the intelligible is acquired only through the intermediary of sensory perception in one way, namely that sensory perception takes the perceptible forms and presents them to the imaginative faculty, and so those forms become subjects of our speculative intellect's activity, and thus there are numerous forms there taken from the perceptible humans. The intellect, then, finds them *varying* in accidents such as it finds *Zayd* particularized by a certain colour, external appearance, ordering of the limbs and the like, while it finds *Omar* particularized by other [accidents] different from those. Thus the intellect receives these accidents, but then it extracts them, *as if* it is peeling away these accidents and setting them to one side, until it arrives at the account in which humans are common and in which there is *no variation* and so acquires knowledge of them and conceptualizes them.[6] (Ibn Sīnā 1956, p. 222; my emphasis).

The intellect can be seen thus as a second order mental act by which understanding is reached when the mind is capable of referring directly or absolutely to the objects and this absoluteness is achieved by identifying an invariant through its accidental variations. Better, as we will discuss in the section *The Epistemic Construction of Numbers*, the point is to grasp an individual as instantiating a given invariant. Hence, the concept of number is therefore based on the identification i.e. the apprehension of an invariant between intentional objects, and, as we will see later on is the result of a construction by iteration from unity. Hence Ibn Sīnā's task in *al-Ilāhiyāt* is to describe the process involved in the generation of numbers from unity. But how can this process be described from experience? How can an invariant whose apprehension is based on experience be conceptually expressed? And how from a particular essence like humanity can the mind conceive the general concept of number. The answers to these questions will be the subject of next sections but let us advance that the process of abstraction should not be thought in the traditional way, but rather as a process by means of which a given individual is seen as instantiating a given concept or invariant that defines the corresponding unity. The point is that plurality and unity must be defined one in terms of the other.

13.2 The Iterative Nature of Number

In Ibn Sīnā's view, essences such as humanity or white cannot be formed by abstraction from perceiving concrete things; since as he argues above, when a specific human being or a white thing is present in the mind or more precisely in imagination what is perceived is the individual being *Zayd* or that white thing. Moreover, how do we

[6] ونقول إنه إنما تكتسب تصور المعقولات بتوسط الحس على وجه واحد، وهو أن الحس يأخذ صور المحسوسات ويسلمها الى القوة الخيالية فتصير تلك الصور موضوعات لفعل العقل النظري الذي لنا، فتكون هناك صور كثيرة مأخوذة من الناس المحسوسين، فيجدها العقل متخالفة بعوارض مثل ما نجد زيدا مختصا بلون وسحنة وهيئة أعضاء، وتجد عمرا مختصا بأخرى غير تلك .فيقبل على هذه العوارض فينزعها فيكون كأنه يقشر هذه العوارض منها ويطرحها من جانب حتى يتوصل الى المعنى الذي يشترك فيه ولا يختلف به، فيحصلها ويتصورها .وأول ما يفتش عن الخلط الذي في الخيال فإنه يجد عوارض وذاتيات، ومن العوارض لازمة وغير لازمة، فيفرد معنى من الكثرة المجتمعة في الخيال ويأخذها الى ذاته.

come to the general notion of number? For this kind of abstraction or rather extraction is not done in the first experience, at least another experience is needed; it is not question of time though temporality is involved. The point is that the concept of number is linked to the operation of building an aggregate:

> Plurality is the aggregate (*mujtama*') of units, we have included unity in the definition of plurality, and we have done something else, we have included the aggregate in its definition. (p. 79, §2)
>
> الكثرة هي المجتمع من وحدات، فقد أخذنا الوحدة في حد الكثرة ثم عملنا شيئا آخر وهو أنا أخذنا المجتمع في حدها.

It looks as if Ibn Sīnā considers *mujtama*' or aggregate to be different from *kathra* i.e. plurality or multiplicity. While plurality, as he explicitly points out, merely means more than one (p. 95, §14), the concept of aggregate or the now commonly used word set suggests instead not only the formation of a single and hence a specific entity from unities, but *points to a process involved for its formation*. The translator is more accurate by translating *mujtama*' as aggregate than set for the Arabic word involves the act of collecting or putting together unities into a single amount or total as Ibn Sīnā explains in his description of the meaning of number ten.

> It must not be said that ten is nothing but nine and one, or five and five, or that is one and one and so on until it terminates with ten. For your statement, "ten is nine and one", is a statement in which you predicated nine of the ten, conjoining the one with it. It would be as though you had said, "ten is black and sweet", where, by the descriptions, the one conjoined to the other must be true of the statement; thus the ten would be nine and also one. If, by the conjunction, you did not intend a definition but instead intend something parallel to the statement, "man is an animal and is rational"-that is, "man is an animal: that animal which is rational"—it would be though as you had said, "ten, is nine, that nine which is one", and this is also impossible. (p. 92, §5).

In this passage Ibn Sīnā points out that it is precisely the conception of numbers as mathematical objects that explains the inadequacy of the grammatical conjunction to render the meaning of elementary arithmetic equations. The point of our author is that the grammatical conjunction, whose function is rather to list, does not explain the composition of a number. In fact, Ibn Sīnā lucidly understands an operation like sum as a transformation of one side of the equation into the other[7]:

[7]It is this grasping of elementary arithmetic processes like sum as operators which will be crucial to his account for repetition by remarkably making it the most basic mathematical operator. It is important to point out that Ibn Sīnā does not conduct this linguistic analysis by chance for he explicitly mentions that the analysis of language is among the tasks of the logician: "because there is some link between word and meaning, the modes of the word can influence the modes of meaning. That's why the logician should also consider the absolute semantic aspect of the word inasmuch as it is not restricted by the language of a certain people to the exclusion of other, except in rare cases" (Ibn Sīnā 1983, p. 131). In *Manṭiq al-Mashriqiyyīn*, he becomes closer to the pragmatist position of the linguists for he begins with this topic in which logical analysis is tightly linked to the discussion of the meaning of sentences. A prominent example is his analysis of Arabic sentences, he explains their structure by admitting the existence of a second type of propositions which are made of just two components and he points out that the copula is not needed to link the subject to the predicate: "إذا كانت القضية غير ثلاثية، إنما هي ثنائية فقط لم تذكر فيها الرابطة استغناء، لأن محمولها كلمة أو اسم مشتق اشتقاقا يتضمن النسبة المذكورة على حسب اللغة." (Ibn Sīnā 1910, p. 67)

> Rather, ten is the sum of nine and one when taken together, they are transformed (صار) to something other than either. (p. 92, §5)
>
> بل العشرة مجموع التسعة والواحد إذا أخذا جميعا فصار منهما شيء غيرهما.

It is crucial to remind that, according to Ibn Sīnā, number is not a *property* of an aggregate but emerges from a process that builds up that aggregate, this makes his approach similar to the view that Frege will express later.[8] Let us now follow Ibn Sīnā's own development. The first stage of the development points out that the elements of an aggregate are separate units, the combination of which yields a number.

> The one by accident consists in saying that something united with another thing is that other and that both are one (p. 74, §2).
>
> والواحد بالعرض هو أن يقال في شيء يقارن شيئا آخر أنه هو الآخر وأنهما واحد.

The same idea is forcefully repeated in book 1 chapter 5:

> If you said 'The reality of A is something and the reality of B is another thing,' this would be sound, imparting knowledge; because in saying this you make the reservation within yourself that the former is something specific, differing from that other latter thing. This would be as if you said, "This is the reality of A, and the reality of B is another reality. p. 25, §11).
>
> إذا قلت: حقيقة آ شيء ما وحقيقة ب شيء آخر، إنما صح هذا وأفاد لأنك تضمر في نفسك أنه شيء آخر مخصوص مخالف لذلك الشيء الآخر، كما لو قلت: إن آ حقيقة وحقيقة ب حقيقة أخرى.
>
> It is like the very unity which is the principle of number – I mean, that which is such that if something else is added to it, the combination of the two becomes number (p. 76, §11).
>
> كنفس الوحدة التي هي مبدأ العدد، أعني التي إذا أضيف إليها غيرها صار مجموعها عددا.

In each of the three examples, he describes what seems to be needed for the formation of number 2 by appealing to a thing other than the previous thing. What is needed now is a specific act of repetition:

> The only meaning of repetition one understands in this is the bringing to be of something numerically other than the first. (p. 253, §14).
>
> ليس يفهم للتكرير فيه معنى إلا إيجاد شيء آخر غير الأول بالعدد.

Ibn Sīnā is thus able to reduce the basic conceptual apparatus needed to define numbers to just two elements: unity and repetition or a first element and a process.

[8] Here is one of the passages in which Frege argues against the view that number is a property of things; "It marks, therefore, an important difference between colour and Number, that a colour such as blue belongs to a surface independently of any choice of ours. [...] The Number 1, on the other hand, or 100 or any other Number, cannot be said to belong to the pile of playing cards in its own right, but at most to belong to it in view of the way in which we have chosen to regard it; and even then not in such a way that we can simply assign the Number to it as a predicate." Frege (1884, p. 29, §22).

13.3 The Epistemic Construction of Numbers

13.3.1 The Formation of Number Two: Combinative Unity and Memory

It is important to note that, in the last quoted passage, Ibn Sīnā chooses to further emphasise his conception of repetition in a polemical context in which he dismisses out of hand "those who generate number through repetition with unity remaining constant for the one," for he argues "if repetition enacts a number and each of the first and the second does not have unity, then unity is not the principle for the composition of number" (p. 253, § 14).[9] His objection is that if by repetition is meant that the same unity is given to each object to be counted, then the result is unity and not number because unity does not play here the role of the principle of the composition of number. He expresses this idea by saying in the above passage that the second does not, that is, it cannot be said to have a real unity. In other words, this conception of repetition confuses the essential (or metaphysical) unity which could be seen as a property of every single object and the numerical unity as the principle of the composition of number.[10] Repetition here fails to function as an operator since its object is not the previous result or in Ibn Sīnā's words unity is not applied as a result of linking the current to the previous experience. And he immediately repeats again what is needed to make a repetition an operator: "if one inasmuch as it is one is unity and the second inasmuch as it is second is unity, then there are two unities" (p. 253, § 14). The number 2 cannot be generated by assigning the same unity to the second object to be counted but by applying another unity, i.e. a unity other than the first, and as he concludes before this is "the only meaning of repetition one understands". Ibn Sīnā seems to present his conception of repetition as corresponding to the underlying counting process performed by an epistemic agent in a phenomenological setting in which repetition is described as the result of a specific intentional act. The motivation of his phenomenological analysis seems to be to elucidate how the specific repetition operation can be actually performed by a human being (or perhaps more generally by an epistemic agent)—different to say from Peano-Dedekind approach to the notion of successor or Frege's concept of hereditary chain. This step might be motivated by Ibn Sīnā's interest in the actual practice of mathematics, thus a proper epistemology characterising the specific act by the means of which numbers are generated is due. The point of the discussion seems to explain his understanding of repetition which consists in applying another unity to the previous one, i.e. why the formation of a number, like 2 for example, requires a unity other than the first. His explanation could also be regarded as an attempt to clarify for those who confuse the numerical

---

[9]This is strikingly similar to Frege's objection that will be discussed later.

[10]In his *Commentary on* al-Ilahiyāt, al-Shirāzī better known as Mulla Ṣadrā (1572–1640) further illuminates what Ibn Sīnā has in mind by this distinction: "the unity which is a principle of the mathematical numbers is other than the unity which can be found in the separate substances for the separate substances do not possess a quantitative number generated by repetition of similar units." (in Avicenne 1978, p. 302).

unity, which is the principle of the composition of number, with the essential unity by showing how the meaning of the first emerges from the latter.

The departure point of the phenomenological understanding of the repetition act is that the new and the former experience are linked with the help of memory or more broadly, imagination. Imagination re-presents the first experience stored in memory, i.e. the first intentional object which is no longer here, the imagination acts thus as a reminder (*munabbih*): the present white thing acts as a *munabbih* of the previous white thing. That is why the intellect apprehends forms through memory. If, due to some fatal mental disease or major mental disorder, for example, imagination fails to re-present the previous experience required for the apprehension of the invariant by the intellect, then there is no *munabbih* i.e. no reminder and consequently no number for the second experience will not be experienced as a second one but as a new one. A series of different experiences is useless if each one cannot be linked to the previous one. This objectively corresponds to the fact that the previous units should not miraculously disappear when putting them together or counting. Likewise, no new number can be generated if an intentional object is or perceived as exactly the same as the former (two simultaneously given identical objects are nevertheless differentiated by their localisation). This is what prevents Ibn Sīnā from defining from the outset number in terms of repetition for the kind of repetition required is not the commonly understood i.e. the same thing or the same unity but the *different same* that can only brought about by another experience. That is why the generation of a number, such as 2 for example, out of a previous one requires the emergence of a new experience of an act that should be linked to the previous one in such a way that the new experience is apprehended and re-cognised as composed by the repetition of the first—again in this example it seems to be very natural to understand this as describing the phenomenological act that corresponds to the *successor* of 1.

> For unity is not repeated unless it would be there in one succession after another. And this succession would have to be either temporal or essential [i.e. *de re*] (ذاتية).. If it is temporal and unity is not annihilated in the intermediary stage, then it would be as it was before, not something that has been repeated. If it ceases to exist and is then brought into being, then what has been brought into being is another individuality. If the succession is essential, then this is more evident.[11] (p. 253, §14)

By grasping repetition as an operator in which time plays little role, Ibn Sīnā explains why it is such a basic and powerful pure mathematical operation in the construction of mathematical concepts. The author of *al-Shifā'* would no doubt agree that repetition is performed in time, but time appears to be external or not *sufficient* to counting as such, for it is not essential to the bringing about of the actual occurrence of another thing let alone linking it to the previous one as a second instance. For if it was the case, time would appear as the cause of the occurrence of the same thing and thus

[11] الوحدة لا تكرر إلا بأن تكون هناك مرة بعد مرة، وهذه المرة إما زمانية أو ذاتية. فإن كانت زمانية ولم تعدم في الوسط فهي كما كانت لا أنها كررت. وإن عدمت ثم أوجدت فالموجدة شخصية أخرى. وإن كانت ذاتية فذلك أبين.

of its mental existence.[12] This is similar to repetition in experimental sciences where repeating the same experience is also performed in time, but the reason for conducting a second experiment is to establish the regularity of the phenomena by confirming the first result. This explains why Ibn Sīnā insists so much on the actual occurrence of a second experience (yet to be re-cognized as such). For the formation of essences requires change in the mind, and change in the mind cannot happen miraculously by itself by some kind of magic but should be brought about by change in intentional objects and we shall see shortly why change is critical in the formation of numbers, otherwise his account would be deemed merely psychological. This is what prevents Ibn Sīnā from defining from the outset number in terms of repetition for the kind of repetition required is not the commonly understood i.e. the same thing or the same unity but the different same that can only brought about by another experience. The formation of numbers such as 2 for example requires therefore another experience that should be linked to the previous one. Ibn Sīnā calls *combinative unity* (الوحدة الإجتماعية) the mental act that combines what is otherwise separated by nature.

> Combinative unity has plurality in act. And as a result there is plurality over which unity is superimposed, but which does not remove plurality from it. (p. 75, §7)
>
> والوحدة الإجتماعية فيها كثرة بالفعل. فهناك كثرة غشيتها وحدة لا تزيل عنها الكثرية.

One of the major objections to the formation of number two raised by the founder of modern logic will be briefly discussed. Ibn Sīnā tackles the issue of how pluralities can be conceived as a unique object: unities are linked into a single mental act—namely, the act of combinative unity—that provides its existence. The achieved unity is thus an intentional object ontologically dependent on the epistemic subject without whom the unity would collapse. Still how is it that linking a series of experiences can form a specific single entity? How can for example the concept of *two humans* be generated by just linking an experience of *Zayd* to that of '*Amr*? The problem is that if we just apprehend the essence of both individuals we will only come to the result that both individuals fall under the concept human and we will not be able to distinguish them as differentiate units. This is precisely Frege's sharp objection against such kind of phenomenological approaches to the foundations of arithmetic.

> For suppose that we do, as Thomae demands, "abstract from the peculiarities of the individual members of a set of things", or "disregard, in considering separate things, those characteristics which serve to distinguish them". In that event we are not left, as Lipschitz maintains, with "the concept of the number of the things considered"; what we get is rather a general concept under which the things considered fall. The things themselves do not in the process lose any of their special characteristics. For example, if I, in considering a white cat and a black cat, disregard the properties which serve to distinguish them, then I get presumably the concept 'cat' [not two cats]. (Frege 1884, §34)

Let us see how our author meets Frege's challenge. Ibn Sīnā first step is to acknowledge that plurality and unity must be understood simultaneously since they are correlative terms though their conception belongs to two different mental acts:

[12] But time seems to be essential not to counting as such but to the actual recognition of identity for it looks as if the mind needs time to be able to effectively objectify a phenomenon. Further discussion of this topic is out of the scope of this paper.

> As for plurality, it is necessarily defined in terms of the one because the one is the principle of plurality, the existence, and quiddity of the latter deriving from it. Moreover, whatever definition we use in defining plurality, we necessarily use in it the one.[13] (p. 79, §2).

> It seems that plurality is better known to our act of imagining; unity, better known to our intellects. It seems that unity and plurality are among the things that we conceive a priori. However, we imagine plurality first, whereas we apprehend unity intellectually, without [a priori] intellectual principle for its conception. Our defining plurality in terms of unity would, then, be an intellectual definition. Here, unity is taken as conceived in itself and as one of the first principles of conception. And our explaining unity in terms of plurality would be a directing of attention wherein the imaginative course is used to indicate an intelligible [unity] which we already have but which we do not conceive to be present in the mind.[14] (p. 80, §4).

According to this view, Ibn Sīnā might meet Frege's challenge in the following way: If I grasp Zayd I grasp him as instantiating a concept that provides the unity and similarly for Amr, what does it mean? Amr and Zayd are grasped as different instantiations of the concept Human, i.e., that they are different elements of the same equivalence class. So what is required now is the notion of being equal in an aggregate, or more precisely, we need a notion of equivalence class. This is a necessary condition for the construction of an aggregate, we will come back to this in the last section.

13.3.2 *Infinity as Repetition Operator Performed by an Epistemic Agent*

But what about the aggregate of numbers itself? They are generated by some kind of inductive definition as described in the preceding sections. Once the mind knows how to form the number two, it can generate the rest of numbers, as elements of an equivalent class, in virtue of the essential feature of the mind to go beyond what is concretely given. The formation of two triggers a specific mental act of process i.e. an indefinite repetition. And what makes this inductive definition to be actually performed by an epistemic subject is memory i.e. namely the act of remembering the grasping of the first individual, then remembering the grasping of grasping the first individual and so on:

> It is within the power of the soul to apprehend intellectually, and [once it has made its first apprehension] to apprehend that it has apprehended, and to apprehend that it has apprehended that it has apprehended, and to construct relations within relations and to make for the one thing different states of relations ad infinitum in potency.

---

[13] وأما الكثرة فمن الضروري أن تحد بالواحد، لأن الواحد مبدأ الكثرة ومنه وخودها وماهيتها؛ ثم أي حد حددنا به الكثرةاستعملنا الواحد بالضرورة.

[14] يشبه أن تكون الكثرة أيضا أعرف عند تخيلنا، والوحدة أعرف عند عقولنا. ويشبه أن تكون الوحدة والكثرة من الأمور التينتصورها بديا، لكن الكثرة نتخيلها أولا والوحدة نعقلها من غير مبدإ لتصورها عقلي بل إن كان ولا بد فخيالي. ثم يكون تعريفنا الكثرة بالوحدة تعريفا عقليا وهنالك تؤخذ الوحدة متصورة بذاتها ومن أوائل التصور، ويكون تعريفنا الوحدة بالكثرة تنبيها يستعمل فيه المذهب الخيالي لنومئ إلى معقول عندنا لا نتصوره حاضرا في الذهن.

> It is within its proximate power to intellectually apprehend this-as, for example, bringing to mind infinite multiple progressions bringing to mind the infinite doubling of numbers and, indeed, the infinite recurrence through doubling of the same relation between a number and one similar to it.[15] (p. 160, §8)

Thus the crux of Ibn Sīnā's answer to Frege's challenge is not that there is a process by means of which we extract an invariant but rather that the point is that there is no other way to grasp an individual as an instance of the invariant that defines the correspondent unity. Another kind of objection is the following: this account which is based on the apprehension of an invariant cannot explain the formation of numbers from concrete different objects which have no species in common like for example Zayd, Bagdad, white. Ibn Sīnā's answer would be: if totally different objects can be accounted at all, it is only in virtue of the apprehension of a non-concrete invariant such as *to be a thing*. In fact, this is a consequence of defining pluralities as instantiating a given unity. Ibn Sīnā's epistemology is symptomatic of the major change brought about by al-Khwārizmī's *Book of Algebra* that drives him to establish that it is in the nature of the mind to go beyond its intentional object by indefinitely relating it to others in variety of ways, and recurrence, i.e. repetition of the same epistemic mental act, turns out to be just one of the simplest relations, for the powerful creativity of the human mind can construct more complex relations by reasoning by induction over relations. It is remarkable that the 11th Arabic Islamic physician is involved in grounding the notion of mathematical induction in a way that strongly reminds us of a similar view defended by the 19th century French mathematician Henry Poincaré for whom mathematical induction: "n'est que l'affirmation de la puissance de l'esprit qui se sait capable de concevoir la répétition indéfinie d'un même acte dès que cet acte est une fois possible." (Poincaré 1902, p. 41)

In the following section, further support will be given of the relation between plurality and unity in the context of Ibn Sīnā's notion of aggregates or equivalence classes.

13.4 Essence as Equivalence Relation

Another major innovation of Ibn Sīnā's mathematical investigations is his attempt to formalize his own conceptual analysis of the formation of numbers within equivalence classes. As a result of his extensive analysis, Ibn Sīnā is able to define numbers as follows:

> Number is formed from similar unities and nothing else. (p. 252, §11)
>
> العدد يحدث من وحدات متشاكلة لا غير.

---

15 إن في قوة النفس أن تعقل، وتعقل أنها عقلت، وتعقل أنها عقلت أنها عقلت، وأن تركب إضافات في إضافات، وتجعل للشيء الواحد أحوالا مختلفة من المناسبات إلى غير النهاية، بالقوة.
فإن ههنا مناسبات في الجذور الصم وفي إضافات الأعداد كلها قريبة المنال من النفس، وليس يلزم أن تكون النفس في حال واحدة تعقل كلها أو أن تكون مشتغلة على الدوام بذلك، بل في قوتها القريبة أن تعقل ذلك مثل إخطار المضلعات التي لا نهاية لها بالبال، ومزاوجة عدد بأعداد لا نهاية لها بالبال، بل بوقوع مناسبة عدد مع مثله مرارا لا نهاية لها بالتضعيف.

The idea of what we called the *different same* is here captured by the more functional similarity relation. This is confirmed by what Ibn Sīnā tells us elsewhere when he discusses the different means by which unity of different objects is achieved.

> Unity is [what is] similar, whereas what is contrary to it is varied, changeable and ramified. Thus, identity consists in the realisation for plurality of an aspect of unity in another respect. This includes that which is accidental and is analogous to the one by accident. For, just as [in the former case] it is said "there is one," here it is said, "[there is] that which is the same." That which is the same in quality is the similar (*shabīh*), that which is the same in quantity is the equal, that which is the same in relation is called the corresponding (*munāsib*). As for the thing which is [the same] in essence, it is present in the things that render the essence subsistent. Thus, what is the same in genus is called homogeneous; what is the same in species is called similar (*mumāthil*), and also that which is the same in properties is called resembling.[16] (p. 237, §2)

Ibn Sīnā's underlying idea is systematically to turn predicates like to "be human" into "similar in humanity", to "be animal" into "similar in animality"; "white" into "similar in colour" or "governance" into "similar in state of affairs (*ḥāl*)". But why to make such conversion? And what we gain by making it? It is in book 5 chapter 1, Ibn Sīnā reminds us of the meaning of abstracted entities like humanity.

> Universality occurs to some nature if such nature comes to exist in mental conception (التصور الذهني)... That aspect of "the human" that is intellectually apprehended in the soul is the universal. But its universality is not due to its being in the soul [i.e. psychologically conceived] but due to its relating (مقيس)[17] to many individuals [i.e. logically related], existent or imagined, that are governed for it by the same governing rule [of instantiation].[18] (p. 159, §5)

This is clearly linked to the view discussed in the preceding sections that humanity has no meaning other than its instantiation by individuals. For Ibn Sīnā then essences cannot really be defined but only subject to instantiation for we can say of each object presented to us whether it is a human or not. And since the apprehension of essences requires at least two objects, predicates become de facto part of the class of relations.

---

[16] الوحدة متشابهة وما يضادها متفنن متغير متشعب. فالهوية هو أن يحصل للكثرة وجه وحدة من وجه آخر؛ فمن ذلك ما بالعرض وهو على قياس الواحد بالعرض. فكما يقال هناك واحد يقال ههنا هو هو. وما كان هو هو في الكيف فهو شبيه، وما كان هو هو في الكم فهو مساو، وما كان هو هو في الإضافة يقال له مناسب. وأما الذي بالذات فيكون في الأمور التي تقوّم الذات. فما كان هو هو في الجنس قيل مجانس، وما كان هو هو في النوع قيل مماثل. وأيضا ما كان هو هو في الخواص يقال له مشاكل.

[17] This general concept of *al-qiyās*, which can be translated as co-relational inference (Young 2016), comes from *al-fiqh* or Islamic jurisprudence, it is an iterative constructive procedure by which the jurists infer a general law of which particular legal judgements appear as instantiations, the same concept is used in similar way by al-Khwārizmī in his landmark book *Algebra and al-Muqābala* to capture the relationship between particular equations and the six canonical equations of which they appear as instantiations (Al-Khwārizmī 2007, p. 51). It is important to point out that Ibn Sīnā, who belongs to the ḥanafite juridical school like al-Khwārizmī, describes the relation of a universal to individuals as one-to-many, he further makes sure to distinguish it from the psychological conception making it in effect a logical-epistemic relation, that is why his construction of the natural numbers can be called a logical-epistemic construction. This is the relation that he captured by "falling under a concept" in his later work *Pointers and Remainders*.

[18] تعرض الكلية لطبيعة ما إذا وقعت في التصور الذهني... فالمعقول في النفس من الإنسان هو الذي هو كلي، وكليته لا لأجل أنه في النفس، بل لأجل أنه مقيس إلى أعيان كثيرة موجودة أو متوهمة حكمها عنده حكم واحد.

As a result, essences can be defined in terms of the conditions which should be satisfied by the relation between objects. And what is the main property involved in effectively constructing an equivalence class? Ibn Sīnā tells us in the case of humanity what does it mean for example for an individual like *Zayd* to be similar to another individual like *'Amr*?

> If the humanity in *'Amr* taken as entity by itself, not in the sense of a definition, exists in *Zayd* then whatever occurs to this humanity in *Zayd* would necessarily occur to it when in *'Amr*. (p. 158, § 4)
>
> إن الإنسانية التي في عمرو إن كانت بذاتها، لا بمعنى الحد، موجودة في زيد؛ كان ما يعرض لهذه الإنسانية في زيد لا محالة يعرض لها وهي في عمرو.

Indeed, one of the consequences of being equal members of an equivalence class is that substitution applies (and substitution is made possible by the symmetry; reflexivity and transitivity of the relation). Three examples have been identified so far in which Ibn Sīnā effectively applies substitution: the first example can be found in the passage in which Ibn Sīnā performs the substitution principle[19] to prove the circularity of Euclid's definition.[20]

Example 1:

Number is a plurality composed of units (Euclid's definition); but

[19]This principle is clearly expressed by Ibn Sīnā's predecessor al-Fārābī in the following passage:

And we can also say in each of the two things in virtue of each one of them leads to the same purpose that they are all the same. We therefore say regarding their plurality *use whatever you wish since they are one and the same*. (Al-Fārābī 1989, §3 p. 38; my emphasis)

Al-Fārābī distinguishes between the two kind of substitutions: the substitution of elements of an equivalence class and the substitution of names that refer to the same object, he calls this identity relation "*wāhid bi al-'adad* i.e. one in number" or more significantly "*wāhid bi'aynihi* i.e. one in itself." One of the interesting examples he provides is the customary use of name and *kunya* to refer to the same person (ibid., p. 41 §6). Ibn Sīnā provides the following specific example in his *al-Ilāhiyāt*: "*Zayd* and *Ibn 'Abdallah* (i.e. *'Abdallah*'s son) are one" (p. 74, §2); many of the chapters of book 3 in which Ibn Sīnā discusses, among other things, the one and many cannot be very well understood without reading al-Fārābī's distinctive *Kitāb al-wāḥid wa al-waḥda* (*On One and Unity*). Ibn Sīnā has not only explicitly acknowledged his debts to al-Fārābī in his autobiography (in Gohlman 1974, pp. 32–34), the latter seems to be the only philosopher that he praised and appreciated among all his predecessors as he declared in the following passage:

> As for Abū Naṣr al-Fārābī, we must have a very high opinion of him and he should not be put on the same group of people (ولا يُجْرى مع القوم في ميدان) for he is all but the most excellent of our predecessors (أفضل من سلف من السلف). May God facilitate the meeting with him, so it shall be useful and beneficial. (in Badawi 1978, p. 122)

[20]One is astounded by those who define number and say, "number is a plurality composed of units or of ones," when plurality is the same as number (والكثرة نفس العدد) and the reality of plurality consists in that it is composed of units. Hence, their statement, "plurality is composed of units," is like their saying, "plurality is plurality (إن الكثرة كثرة)." For plurality is nothing but a *name* for that which is composed of units (فإن الكثرة ليست إلا اسما للمؤلف من الوحدات). (p. 80, §5; my emphasis)

> Plurality is composed of units, and Plurality is the same as number (والكثرة نفس العدد);[21] hence he concludes

Plurality is plurality (الكثرة كثرة)

The other two examples in which the substitution is used are identified by Hodges and Moktefi.

Example 2 (*al-Qiyās* 472.15f)[22]:

Zayd is this person sitting down, and
This person sitting down is white
So *Zayd* is white

Example 3 (*al-Qiyās* 488.10)[23]:

Pleasure is B
B is the good
Therefore pleasure is the good

For examples of the transitivity relation[24]:

> If A is equal to B and B is equal to C, then A is equal to C[25]

Ibn Sīnā also states (and even attempts to prove) the transitivity of the parallel relation in his *Danesh Name* or *The Book of Scientific Knowledge*: "We say when a line is parallel to a second line which is parallel to a third, it follows that the first is parallel to the third" (Avicenne 1986, II p. 97), i.e.

> If $L_1 \parallel L_2$ and $L_2 \parallel L_3$ then $L_1 \parallel L_3$

---

[21]Ibn Sīnā explicitly states here an equality relation between the two members, hence we can substitute whatever term we like in the first proposition. A similar proof of a circular definition using substitution is expressed in the following passage:

> If you say 'The thing is that about which (ما) it is valid to give an informative statement (الشيء هو ما يصح الخبر عنه),' it is as if you have said, 'The thing is the thing about which (الذي) it is valid to give an informative statement (الشيء هو الشيء الذي يصح الخبر عنه) because the meaning of 'whatever' (ما),, 'that which' (الذي) and the 'thing' (الشيء) is one and the same. You would have then included 'the thing' in the definition of 'the thing'. (p. 24, §7)

[22]Hodges and Moktefi (2013, p. 79).

[23]Ibid., p. 95.

[24]It is interesting to point out that al-Jurjānī (1340-1413) knows the transitivity relation that he calls the rule of equivalence (*qiyās al-musāwāt*). In his famous *Mu'jam al-ta'rīfāt* or *Dictionary of Definitions*, he provides two kinds of example to illustrate his definition. The first represents the class of relations which are transitive like equality, and the second those relations which are not transitive like "to be half (*niṣf*)" as he explains: "A is half of B and B is half of C, it is not true (*falā yaṣduqu*) to infer that A is half of C since half of half is not a half but a quarter." (Al-Jurjānī 2004, p. 153).

[25]*Al-Qiyās* II. 4 in Hodges and Moktefi (2013, p. 36).

The problem might now be that if the underlying equivalence classes are to be taken as the essence of the individuals that instantiate that class what about the accidental properties? Perhaps one can see them as the predicates that are defined on the equivalence class. Such a strategy would assume that we distinguish two differentiate predication acts that correspond to two different types: the underlying set and the propositional functions built on that set. The predication *a is B* is ambiguous then, it might be understood *as a instantiates the aggregate B (a is an element of the set B)* and as the proposition *Ba is true* where *a* instantiates the aggregate *A* (*Ba is true and a is an element of the set A*). In his discussion of a wide range of examples of building equivalence classes in book 3 chapter 2, Ibn Sīnā presents us with the following case (for a formal study of this and similar cases see Rahman and Salloum (2013):

> Regarding the one in terms of equality, this is by virtue of some similar correspondence (*munāsaba*), in that the state of the ship, for example, with respect to the captain and the state of the city with respect to the king are one. For these are two states of things that coincide. (p. 78, § 17)
>
> وأما الواحد بالمساواة فهو بمناسبة ما، مثل أن حال السفينة عند الربان وحال المدينة عند الملك واحدة، فإن هاتين حالتان متفقتان.

And more generally:

> Regarding the things that are numerically many, they are only spoken of as one in another respect by reason of a *coincidence they have in meaning*. Their coincidence would be either in some relation, in some predicate other than relation or in some subject. The predicate would be either genus, species, differentia or accident. (p. 78, §19; my emphasis)
>
> أما الأشياء الكثيرة بالعدد فإنما يقال لها من جهة أخرى واحدة لاتفاق بينها في معنى. فإما أن يكون اتفاقها في نسبة أو في محمول غير النسبة، وإما في موضوع. والمحمول إما جنس وإما نوع وإما فصل وإما عرض.

This example confirms our analysis that numbers and more generally class of objects, which can be of completely different essences like ship and kingdom, can nevertheless be formed by the apprehension of a nonconcrete invariant namely the governance relationship functioning as the similarity criterion, and as a result the ordered pair (captain, ship) and (king, city) forms an equivalence class. As mentioned in the first quote of this section, equality is seen as numerical equality. Hence, it seems then that equality as a specific relation is defined in terms of the more general similar relation. The extensional contemporary definition is to identify equality as the smallest equivalence class. However, extensionality does not seem to be the conceptual frame of Ibn Sīnā. Perhaps the best way to think about it is establishing some kind of bijective function between two equivalence aggregates.

References

Primary Sources

Avicenna. (2009). *The Physics of* The Healing (الشفاء: السماع الطبيعي). A Parallel English-Arabic text translated, introduced and annotated by Jon McGinnis, Islamic Translation Series, Birgham Young University Press, Prove, Utah.

Avicenna. (2005). *The Metaphysics of* The Healing (الشفاء: الإلهيات). A Parallel English-Arabic text translated, introduced and annotated by Michael E. Marmura, Islamic Translation Series, Birgham Young University Press, Prove, Utah.

Ibn Sīnā. (1992). *Kitāb al-Ishārāt wa aṭ-Ṭanbīhāt with a Commentary by Nasīr al-Dīn al-Ṭūsī*, Second Part, ed. S. Dunyā, Dār al-Ma'ārif, Cairo.

Avicenne. (1986). *Le Livre de Science*, Traduit du persan par Mohammad Achena et Henri Massé, Deuxième édition revue et corrigée, Les Belles Lettres – UNESCO.

Ibn Sīnā. (1983). *Kitāb al-Ishārāt wa aṭ-Ṭanbīhāt with a Commentary by Nasīr al-Dīn al-Ṭūsī*, First Part One, ed. S. Dunyā, Dār al-Ma'ārif, Cairo.

Avicenne. (1978). *La Métaphysique du Shifā', Livres I à V*, Introduction, traduction et notes par Georges C. Anawati, J. Vrin, Paris.

Ibn Sīnā. (1968). *Kitāb al-Ishārāt wa aṭ-Ṭanbīhāt with a Commentary by Nasīr al-Dīn al-Ṭūsī*, Third Part, ed. S. Dunyā, Dār al-Ma'ārif, Cairo.

Ibn Sīnā. (1956). *Al-Shifā', al-Manṭiq*, vol. V: *al-Burhān*, ed. Afifi, directed by Ibrāhīm Madhkour, al-Maṭba'a al-Amīrīa, Cairo.

Avicenne. (1951). *Livre des directives et remarques. Kitāb al-Ishārāt wa L-Tanbīhāt* (*Pointers and Reminders*), Traduction, introduction et notes par A.-M. Goichon, Paris-Beyrouth, Editions Unesco, Vrin.

Ibn Sīnā. (1910). *Manṭiq al-mashriqiyyīn*, al-Maktaba al-salafia, Cairo.

Gohlman, W. E. (1974). *The life of Ibn Sīnā. A Critical Edition and Annotated Translation*. State University of New York Press: Albany, NY.

Al-Fārābī. (1989). *Kitāb al-wāḥid wa al-waḥda*, Arabic Text Edited with Introduction and Notes by M. Mahdi, Ṭūbqāl li al-Nashr, Casablanca.

Al-Jurjānī. (2004). *Mu'jam al-ta'rīfāt*, ed. Muhammad Siddiq al-Minshawi, Cairo, Dār al-Faḍīla.

Al-Khwārizmī. (2007). *Kitāb al-jabr wa al-muqābala* in *Le commencement de l'algèbre*, texte établi, traduit et commenté par R. Rashed, Albert Blanchard, Paris; for the English version, Al-Khwārizmī: *The Beginnings of Algebra*, Edited with translation and commentary by Roshdi Rashed, Bilingual edition, Saqi Books, 2009.

Al-Kindī. (1950). *Rasā'il al-Kindī al-falsafiyya*, ed. Mu ammad 'Abd al-Hādī Abū Rīda, Cairo.

Frege, G. (1884). *The Foundations of Arithmetic, A Logico-Mathematical Enquiry into The Concept of Number*, Translated by J. L. Austin, Second Revised Edition, Harper Torchbooks, The Science Library Harper & Brothers, New York.

Poincaré, H. (1902). *La science et l'hypothèse*. Paris: Flammarion.

Secondary Literature

Badawi, A. (1978). *Aristū 'inda al-'arab*, al-Kuwait, Wakālat al-Maṭbū'at.

Hodges, W. \& Moktefi, A. (2013). *Ibn Sīnā on Logical Analysis*, (Draft), available at Hodges's website: http://wilfridhodges.co.uk/arabic29.pdf.

Rashed, R. (2015). *Classical Mathematics from Al-Khwārizmī to Descartes*, translated by M. H. Shank, Centre for Arab Unity Studies, London and New York: Routledge.

Tahiri, H. (2018). When the present misunderstands the past. How a modern arab intellectual reclaimed his own heritage. *Arabic Sciences and Philosophy*, *28*(1), 133–158.
Rahman, S. \& Salloum, Z. (2013). "Quantité, unité et identité dans la philosophie d'Ibn Sīnā". In H. Tahiri (Ed.), *La périodisation en histoire des sciences et de la philosophie. La fin d'un mythe*, Cahiers de Logique et d'Epistémologie (Vol. 13, pp. 51–61). London: College Publications.
Young, W. E. (2016). *The Dialectical Forge. Juridical Disputation and the Evolution of Islamic Law*. Dordrecht: Springer.

Author Index

H. Tahiri (ed.), *The Philosophers and Mathematics*, Logic, Epistemology, and the Unity of Science 43, https://doi.org/10.1007/978-3-319-93733-5

Subject Index

H. Tahiri (ed.), *The Philosophers and Mathematics*, Logic, Epistemology, and the Unity of Science 43, https://doi.org/10.1007/978-3-319-93733-5

Printed in the United States
By Bookmasters